"十三五"江苏省高等学校重点教材
（编号：2020-2-081）

工程制图与数字化表达

（双语）

Engineering Drawing and Digital Expression

主　编　赵海燕　杜　洁

副主编　郭南初

参　编　徐培炘　张亚琴

机 械 工 业 出 版 社

数字化是智能制造的基础。本书是为了满足智能制造对设计表达的数字化要求，根据现行的《技术制图》和《机械制图》国家标准编写而成的。本书吸收高等职业教育工程制图多年教学改革的经验，对传统教学内容进行了必要增减和优化整合，根据工作过程系统化和创新能力培养要求，将教学内容划分为7章，分别是产品设计与表达、计算机绘图、制图基础、机械制图标准、常用机件及结构要素的表示、零件图的识读与绘制、装配图的识读与绘制。

本书以培养学生的读图能力、数字化表达能力和创新能力为目标，提供大量带尺寸的立体图，有利于将知识的学习和技能的训练融为一体，实现学习和练习一体化、教材和习题集一体化，并通过在线开放课程网站提供补充习题，以满足不同学生、不同学校的需要。

本书可作为高等职业院校机械类及相关专业学生的教材。

Digitalization is the foundation of intelligent manufacturing. According to the current edition of *Technical Drawings* and *Mechanical Drawings* by national standard, this book aims to meet the digital requirements of intelligent manufacturing for design expression by utilizing experiences of higher vocational education educational reforms of Engineering Drawing. This book makes necessary adjustments and improvements to traditional teaching contents of mechanical drawing. To reflect and satisfy the requirements of systematic work process and innovative ability cultivation, this book divides its learning materials into seven chapters: product design and expression, computer graphics, fundamentals of drawing, standards of mechanical drawing, the representation of commonly-used machine parts and structural elements, the reading and drawing of the detail drawings, the reading and drawing of the assembly drawings.

This book aims to cultivate students' ability of drawing reading, digital expression and innovation. By providing a large number of 3D drawings labeled with sizes, it is beneficial to integrate the knowledge learning and skill training, and realizes the integration of textbook and problem set, as well as learning and practice. Moreover, the additional exercises are supplemented on the online open course website to meet the needs of different students and schools.

This book can be used as a textbook for students majoring in mechanical engineering and similar majors in higher vocational colleges.

图书在版编目（CIP）数据

工程制图与数字化表达：汉英对照 / 赵海燕，杜洁主编．—北京：机械工业出版社，2021.9（2024.9 重印）

“十三五”江苏省高等学校重点教材

ISBN 978-7-111-69240-9

Ⅰ．①工… Ⅱ．①赵… ②杜… Ⅲ．①数字技术—应用—工程制图—高等职业教育—教材—汉、英 Ⅳ．① TB237

中国版本图书馆 CIP 数据核字（2021）第 200063 号

机械工业出版社（北京市百万庄大街 22 号　邮政编码 100037）

策划编辑：曹帅鹏　责任编辑：曹帅鹏　车　忱

责任校对：李　伟　责任印制：常天培

固安县铭成印刷有限公司印刷

2024 年 9 月第 1 版第 3 次印刷

184mm × 260mm · 13.25 印张 · 328 千字

标准书号：ISBN 978-7-111-69240-9

定价：59.00 元

电话服务　　　网络服务

客服电话：010-88361066　机　工　官　网：www.cmpbook.com

010-88379833　机　工　官　博：weibo.com/cmp1952

010-68326294　金　书　网：www.golden-book.com

封底无防伪标均为盗版　机工教育服务网：www.cmpedu.com

前 言

根据高等职业教育改革的发展方向、智能制造对技术技能型人才的培养要求和党的二十大提出的推动共建“一带一路”高质量发展的精神，本书从高等职业教育的特点出发，以工作过程系统化为原则，强调识图、数字化表达和创新等基本能力的培养，依托江苏省外国留学生英文授课精品课程“Engineering Drawing and Digital Expression”和智能控制技术专业的国家资源库子项目“工程制图与数字化表达”，采用中、英文双语进行编写，以便“一带一路”留学生和中外合作学生学习，也可以作为普招学生的专业英语教材，提升各类学生的双语表达能力。

本书根据现行的《技术制图》《机械制图》等有关国家标准、职业和专业标准，引入和利用现代计算机技术为制图课程服务，以产品（机器部件）为主线，从系统的角度和学生的认知规律考虑教材内容的编排，使教学内容由浅入深、循序渐进，符合认知规律。

全书内容包括绘图方法、制图基础、机械制图标准、常用机件及结构要素的表示、零件图的识读与绘制和装配图的识读与绘制，形成了一条完整、系统的空间思维能力和创新能力的培养链，整个过程贯穿现代成图技术的使用，也确保了课程章节顺序符合学生的认知规律。教材将制图课程与机械关联知识有效融合，注重学科间的关联，强调国家标准规定画法的贯彻执行，重视将学生的设计思想表现为工程图的能力训练。

本书由赵海燕、杜洁担任主编，郭南初担任副主编。赵海燕编写第 3 章、第 7 章和 4.1、4.3 节，杜洁编写第 1 章、第 6 章和 4.2 节，郭南初编写第 5 章，徐培炘编写 2.1、2.2 节并负责英文校稿，张亚琴负责中文校稿。

本书是高职教育教材编写的一种探索和尝试，限于编者水平，疏漏及不当之处在所难免，恳请读者批评指正。意见或建议反馈可发送至邮箱 zhy@jssvc.edu.cn。

编者

Preface

According to the development of higher vocational education reform, the requirements of intelligent manufacturing for the training of technical and skilled talents, and the spirit of promoting high-quality development of “The Belt and Road Initiative”proposed by the 20th CPC National Congress, this book starts from the characteristics of higher vocational education, and emphasizes the principle of systematic work process. The cultivation of basic abilities such as drawing, digital expression and innovation is based on the excellent English-taught course “Engineering Drawing and Digital Expression" for foreign students in Jiangsu Province and the sub-project of the national resource bank of intelligent control technology "Engineering Drawing and Digital Expression". It is written in Chinese and English to facilitate the study of "The Belt and Road Initiative" international students and Chinese-foreign cooperative students. It can also be used as a professional English material for general enrollment students. It can also improve the bilingual expression ability of various kinds of students.

According to the current edition of *Technical Drawing*, *Mechanical Drawing* and other relevantnational standard, vocational and professional standards, this book introduces and utilizes modern computer technology to serve the course of cartography, and takes products (machine components)the main line of the course system, and considers the arrangement of teaching material content from the point of view of system and students' cognitive regulation, so that the teaching content is in line with the cognitive regulations step by step.

The teaching materials form an integrated system of spatial thinking ability and innovation ability from drawing methods, drawing foundations, mechanical drawing standards, representation of common parts and structural elements, reading and drawing of detail drawings, and reading and drawing of assembly drawings. Modern photographing technology is employed throughout the whole process and training chain,which also ensure that the sequence of the course chapters conforms to the students' cognitive regulation.The teaching material effectively integrates the drawing course with the knowledge of mechanical association, pays attention to the association between disciplines, emphasizes the implementation of the drawing method stipulated by the national standard, and pays attention to the training of students' ability to express their design ideas into engineering drawings.

The book was co-written by several authors, with Zhao Haiyan and Du Jie as chief authors and Guo Nanchu as deputy chief author. Zhao Haiyan wrote Chapter 3, Chapter 7 and Section 4.1, Section 4.3. Du Jie wrote Chapter 1,Chapter 6 and Section 4.2.Guo Nanchu wrote Chapter 5. Xu Peixin wrote Section 2.1, Section 2.2 and made English proofreading; Zhang Yaqin made Chinese proofreading.

This book is an exploration and attempt to compile teaching material for higher vocational education. However, it is limited to the knowledge level of editors, and the omissions and improprieties are inevitable. Readers who use this book are earnestly requested to criticize and correct. Comments or feedback can be sent to E-mail:zhy@jssvc.edu.cn.

Authors

Contents 目　录

前言　Preface ······ III

第 1 章　产品设计与表达　Chapter 1　Product Design and Expression ······ 1

1.1　产品表达方法　1.1　Representation of Products ······ 1
1.2　工程图的作用　1.2　Functions of Engineering Drawings ······ 2
1.3　课程任务　1.3　Course Tasks ······ 3
1.4　学习方法　1.4　Learning Methods ······ 4
[本章习题]　[Chapter exercises] ······ 5

第 2 章　计算机绘图　Chapter 2　Computer Graphics ······ 6

2.1　计算机绘图概述　2.1　Computer Graphics Summary ······ 6
2.2　体验 Inventor　2.2　Experience Inventor ······ 7
[本章习题]　[Chapter exercises] ······ 9

第 3 章　制图基础　Chapter 3　Fundamentals of Drawing ······ 11

3.1　正投影基本原理　3.1　Principle of Orthographic Projection ······ 11
3.1.1　投影的概念及分类　3.1.1　Concept and Types of Projection ······ 11
3.1.2　物体的三视图　3.1.2　Three-view Drawings ······ 12
3.2　平面图形的分析　3.2　Analysis of Plane Figures ······ 15
3.2.1　平面图形的尺寸分析　3.2.1　Dimensional Analysis of Plane Figure ······ 16
3.2.2　平面图形的线段分析　3.2.2　Line Segment Analysis of Plane Figure ······ 17
3.3　组合体的构形与分析　3.3　Configuration and Analysis of Composite Object ······ 18
3.3.1　基本体与组合体　3.3.1　Basic Object and Composite Object ······ 19
3.3.2　组合体的构形设计　3.3.2　Configuration Design of the Composite Object ······ 21
[本章习题]　[Chapter exercises] ······ 31

第 4 章　机械制图标准　Chapter 4　Standards of Mechanical Drawing ······ 33

4.1　机械制图国家标准的一般规定　4.1　General Provisions of National Standards of Mechanical Drawing ······ 33
4.1.1　图纸幅面和格式（GB/T 14689—2008）　4.1.1　Drawing Sheet Size and Layout（GB/T 14689—2008）······ 33
4.1.2　比例（GB/T 14690—1993）　4.1.2　Scales（GB/T 14690—1993）······ 34
4.1.3　字体（GB/T 14691—1993）　4.1.3　Font（GB/T 14691—1993）······ 35
4.1.4　图线及画法（GB/T 4457.4—2002）　4.1.4　Lines and Drawing Techniques（GB/T 4457.4—2002）······ 36

4.1.5 尺寸注法（GB/T 4458.4—2003） 4.1.5 Dimensioning (GB/T 4458.4—2003) …… 37
4.2 机件表达方法 4.2 Representation of Machine Parts …… 42
4.2.1 视图 4.2.1 View …… 43
4.2.2 剖视图 4.2.2 Section View …… 53
4.2.3 断面图 4.2.3 Cross-section View …… 72
4.2.4 其他表达方法 4.2.4 Other Representation …… 76
4.3 技术要求 4.3 Technical Requirements …… 80
4.3.1 表面结构的表示法 4.3.1 The Representation of the Surface Structure …… 81
4.3.2 极限与配合 4.3.2 Limits and Fits …… 85
4.3.3 几何公差 4.3.3 Geometric Tolerance …… 92
[本章习题] [Chapter exercises] …… 95

第5章 常用机件及结构要素的表示 Chapter 5 The Representation of Commonly-used Machine Parts and Structural Elements …… 97

5.1 螺纹 5.1 Threads …… 97
5.1.1 螺纹概述 …… 97
5.1.1 Threads Summary …… 97
5.1.2 螺纹的种类和标注 5.1.2 Types and Labeling of the Thread …… 101
5.2 螺纹紧固件 5.2 Thread Fastener …… 105
5.2.1 螺纹紧固件的标记 5.2.1 Labeling of Thread Fastener …… 105
5.2.2 螺纹联接 5.2.2 Thread Connection …… 107
5.3 键、销联接 5.3 Key and Pin Connection …… 109
5.3.1 键联接 5.3.1 Key Connection …… 109
5.3.2 销联接 5.3.2 Pin Connection …… 111
5.4 齿轮 5.4 Gear …… 113
5.4.1 直齿圆柱齿轮各部分名称和主要参数 5.4.1 Names and Major Parameters of Parts of Straight Toothed Spur Gear …… 114
5.4.2 各部分尺寸计算公式 5.4.2 Computational Formula of Part Dimensions …… 116
5.4.3 齿轮规定画法 5.4.3 Stipulated Drawing Method of the Gear …… 116
5.5 滚动轴承 5.5 Rolling Bearing …… 118
5.5.1 滚动轴承的种类 5.5.1 Types of Rolling Bearing …… 118
5.5.2 滚动轴承的代号（GB/T 272—2017） 5.5.2 Codes of Rolling Bearing (GB/T 272—2017) …… 119
5.6 弹簧 5.6 Spring …… 122
5.6.1 圆柱螺旋压缩弹簧各部分名称及尺寸计算 5.6.1 Names and Dimension Calculation of Different Parts of Cylindrical Helical Compression Spring …… 122

5.6.2 圆柱螺旋压缩弹簧的绘制 5.6.2 Drawing of Cylindrical Helical Compression Spring ······ 124
[本章习题] [Chapter exercises] ······ 124

第 6 章 零件图的识读与绘制 Chapter 6 The Reading and Drawing of the Detail Drawings ······ 125

6.1 零件图的作用和内容 6.1 Function and Contents of Detail Drawings ······ 125
6.1.1 零件图的作用 6.1.1 Function of Detail Drawings ······ 125
6.1.2 零件图的内容 6.1.2 Contents of Detail Drawings ······ 125
6.2 零件图的视图选择 6.2 Choosing Views of Detail Drawings ······ 126
6.2.1 主视图的选择 6.2.1 Choosing the Front View ······ 127
6.2.2 其他视图的选择 6.2.2 Choosing Other Views ······ 130
6.2.3 典型零件分析 6.2.3 Analysis of Typical Parts ······ 131
6.3 零件图的尺寸标注 6.3 Dimensioning of Detail Drawings ······ 138
6.3.1 尺寸基准的选择 6.3.1 Selection of Dimensioning Datum ······ 139
6.3.2 尺寸标注方法 6.3.2 Dimensioning Method ······ 141
6.3.3 零件上常见结构的尺寸标注 6.3.3 Dimensioning of Common Structures on Parts ······ 144
6.4 零件上常见的工艺结构 6.4 Common Process Structure on Parts ······ 145
6.4.1 铸造结构 6.4.1 Casting Structure ······ 145
6.4.2 机械加工零件的工艺结构 6.4.2 Process Structure of Machined Parts ······ 149
6.5 读零件图 6.5 Interpreting Detail Drawings ······ 152
6.5.1 概述 6.5.1 Summary ······ 153
6.5.2 分析视图 6.5.2 View Analysis ······ 154
6.5.3 分析尺寸 6.5.3 Analyzing Dimensions ······ 155
6.5.4 了解技术要求 6.5.4 Understanding Technical Requirements ······ 156
6.5.5 综合分析 6.5.5 Comprehensive Analysis ······ 156
6.6 绘制零件图 6.6 Drawing Detail Drawings ······ 157
[本章习题] [Chapter exercises] ······ 157

第 7 章 装配图的识读与绘制 Chapter 7 The Reading and Drawing of the Assembly Drawings ······ 163

7.1 装配图的内容 7.1 Contents of Assembly Drawings ······ 163
7.2 装配图的尺寸标注、技术要求和零件编号 7.2 Dimensioning of Assembly Drawings, Technical Requirements and Part Numbers ······ 165
7.2.1 装配图的尺寸标注 7.2.1 Dimensioning of Assembly Drawings ······ 165
7.2.2 装配图的技术要求 7.2.2 Technical Requirements of Assembly Drawings ······ 166
7.2.3 装配图的零件序号和明细栏 7.2.3 Part Numbers and Part Lists of Assembly Drawings ······ 166

7.3 装配图的表达方法 7.3 Representation Methods of Assembly Drawings ………… 173
7.3.1 规定画法 7.3.1 Conventional Representation ………… 173
7.3.2 特殊表达方法 7.3.2 Special Representation Methods ………… 175
7.4 装配图的绘制 7.4 Assembly Drawing ………… 176
7.4.1 部件装配关系分析 7.4.1 Analysis of Subassembly Assembly Relationship … 177
7.4.2 部件装配 7.4.2 Subassembly Assembly………… 178
7.4.3 装配图的表达方案 7.4.3 Representation Method of Assembly Drawing …… 183
7.4.4 绘制装配图 7.4.4 Drawing of Assembly Drawing ………… 186
7.5 装配结构的合理性 7.5 Rationality of Fitting Structure ………… 188
7.5.1 相邻两零件的接触面 7.5.1 Contact Surface of Two Adjacent Parts ………… 188
7.5.2 轴与孔配合 7.5.2 Fitting of Shafts and Holes ………… 189
7.5.3 拆装空间 7.5.3 Space for Disassembly ………… 190
7.5.4 防松结构 7.5.4 Anti-loosening Structure ………… 190
7.5.5 轴向固定结构 7.5.5 Axial Fixed Structure ………… 191
7.6 读装配图 7.6 Reading Assembly Drawings ………… 192
7.6.1 读装配图的方法和步骤 7.6.1 Methods and Steps of Reading Assembly Drawings ………… 193
7.6.2 由装配图拆画零件图 7.6.2 Extracting Part Drawing from Assembly Drawing ………… 197
[本章习题] [Chapter exercises] ………… 200
参考文献 References ………… 204

第 1 章　产品设计与表达

Chapter 1　Product Design and Expression

1.1　产品表达方法

产品构思与产品表达是产品设计阶段中的重要组成部分。产品构思（即构形）时，设计者必须把自己的构思用语言、文字、图样等形式表达出来，而用语言或文字来表达物体的形状和大小是很困难的。因此，用三维模型或二维工程图形形象地表达物体形状和大小的图样，就成为生产中不可缺少的技术文件。在工程上，为了准确地表达机械设备、仪表、仪器等的形状、结构和大小，根据投影原理、国家标准和有关规定画出的图样，叫作工程图，它是工程界的技术语言。如图 1-1 所示为轴承拆卸器零件横梁的三维模型，图 1-2 所示为横梁的二维工程图。

1.1　Representation of Products

1.1

Envisioning and expressing the product are essential to product design. While designers can express their ideas with either words or graphs, the latter are clearly more desirable for direct representations of shapes and sizes. 3D models and 2D drawings are indispensable to production. Engineering drawings are such drawings that accurately express the shape, structure and size of an equipment and adhere to the projection principle, national standards and related regulations (Fig.1-1, Fig.1-2). Engineers around the world communicate through engineering drawings.

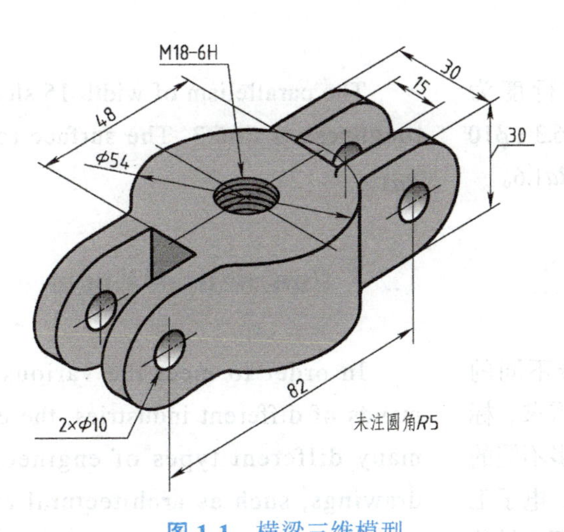

图 1-1　横梁三维模型

Fig.1-1　3-D model of a crossbeam

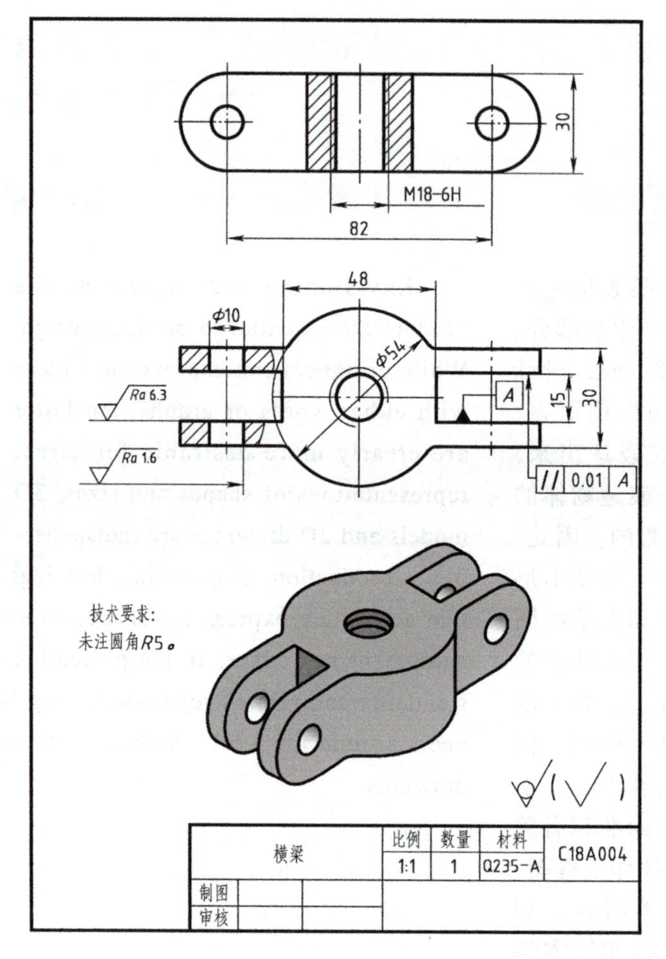

图 1-2 横梁工程图

Fig.1-2 Engineering drawing of a crossbeam

宽度为 15 的槽平行度为 0.01，表面粗糙度为 *Ra*6.3；ϕ10 的圆孔内表面粗糙度为 *Ra*1.6。

The parallelism of width 15 slot is 0.01 and the surface roughness is *Ra*6.3. The surface roughness of ϕ10 hole is *Ra*1.6.

1.2 工程图的作用

1.2 Functions of Engineering Drawings

由于不同行业有着不同的表达内容、表达方式、需求、标准等，故工程图也有很多不同的类型，例如建筑工程图、电子工程图、化工工程图等。而机械的设计、制造、维修中使用的工程

In order to meet the various demands of different industries, there are many different types of engineering drawings, such as architectural engineering drawing, electrical engineering drawing, chemical engineering draws,

图，称为机械工程图。

设计者通过工程图来表达设计意图；制造者通过工程图来了解设计要求，并依据图样来制造机器；使用者通过工程图来了解机器的结构和使用性能；在各种技术交流活动中，工程图也是不可缺少的。机械工程图的作用如图 1-3 所示。

etc. Mechanical engineering drawings are engineering drawings used in the design, manufacture and maintenance of machinery.

The designer expresses the design intention with engineering drawing. The manufacturer interprets the drawing and produces the equipment accordingly. Engineering drawing helps the user understand the structure and functions of the equipment (Fig.1-3). In a word, engineering drawing is essential to technical communication.

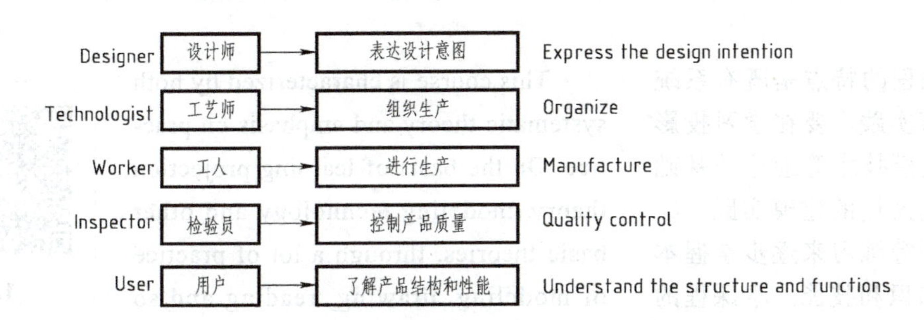

图 1-3　机械工程图的作用

Fig.1-3　Functions of mechanical engineering drawing

1.3　课程任务

本课程研究产品表达规律及方法，内容包括创建、绘制、阅读三维与二维工程图。绘制图样是将实物或头脑中的三维形体用三维建模技术或根据投影原理采用适当的表达方法表达出来。阅读图样是查看三维模型或采用形体分析法逆向思考，将二维工程图转化为头脑中的三维模型。本课程的学习任务主要有以下几点。

（1）学习正投影理论和图学思维方法，掌握工程形体的构成及表达方法。

（2）培养工程意识，贯彻、执行工程图的国家标准。

1.3　Course Tasks

This course studies the rules and methods of product expression, including creating, drawing and reading 3D and 2D engineering drawings. Drawing is the use of 3D modeling techniques or projection principles to express 3D shapes or images using appropriate expression methods. Reading drawings is to view the 3D model or to use the shape analysis to transform the 2D engineering drawing into a 3D model. The main tasks of this course are as follows.

（1）Learning orthographic projection methods and graphic thinking methods; mastering the structure and expression methods of engineering form.

（2）Cultivating engineering consciousness; implementing the national standard of engineering drawing.

（3）能绘制、阅读中等复杂程度的机械工程图样，可正确表达自己的设计意图。

（4）培养创新精神和实践能力，自主学习能力，分析问题和解决问题的能力。

（5）培养严谨求实、认真负责的工程素养。

（3）Being able to draw and read the mechanical engineering drawings of medium complexity to express the design intention correctly.

（4）Developing innovative and practical skills, independent learning, analytical, and problem-solving skills.

（5）Being rigorous and realistic, serious and responsible.

1.4 学习方法

本课程的特点是既有系统理论又有实践。要在学习投影理论、建模技术等理论的基础上，通过大量的建模实践、绘图和读图等练习来逐步掌握本课程的知识和技能。本课程的学习方法如下。

（1）以“图”为中心，围绕“图”进行学习和训练，通过“物体”与“图形”的相互转化训练，提高空间思维能力和空间想象力。即“由物得图、由图想物”。

（2）通过学习微课和相关资料完成课程的课前学习，并保质保量地完成相应部分的习题，才能使所学知识得以巩固。在课堂讨论时应积极主动地思考，课后应及时进行练习，以加深对所学内容的理解，并巩固所学内容。即采用“翻转课堂”的形式进行教学。

（3）运用所学知识和方法，观察、分析物体，并解决实际问题，以实现理论知识向能力的转化。即利用所学知识和能力进行“探索、创新和运用”。

1.4 Learning Methods

1.4

This course is characterized by both systematic theory and emphasis on practice. On the basis of learning projection theory, modeling technology and other basic theories, through a lot of practice of modeling, drawing, reading and so on, students will gradually grasp the knowledge and skills of this course. The learning methods of this course are as follows.

（1）In order to improve the spatial thinking ability and space imagination, the “figure” is the center of the study and training, and students learn to transform the “object” and “figure” to each other. That is “obtain image from object, think of object from image.”

（2）In order to consolidate knowledge, students must complete studying micro-courses and related materials before class, and complete the corresponding parts of the exercises after class. Students should participate in class discussion actively, and practice after class to deepen the understanding and consolidate the content of the study. This is the “Flipped Classroom ” form for teaching and learning.

（3）Students should observe and analyze objects and solve practical problems by using the knowledge and methods learned, and practice to realize the transformation of theoretical knowledge to ability. That is, the use of the knowledge and ability to carry out “exploration, innovation and use”.

（4）严格要求自己，随时注重对严谨、认真、负责、细致等优秀工程素养的培养。即“做事先做人”。

（4）Students should be strict with themselves. They should focus on developing rigorous, serious, responsible, meticulous and other outstanding engineering quality. That is “To do things one must first be a man”.

[本章习题]

1. 机械工程图的作用是什么?
2. 如何学好这门课程?

[Chapter exercises]

1. What is the function of mechanical engineering drawing?
2. How to learn this course well?

第 2 章　计算机绘图

Chapter 2　Computer Graphics

2.1　计算机绘图概述

2.1　Computer Graphics Summary

计算机绘图（Computer Graphics，简称 CG）是应用计算机软件及计算机硬件来处理图形信息，从而实现图形的生成、显示及输出的计算机应用技术，是工程技术人员必须掌握的基本技能之一。在新产品设计时，除了必要的计算外，绘图就占用了大量时间。采用计算机绘图可大大缩短产品开发周期，促进产品设计的标准化、系列化。计算机绘图是计算机辅助设计（Computer Aided Design，简称 CAD）的最重要组成部分。

Inventor 是美国 Autodesk 公司推出的一款三维可视化实体模拟软件，它包括零件造型、钣金、装配、表达视图和工程图等设计模块。其主要特点是简单易用，二、三维数据可无缝转换等。

工程图是设计意图实现中设计信息最主要的携带和表达者。Inventor 提供了二维与三维数据相关联的功能，让用户方便地实现从三维模型开始，按照平行正投影规则得到合格的工程图。

Computer Graphics (CG) is the technology of applying computer software and hardware to graph analysis to achieve graph generation, display and output. It is one of the basic skills that engineering and technical personnel must master. When designing a new product, in addition to the necessary calculation, the drawing takes up a lot of time. The use of computer graphics can greatly shorten the product development cycle and promote the standardization of product design and serialization. This is the most important part of Computer Aided Design (CAD).

2.1

Inventor is a 3D virtual simulation software introduced by the Autodesk Corporation in the US. It includes parts modeling, sheet metal, assembly, expression view and engineering drawing design modules. It is simple and easy to use, and provides two or three-dimensional data seamless conversion.

Engineering drawing is the most important carrier and expression of design information in design intention realization. Inventor provides two-dimensional and three-dimensional data-related functions, allowing users to easily implement from the three-dimensional model, according to the parallel positive projection rules to obtain qualified engineering drawings.

2.2 体验 Inventor

1. Inventor 的组成

Inventor 中与“先进成图技术”有关的模块包括以下几个部分。

零件造型——通过绘制草图，利用相应的定位特征（参考元素）来生成零件特征，再加上适当地放置特征，最终生成所需的零件。

钣金设计——具有处理钣金冲压等功能，并在此基础上完成展开。

装配——在部件环境中，可以将多个零部件装配起来建立一个部件。当然也可以在此环境中放置已存在的零部件，还可以在位创建零部件。

表达视图——为了向他人直观地表达自己的设计想法等，对已经装配完成的部件进行分解，生成的分解视图就称为表达视图，也称为爆炸图。“表达视图”可以输出 AVI、WMV 等格式的动画文件。

工程图——通过对已经创建的三维模型（如零件或者部件），创建工程视图，从而将三维模型转化为二维视图，以方便企业生产者、工程师与设计师之间的交流。

2. 基本成图技术

单击图标即可启动 Inventor 界面，如图 2-1 所示。

单击图 2-1 左上角的“新建”按钮，出现“新建文件”对话框，如图 2-2 所示。该对话框中

2.2 Experience Inventor

1. Composition of Inventor

Inventor's“Advanced Mapping Technology” related module includes the following sections.

Part modeling — by drawing sketches, using the corresponding positioning features (reference elements) to generate part features, and then combined with the appropriate placement characteristics to produce the necessary parts.

Sheet metal design— the ability to handle sheet stamping pressure and on this basis to complete the expansion.

Assembly — in a component environment, multiple parts, both existing and created, can be assembled to create a component.

Expression view — In order to visually express your own design ideas and such, the parts that have already been assembled are decomposed, and the exploded view is called an expression view, also known as an explosion diagram. The expression view can be exported to animated files in formats such as AVI, WMV, and so on.

Engineering drawings — by creating engineering views of 3D models such as parts or components, the 3D model is transformed into a 2D view to facilitate communication between enterprise producers, engineers and designers.

2. Basic drawing techniques

Click on icon to start the Inventor interface, as shown in Fig.2-1.

Click the “New” button in the upper-left corner of Fig.2-1 to see the “New File” dialog box, shown in Fig.2-2, which contains 4 categories of 7 working modes. In this section we will complete the “gasket” part diagram shown in Fig.2-3: ① create the three-dimensional model of the “gasket” ; ② generate the engineering drawing according to the

包含4类7种工作模式。在本节我们要完成图2-3所示的“垫片”工程图：①完成“垫片”的三维模型创建；②依据垫片三维模型生成其工程图。

建模方法：①创建草图，如图2-3所示；②这里假设垫片的厚度为1mm，对草图进行拉伸，生成三维模型，如图2-4所示。

工程图生成方法：①生成基础视图，如图2-5所示；②标注中心线，如图2-6所示；③标注尺寸，如图2-7所示。

three-dimensional model of the gasket.

The modeling method is as follows: ① Create a sketch, as shown in Fig.2-3; ② Assumes that the gasket thickness is 1mm, and the sketch is stretched to produce a 3D model, as shown in Fig.2-4.

The engineering drawing generation method are as follows: ① generate the base view, as shown in Fig.2-5; ② mark center line, as shown in Fig.2-6; ③ dimensioning, as shown in Fig.2-7.

图 2-1 Inventor 启动界面

Fig.2-1 Inventor start interface

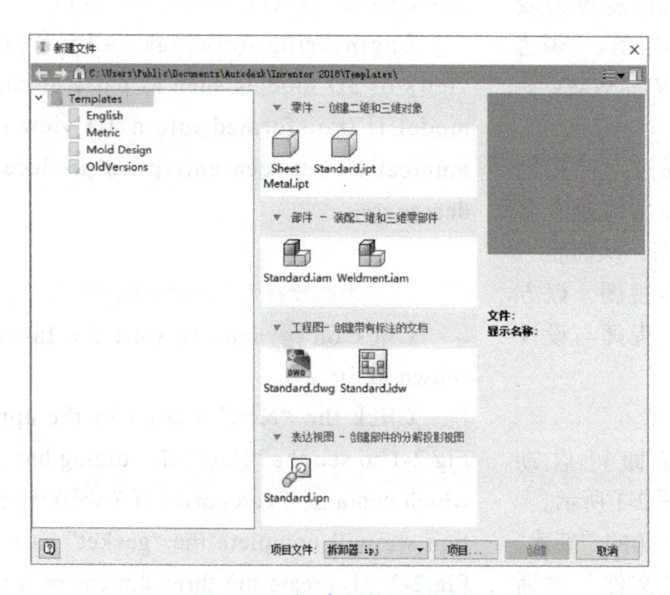

图 2-2 “新建文件”对话框

Fig.2-2 “New File” dialog box

[本章习题]

完成图 2-8、图 2-9 手柄三维模型的创建和工程图的生成。

[Chapter exercises]

Complete 3D model and engineering drawing of the handle, as shown in Fig.2-8, Fig.2-9.

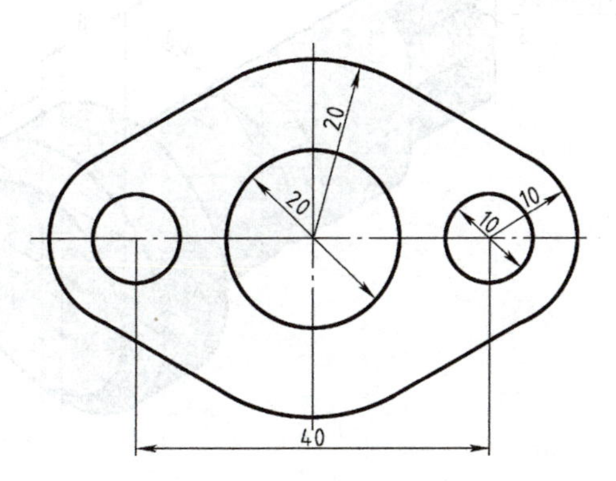

图 2-3　垫片草图

Fig.2-3　Gasket sketch

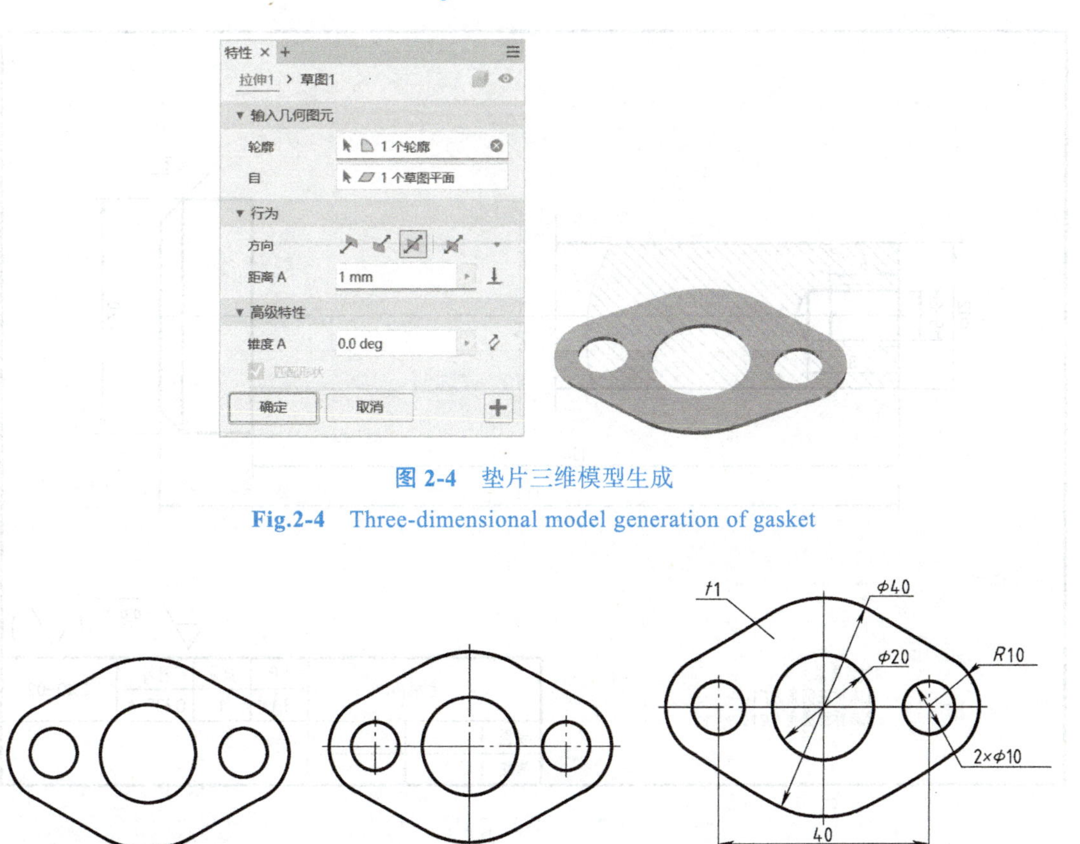

图 2-4　垫片三维模型生成

Fig.2-4　Three-dimensional model generation of gasket

图 2-5　垫片视图

Fig.2-5　Gasket view

图 2-6　标注中心线

Fig.2-6　Mark center line

图 2-7　标注尺寸

Fig.2-7　Dimension

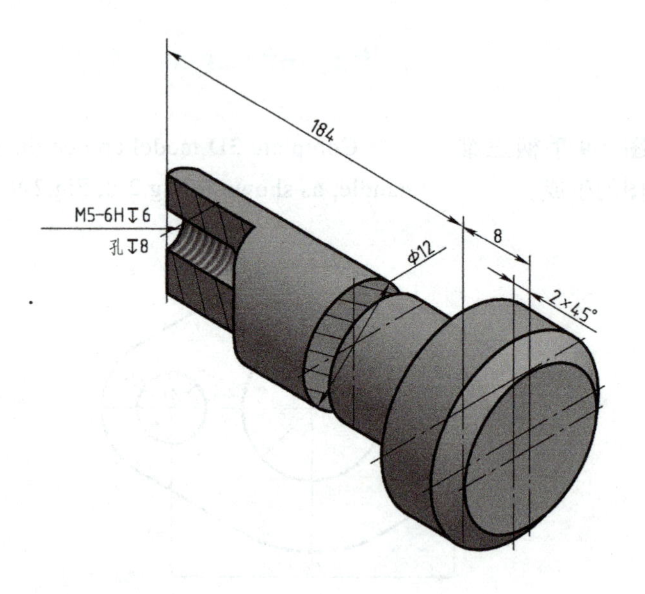

图 2-8　手柄三维模型

Fig.2-8　3D model of the handle

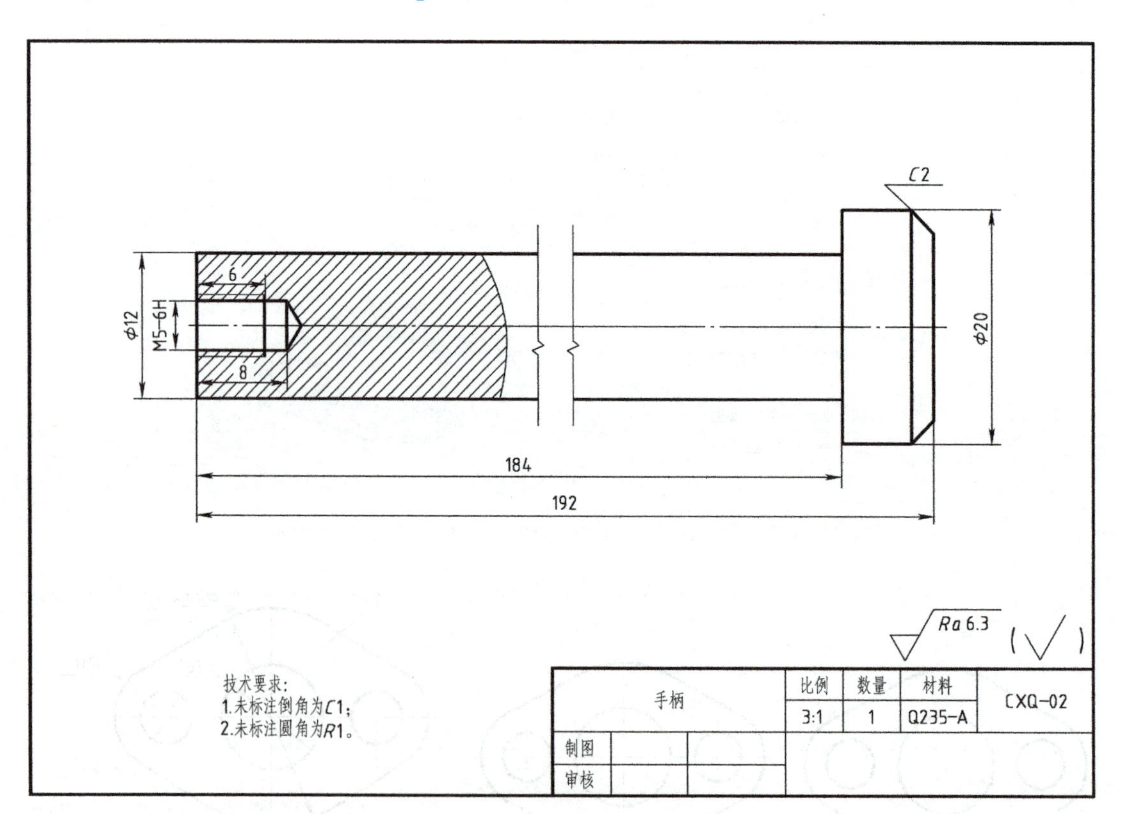

图 2-9　手柄工程图

Fig.2-9　Engineering drawing of the handle

第 3 章 制图基础

Chapter 3 Fundamentals of Drawing

3.1 正投影基本原理

3.1 Principle of Orthographic Projection

3.1.1 投影的概念及分类

1. 投影的概念

当灯光或太阳光照射物体时，在地面或墙上就会产生与原物体相同或相似的影子。人们根据这个自然现象，把能够产生光线的光源称为投影中心，光线称为投射线，承接影子的平面称为投影面。投射线通过物体，向选定的面进行投射，并在该面上得到图形的方法称为投影法，根据投影法所得到的图形称为投影。

2. 投影的分类

根据投射线之间的相对位置，可将投影分为中心投影和平行投影。投射线交汇到一点的投影，称为中心投影，如图 3-1a 所示。投射线相互平行的投影，称为平行投影，平行投影又可分为斜投影和正投影。投射线不垂直于投影面时的投影为斜投影，如图 3-1b 所示；投射线垂直于投影面时的投影为正投影，如图 3-1c 所示。

正投影图能准确表达物体的形状，度量性好，作图方便，所以在工程上得到广泛应用。机械图样主要是用正投影法绘制的，因此，正投影法的基本原理是识读和绘制机械图样的理论基础，也是本课程的核心内容。

3.1.1 Concept and Types of Projection

1. Concept of Projection

3.1.1

When the lamplight or sunlight irradiates an object, there will be a same or similar shadow of the original object on the ground or the wall. According to this natural phenomenon, the light source that can generate light ray is called the center of projection, and the light ray is called the line of projection. The plane of the shadow is called the plane of projection. When the line of projection passes through an object and then projects in the selected plane, there will be a graphic produced on the plane. This method is called a projection. The graphic on the plane is called the projection.

2. Types of Projection

According to the relative position between the lines of projection, the projection can be divided into central projection and parallel projection. When the lines of projection meet at a point, it is called the central projection, as shown in Fig.3-1a. When the lines of projection are parallel with each other, it is called the parallel projection. Parallel projection includes oblique projection and orthographic projection. When the lines of projection are not orthogonal to the plane of projection, it is called oblique projection, as shown in Fig.3-1b; when the lines are orthogonal to the plane, it is called orthographic projection, as shown in Fig.3-1c.

Orthographic projection can express the shape of an object accurately and draw conveniently, so it is widely used in engineering drawings. The mechanical drawing is mainly drawn with the orthographic projection method. Therefore, the principle of the orthographic projection method is the theoretical basis for reading and drawing mechanical drawings, and it is also the core of this course.

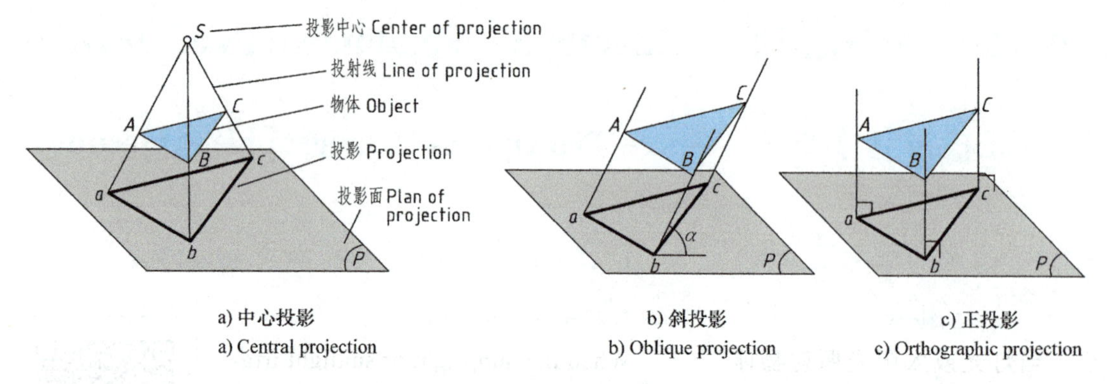

a) 中心投影
a) Central projection

b) 斜投影
b) Oblique projection

c) 正投影
c) Orthographic projection

图 3-1 投影的分类
Fig.3-1 Types of Projection

3.1.2 物体的三视图

1. 三投影面体系的建立

如图 3-2 所示，一般情况下，一个投影不能确定物体的形状。为了准确地表达物体的形状特征，一般选取互相垂直的三个投影面来构成三投影面体系，如图 3-3 所示。

三投影面体系：正立投影面，简称正面，用 *V* 表示；水平投影面，简称水平面，用 *H* 表示；侧立投影面，简称侧面，用 *W* 表示。三个投影面的交线 *OX*、*OY*、*OZ* 称为投影轴，分别代表长、宽、高三个方向。三根投影轴交于一点 *O*，称为原点。

2. 三视图的形成

用正投影法绘制出物体的图形称为视图。三视图是从三个不同方向对同一个物体进行投影的结果，能较完整地表达物体的形状特征，如图 3-4a 所示。

3.1.2 Three-view Drawings

1. The establishment of a three-projection plane system

As shown in Fig.3-2, in general, one projection cannot determine the shape of an object. In order to represent the shape features of an object accurately, three projection planes that are orthogonal from each other are generally selected to form a three-projection planes system, as shown in Fig.3-3.

3.1.2

Three-projection system: A frontal vertical plane is called V-Plane in short; A horizontal projection plane is called H-Plane in short; a profile plane is called W-plane in short. The intersection lines of the three projection planes are called the axes of projection, which are termed as OX-axis, OY-axis and OZ-axis and represent length, width and height respectively. These axes converge at a point which is called the origin.

2. Creating three-view drawings

A drawing produced through orthographic projection is called view. The three-view drawing is the result that the same object is projected from three different directions, which can express the shape features of the object fully, as shown in Fig.3-4a.

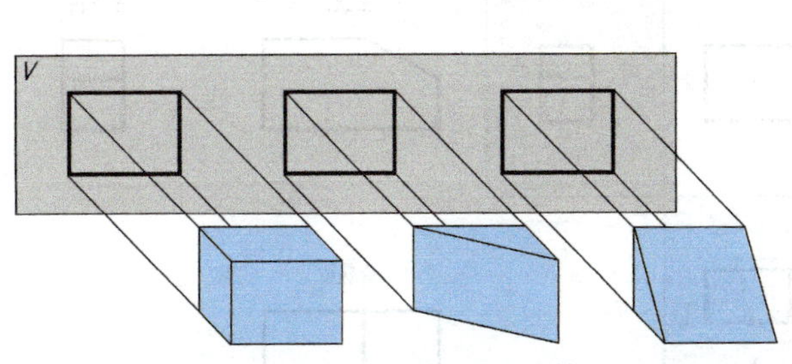

图 3-2 一个视图不能确定物体形状

Fig.3-2 One view cannot determine the shape of an object

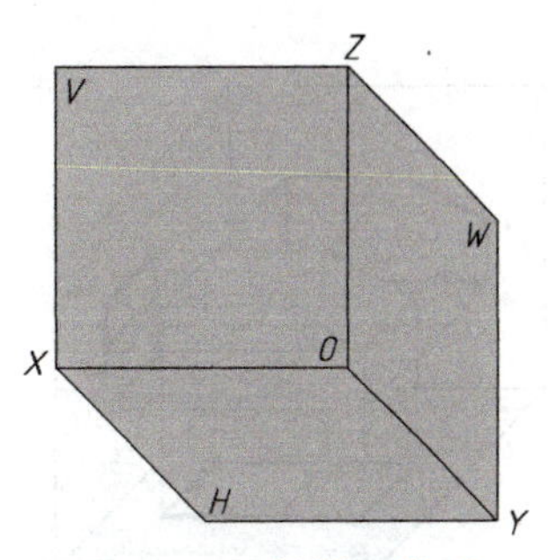

图 3-3 三投影面体系

Fig.3-3 Three projection system

为了看图、画图的方便，必须将三个相互垂直的投影面展开到同一个平面上，如图 3-4b 所示。三视图的展开过程是：保持 *V* 面不动，将 *H* 面绕 *OX* 轴向下旋转 90°，将 *W* 面绕 *OZ* 轴向后旋转 90°。在画视图时，投影面的边框不必画出，三个视图的相对位置不能变动，如图 3-4c 所示。

从物体的前面向后面投射，在 *V* 面所得的视图称为主视图，反映物体的上、下和左、右位置关系；

从物体的上面向下面投射，在 *H* 面所得的视图称为俯视图，反映物体的左、右和前、后位置关系；

从物体的左面向右面投射，在 *W* 面所得的视图称为左视图，反映物体的上、下和前、后位置关系。

3. 三视图之间的投影关系

任何物体均有长、宽、高三个方向尺寸，规定正对主视图（*V* 面）的水平方向为物体的长度（*X* 轴）方向，其宽度（*Y* 轴）和高度（*Z* 轴）方向自然就确定下来了。

For the convenience of reading and drawing, three mutually orthogonal projection planes must be unfolded on the same plane, as shown in Fig.3-4b. The process of unfolding the three-view has three steps. The first one is to keep the V-plane fixed. Then turn the H-plane 90° downwards around the OX-axis. Finally, the W-plane 90° is turned rightwards around the OZ-axis. It is not necessary to draw the plane frame of projection when drawing a view. Fig.3-4c shows that the relative positions of the three-view cannot be changed..

Projecting from the front of the object to the back, the view on the V-plane is called the front view, which shows up up/down and left/right relations of drawing features.

Projecting from the top of the object to the bottom, the view on the H-plane is called the top view, which shows up left/right and front/back relations of drawing features.

Projecting from the left of the object to the right, the view on W-plane is called the left view, which shows up up/down and front/back relationships of drawing features.

3. The projection relationship between the three-view

Any object has dimensions of length, width and height. It is specified that the horizontal direction of the front view (V-plane) is the length of the object (X-axis), then the width (Y-axis) and height (Z-axis) of the object are naturally determined.

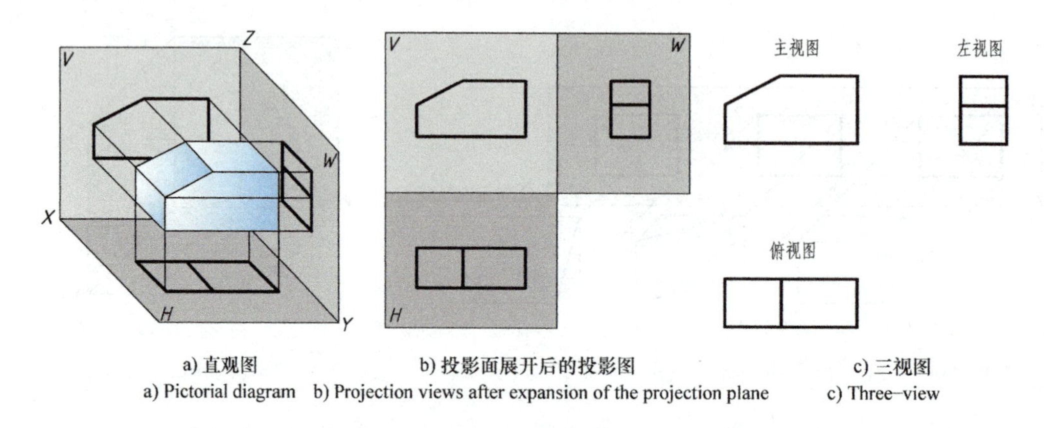

a) 直观图
a) Pictorial diagram
b) 投影面展开后的投影图
b) Projection views after expansion of the projection plane
c) 三视图
c) Three-view

图 3-4 三视图的形成

Fig.3-4 Formation of three-view

主视图反映物体的长度和高度；俯视图反映物体的长度和宽度；左视图反映物体的高度和宽度。

主、俯视图反映了物体的长度（等长）；主、左视图反映了物体的高度（等高）；俯、左视图反映了物体的宽度（等宽）。如图 3-5 所示。三视图的投影规律归纳如下。

主、俯视图**长对正**（等长）。

主、左视图**高平齐**（等高）。

俯、左视图**宽相等**（等宽）。

由三视图的投影规律，根据图 3-6 挂钩立体图绘制的挂钩三视图如图 3-7 所示。

The front view reflects the length and height of the object; the top view reflects the length and width of the object; the left view reflects the height and width of the object.

The front and top views reflect the length of the object (length alignment); the front and left views reflect the height of the object (height alignment); the top and left views reflect the width of the object (width equality). The three-view projection rules are summarized as follows (Fig.3-5).

The length of the front and the top views should be aligned and equal (length alignment).

The height of the front and the left views should be aligned and equal (height alignment).

The width of the top and left views should be aligned and equal (width alignment).

On the basis of the projection rules of the three-view, the hook's three-view (Fig.3-7) is drawn according to the stereogram view (Fig.3-6).

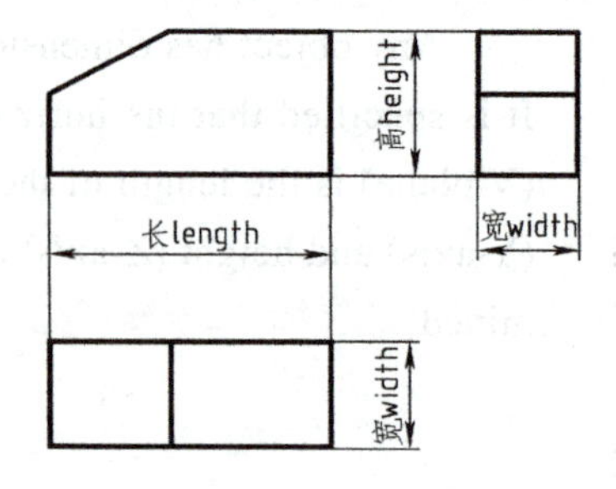

图 3-5 三视图的投影对应关系

Fig.3-5 Three-view projection relations

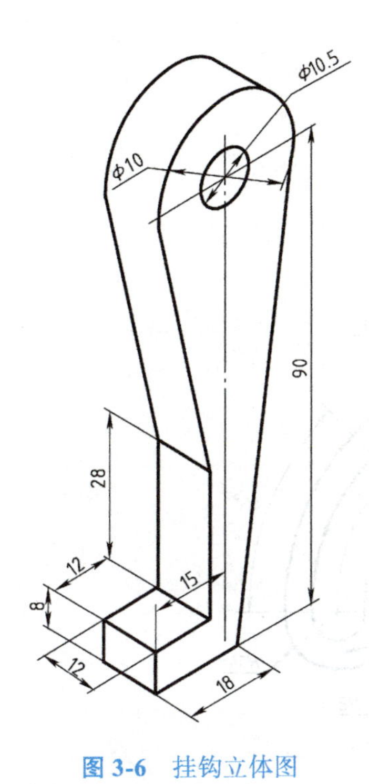

图 3-6 挂钩立体图
Fig.3-6 Perspective view of a hook

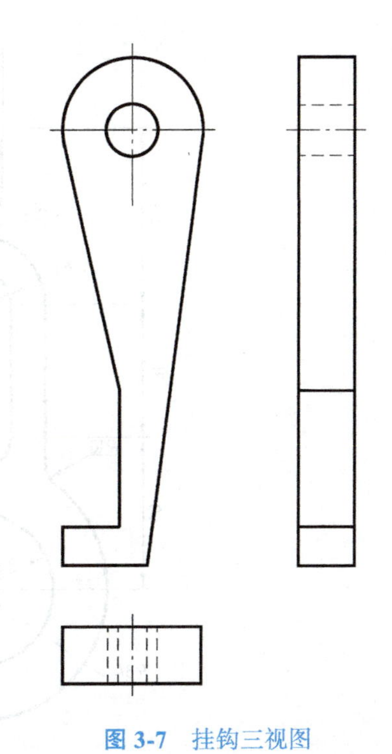
图 3-7 挂钩三视图
Fig.3-7 Three-view of a hook

3.2 平面图形的分析

计算机三维模型的基础就是草图，即平面图形。平面图形由若干直线和曲线封闭连接组合而成。在平面图形中，有些线段可以根据所给定的尺寸直接画出，而有些线段则需利用线段连接关系，找出潜在的补充条件才能画出。要处理好这方面的问题，就必须首先对平面图形中各尺寸的作用、各线段的性质，以及它们之间的相互关系进行分析，在此基础上才能确定正确的画图步骤并正确、完整地标注尺寸。现以图 3-8 所示的连接板为例，介绍平面图形的分析与画法。

3.2 Analysis of Plane Figures

The foundation of a three-dimensional(3D) computer model is a sketch（plane figure）. A plane figure is composed closely of straight lines and curves. In a plane figure, some line segments can be directly drawn according to the given dimension, and some line segments need to use the connection relationship of line segment to find potential supplementary conditions to draw. To deal with this problem, first, we must analyze the role of each dimension in the plane figure, the properties of each line segment, and the relationship between them. Based on this, we can determine drawing steps and dimension correctly and completely. Take connection board as shown in Fig.3-8 as an example to introduce the analysis and drawing of plane figure.

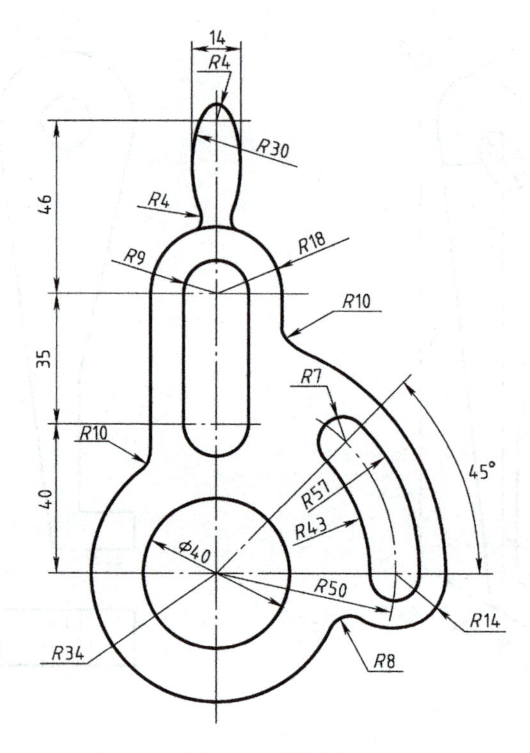

图 3-8　连接板

Fig.3-8　Connection board

3.2.1　平面图形的尺寸分析

平面图形的尺寸分析，主要是分析图中尺寸的基准和各尺寸的作用，以确定画图时所需要的尺寸数量，并根据图中所注的尺寸，来确定画图的先后顺序。

1. 尺寸基准

标注尺寸的起点称为尺寸基准，平面图形中有水平和垂直两个方向的尺寸基准。通常将对称图形的对称线、较大圆的对称中心线及主要轮廓线等作为尺寸基准。当图形在某个方向上存在多个尺寸基准时，应以一个为主（称为主要基准），其余的则为辅（称为辅助基准）。如图 3-8 中 $\phi40$ 轴孔的一对对称中心线分别为该平面

3.2.1　Dimensional Analysis of Plane Figure

The dimensional analysis of plane figures is mainly to analyze the dimensional datum of the figure and the role of each dimension, which could determine the number of dimensions for drawing and the order of drawing according to the dimensions in the figure.

1. Dimension datum

The origin of dimension is called the dimension datum, and the plane figure has dimension references in both the horizontal and vertical directions. The symmetry line of the symmetrical graph, the symmetrical center line of the larger circle, and the main outline are generally used as the dimension datum. When a graph has multiple dimension datum in a certain direction, it should select one as the main datum, and the rest should be the auxiliary datum. A pair of symmetrical center lines of the $\phi40$ shaft hole in Fig.3-8 are the dimension datum of the plane figure in the horizontal and vertical directions (main datum), and also a pair of main da-

图形水平和垂直方向的尺寸基准（主要基准），也是画图时必须首先画出的一对主要基准线。

tum lines which must be drawn first when drawing.

2. 尺寸的作用及其分类

2. The role and classification of dimension

平面图形中的尺寸，按其作用可分为定形尺寸和定位尺寸两类。

The dimensions in the plane figure can be divided into two types according to their functions: size dimension and location dimension.

（1）定形尺寸　用以确定平面图形中各线段（或线框）形状大小的尺寸，称为定形尺寸，如直线段的长度、圆及圆弧的直径或半径、角度的大小等。在图 3-8 中，ϕ40、*R*34、*R*9 均属于定形尺寸。

(1) Size dimension. Dimensions used to determine the shape of each line segment (or wireframe) in a plane figure are called size dimensions, such as the length of a straight line segment, the diameter or radius of a circle and an arc, the size of an angle, etc. In Fig.3-8, ϕ40, *R*34 and *R*9 are all size dimensions.

（2）定位尺寸　用以确定平面图形中各线段（或线框）间相对位置的尺寸，称为定位尺寸。如图 3-8 中的 R50、46、35、40、45° 等均属于定位尺寸。

(2) Location dimension. Dimensions used to determine the relative position between line segments (or wire frames) in a plane figure are called location dimensions. In Fig.3-8, *R*50, 46, 35, 40, 45°, etc, are all location dimensions.

有时某些尺寸既是定位尺寸，又是定形尺寸，如图 3-8 中的 14。尺寸基准也只有在确定线段间的相对位置时才有意义。定位尺寸是图形某一方向主要基准与辅助基准间相互联系的尺寸。

Sometimes some dimensions belong to size dimension as well as location dimension, as shown in Fig.3-8 for 14. Dimension datum only makes sense when determining the relative position between segments. Location dimension is the dimension of the main datum and the auxiliary datum in the certain direction of the drawing.

3.2.2　平面图形的线段分析

3.2.2　Line Segment Analysis of Plane Figure

确定平面图形中任一线段（或线框）一般需要三个条件（两个定位条件，一个定形条件）。例如确定一个圆，应有圆心的坐标（x，y）及直径。凡具备以上条件的线段可直接画出，否则要利用线段连接关系找出潜在的补充条件才能画出。因此，平面图形中的线段一般可分为三种不同性质的线段，现具体分析如下。

Determining any line segment (or wire frame) in a plane figure generally requires three conditions (two location conditions, one shape condition). For example, to determine a circle, there should be coordinates (x, y) of the center and diameter of the circle. Line segments that have three conditions can be drawn directly; otherwise the line segment connection relationship should be used to find potential supplementary conditions to draw. Therefore, the line segments in the plane figure can be generally divided into three segments based on its properties. The specific analysis is as follows.

1. 已知线段

凡是定位尺寸和定形尺寸均齐全的线段，称为已知线段（圆弧）。已知线段（圆弧）能直接画出，如图 3-8 中的 $\phi40$、*R*34、*R*43、*R*57 均为已知线段。画图时应先画出已知线段（圆弧）。

2. 连接线段

只有定形尺寸，而无定位尺寸的线段，称为连接线段（圆弧）。连接线段必须根据与相邻中间线段或已知线段的连接关系才能画出，如图 3-8 中 *R*9 的键槽孔中两段直线，右下角与 *R*14 圆弧相切的大圆弧，均为连接线段。连接线段（圆弧）须最后画出。

3. 中间线段

定形尺寸齐全，但定位尺寸不齐全的线段，称为中间线段（圆弧），如图 3-8 中的 *R*8、*R*30。中间线段必须根据与相邻已知线段的连接关系才能画出。中间线段（圆弧）在其相邻的已知线段画完后才能画出。

1. Given line segment

A line segment with a complete location and shape dimension is called a given line segment (arc). It is known that given line segments (arcs) can be drawn directly, as shown in Fig.3-8, $\phi40$, *R*34, *R*43 and *R*57 are all given line segments. The given line segment (arc) should be drawn first when drawing.

2. Connecting line segment

A line segment that has only a shape dimension and no-location dimension is called a connecting line segment (arc). The connecting line segment must be drawn according to the connection relationship with the adjacent intermediate line segment or the given line segment. Two straight lines in the key slot of *R*9 in Fig.3-8, and the large arc in the lower right corner tangent to the *R*14 arc, are connecting line segments. The connecting line segments (arcs) need to be drawn last.

3. Intermediate line segment

A line segment that has a complete set of size dimensions but incomplete location dimensions is called an intermediate segment (arc), as shown in Fig.3-8 for *R*8 and *R*30. The intermediate line segment must be drawn according to the connection relationships with the adjacent given line segments. The intermediate line segments (arcs) must be drawn after its adjacent given line segments have been drawn.

3.3 组合体的构形与分析

任何机器或部件都是由若干零件按一定的装配连接关系和技术要求装配起来的。如图 3-9 所示的拆卸器就是由 6 个普通零件和两类标准零件按一定的装配连接关系和技术要求装配起来的。

3.3 Configuration and Analysis of Composite Object

Any machines or components are assembled by parts according to the assembly connections and technical requirements. As shown in Fig.3-9, the detacher is assembled by 6 common parts and 2 types of standard parts according to certain assembly connections and technical requirements.

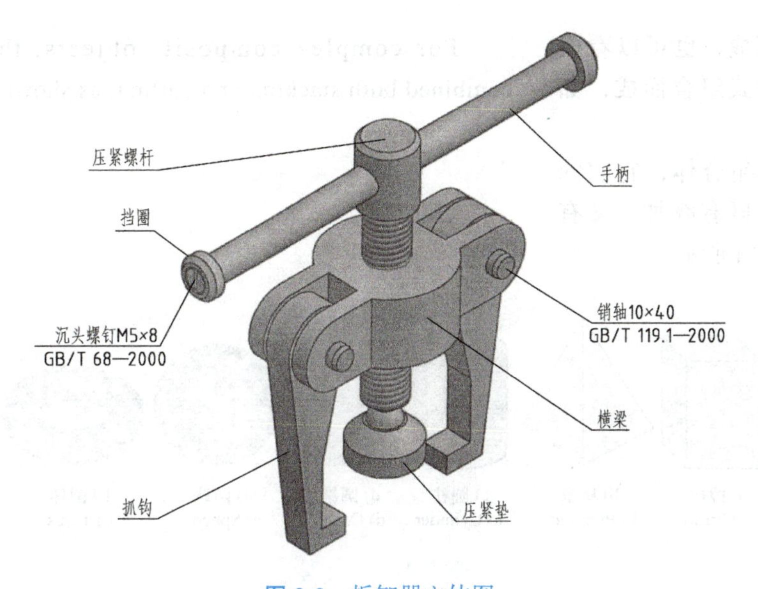

图 3-9　拆卸器立体图

Fig.3-9　Stereo view of the detacher

零件是构成机器或部件的最小单元。零件从几何构形的角度来看，大多由棱柱、棱锥、圆柱、圆锥、球等立体组成。按照立体构成的复杂程度，可将立体分为简单几何体和复杂几何体。

The machine or component consists of the parts, which is the minimum unit. From the perspective of geometric configuration, parts are mostly composed of prisms, pyramids, cylinders, cones, balls and so on. According to the complexity of configuration, the geometry can be divided into simple geometry and complex geometry.

3.3.1　基本体与组合体

3.3.1　Basic Object and Composite Object

简单几何体即基本体。如图 3-10 所示为常见的基本体，图 3-10a、b 称为平面立体，图 3-10c~f 称为曲面立体。图 3-11 为由基本体组成的机件。

复杂几何体即组合体。由若干个基本体按照一定的相对位置和组合方式有机组合而形成的较为复杂的形体称为组合体。

组合体一般有两种基本组合方式：叠加、切割，如图 3-12 所示。叠加与切割具有相对性，有些形体既可以看成是通过叠

Simple geometry is also called basic object. Fig.3-10 reports the common basic object. Fig.3-10a and b are called plane solids, and Fig.3-10c~f are called curved solids. Fig.3-11 shows that a machine is made up of basic objects.

Complex geometry is also called composite object. According to the relative position and combination, a relatively complex geometry formed by several basic objects is called composite object.

There are two basic combinations: stacking and cutting, as shown in Fig.3-12. Stacking and cutting are relative. Some objects can be seen as a combination of stacking or as a combination of cutting, as shown in Fig.3-13.

加方式组合而成，也可以看成是通过切割方式组合而成，如图 3-13 所示。

对于复杂组合体，它们的组合方式往往既有叠加，又有切割，如图 3-14 所示。

For complex composite objects, they are often combined both stacking and cutting, as shown in Fig.3-14.

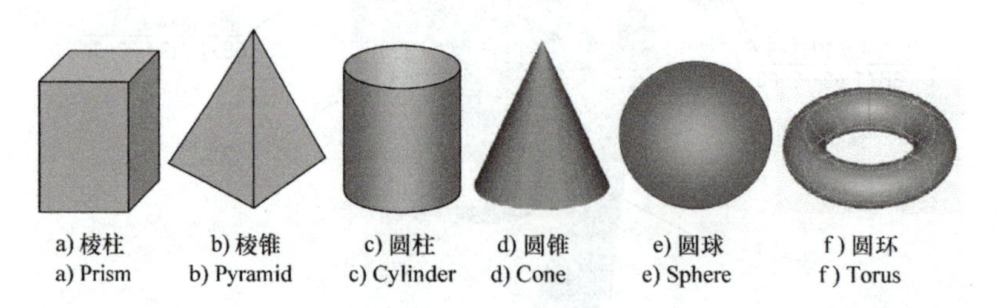

图 3-10 基本体

Fig.3-10 Basic objects

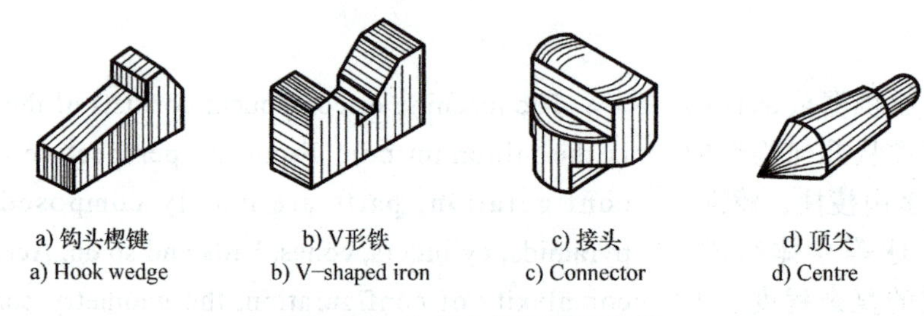

图 3-11 由基本体组成的机件

Fig.3-11 Components consisting of basic objects

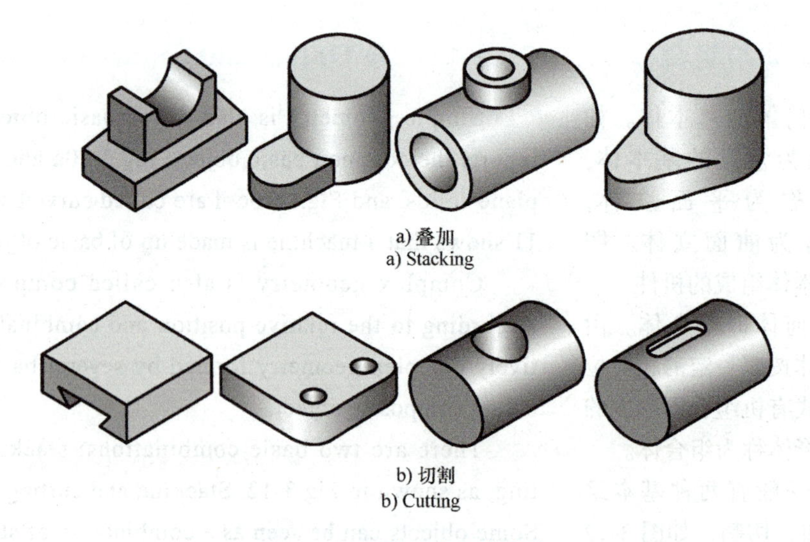

图 3-12 两种基本组合方式

Fig.3-12 Two basic combinations

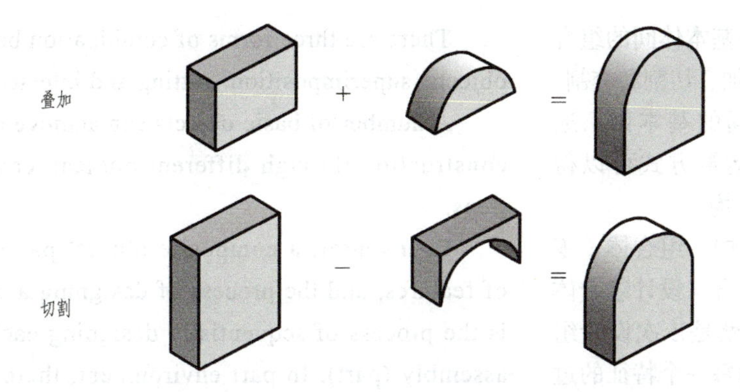

图 3-13　叠加与切割的相对性

Fig.3-13　Relativity of stacking and cutting

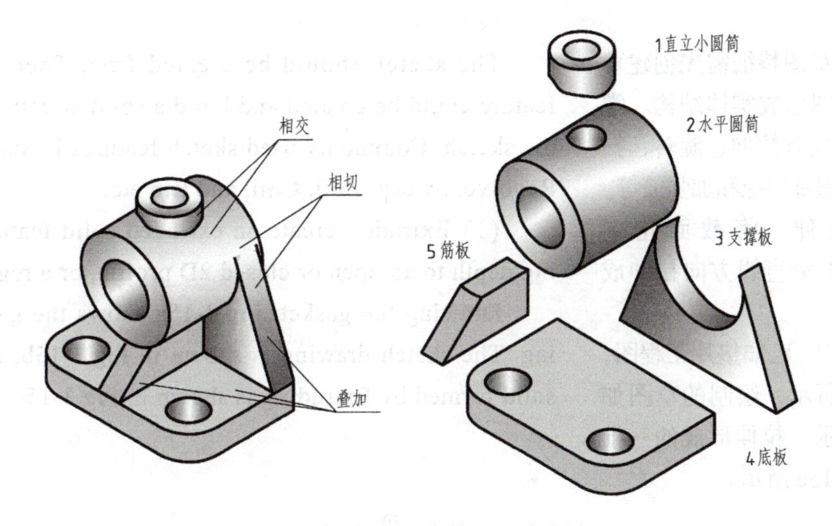

图 3-14　复杂组合体的组合方式

Fig.3-14　Combination of complex composite object

3.3.2　组合体的构形设计

根据已知条件，构思组合体的形状、大小并表达成图的过程称为组合体的构形设计。

利用 CAD 软件构形的基本方法：先用拉伸、旋转、扫掠和放样等命令生成基本形体实体，再通过“并（∪）”、“差（−）”或“交（∩）”等布尔运算构成组合体。

3.3.2　Configuration Design of the Composite Object

According to the known conditions, conceiving the shape, size, and representation design process of the composite object is called the configuration of the composite object.

The basic method of using CAD software configuration is: use the commands of extrude first, revolve, sweep and loft to generate basic objects, and then use Boolean operations such as “union (∪)”, “subtract (−)” or “intersect(∩)”, to form a combination.

组合体中各基本体间的组合方式有三种：叠加、切割和交割。

若干个相同的基本体，通过不同的布尔运算方式可以得到不同的形体结构。

在 Inventor 中，组合体（零件）是特征的集合，设计组合体（零件）的过程就是依次设计组合体（零件）的每一个特征的过程。在零件环境中主要有草图特征、放置特征和定位特征三类。

1. 草图特征

要创建草图特征需先创建草图，依照草图形成实体结构。常用的草图特征有拉伸、旋转、扫掠、放样、螺旋扫掠和加强筋等。

（1）拉伸　将截面轮廓（二维草图）沿直线方向拉伸成实体特征。

绘制垫片：已知垫片工程图，如图 3-15a 所示，绘制的草图如图 3-15b 所示，拉伸形成的三维实体如图 3-15c 所示。

There are three forms of combination between the basic objects: superimposition, cutting and intersection.

A number of basic objects can achieve different objects construction through different boolean calculation operations.

In Inventor, a composite object (part) is a collection of features, and the process of designing a composite (part) is the process of sequentially designing each feature of the assembly (part). In part environment, there are three types of features: sketch features, placed features and positioning features.

1. Sketch feature

The sketch should be created first. Then the sketch feature could be created and build a solid structure based on the sketch. Commonly used sketch features include Extrude, Revolve, Sweep, Loft, Coil, and Rib etc.

(1) Extrude　create an extruded solid feature by adding depth to an open or closed 2D profile, or a region.

Drawing the gasket: Fig.3-15a shows the gasket drawing. The sketch drawing is shown in Fig.3-15b, and the 3D solid formed by Extruding is shown in Fig.3-15c.

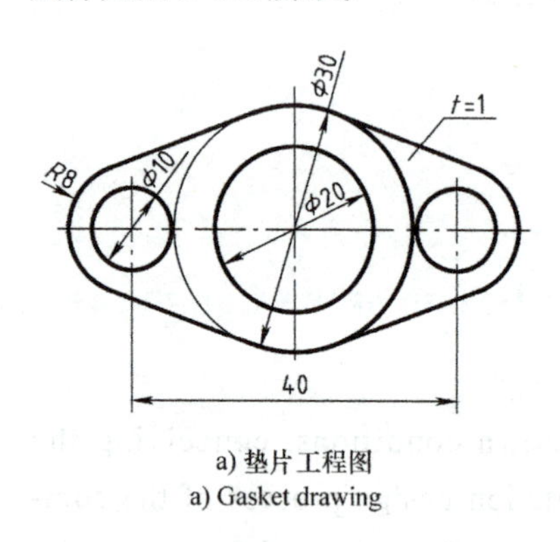

a) 垫片工程图
a) Gasket drawing

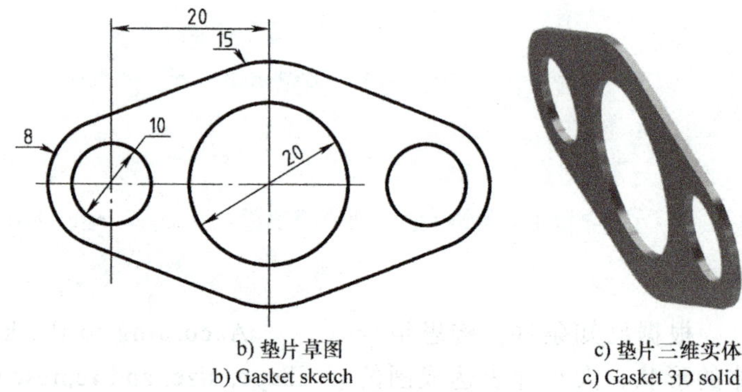

b) 垫片草图
b) Gasket sketch

c) 垫片三维实体
c) Gasket 3D solid

图 3-15　拉伸形成实体

Fig.3-15　Extrude to form a solid

（2）旋转　将截面轮廓（二维草图）绕轴线旋转成实体特征。

绘制手柄：已知手柄工程图，如图 3-16a 所示，将手柄

(2) Revolve　revolving the profile (2D sketch) around axis into solid feature.

Drawing handle: Fig.3-16a shows a handle drawing. Revolving the 2D sketch of the handle around the central

二维草图绕中心轴线旋转成三维实体，如图 3-16b 所示。这里需要说明的是，由于手柄工程图关于中心轴线对称，所以绘制草图时只需要绘制工程图的一半即可，如图 3-16b 立体图中显示的草图。

（3）扫掠 沿某一路径移动截面轮廓（二维草图）形成实体特征，如图 3-17 所示。

绘制 L 型圆棒：已知 L 型圆棒工程图，如图 3-17a 所示，将直径为 15 的圆通过中心线路径扫掠成三维实体，如图 3-17b 所示。

axis to build a 3D solid, as shown in Fig.3-16b. It should be noted that since the handle drawing is symmetrical with the central axis, so only half of the engineering drawing needs to be drawn when sketching, as shown in Fig.3-16b.

（3）Sweep sweep is moving profile contour by one path to build solid feature（Fig.3-17）.

L-shaped round bar: Fig.3-17a shows L-shaped round bar drawing. Sweeping a circle with a diameter of 15 along a centerline to build a 3D solid, as shown in Fig.3-17b.

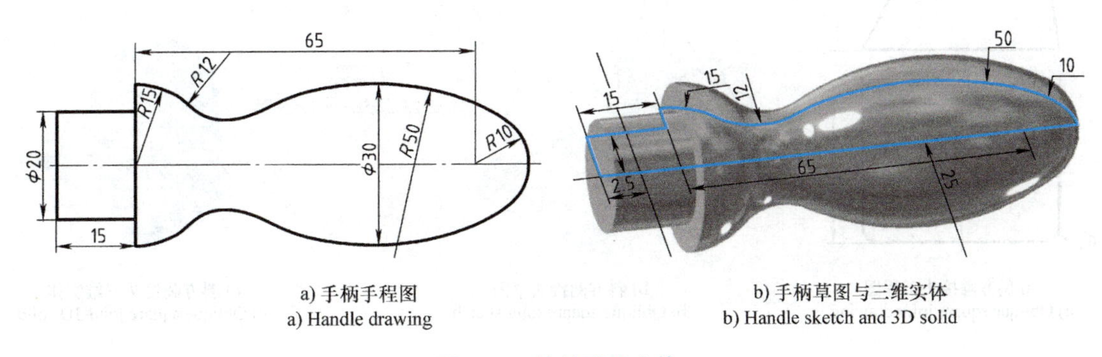

a) 手柄手程图
a) Handle drawing

b) 手柄草图与三维实体
b) Handle sketch and 3D solid

图 3-16 旋转形成实体

Fig.3-16 Revolving to form a solid

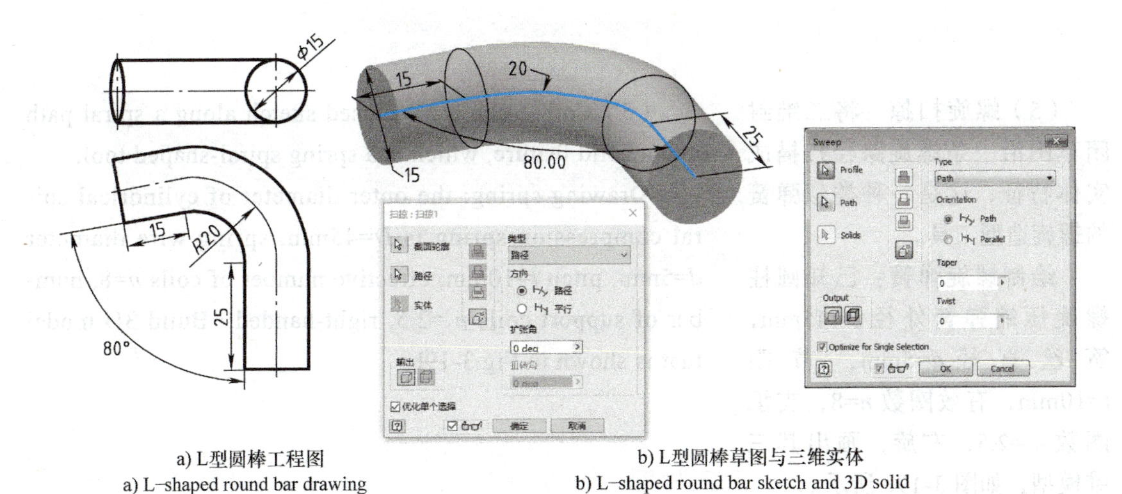

a) L型圆棒工程图
a) L-shaped round bar drawing

b) L型圆棒草图与三维实体
b) L-shaped round bar sketch and 3D solid

图 3-17 扫掠形成实体

Fig.3-17 Sweeping to build an entity

（4）放样　在两个或多个封闭截面（二维草图）之间生成光滑实体特征，如图 3-18 所示。

绘制斜方圆接头：已知斜方圆接头工程图，如图 3-18a 所示，通过直径为 25 的圆和 40×60 的矩形，放样成三维实体，如图 3-18b、c 所示。注意：圆和矩形在两个不同的平面上，两平面相距 50mm。

（4）Loft　generate smooth solid features between two or more closed sections (2D sketch) as shown in Fig.3-18.

Drawing the oblique square joint: Fig.3-18a shows an oblique square joint drawing. Drawing a circle with a diameter of 25 and a rectangle of 40×60 then loft to form a 3D solid, as shown in Fig.3-18b、c. Note: the circle and rectangle are on two different planes, and the distance of two planes is 50mm.

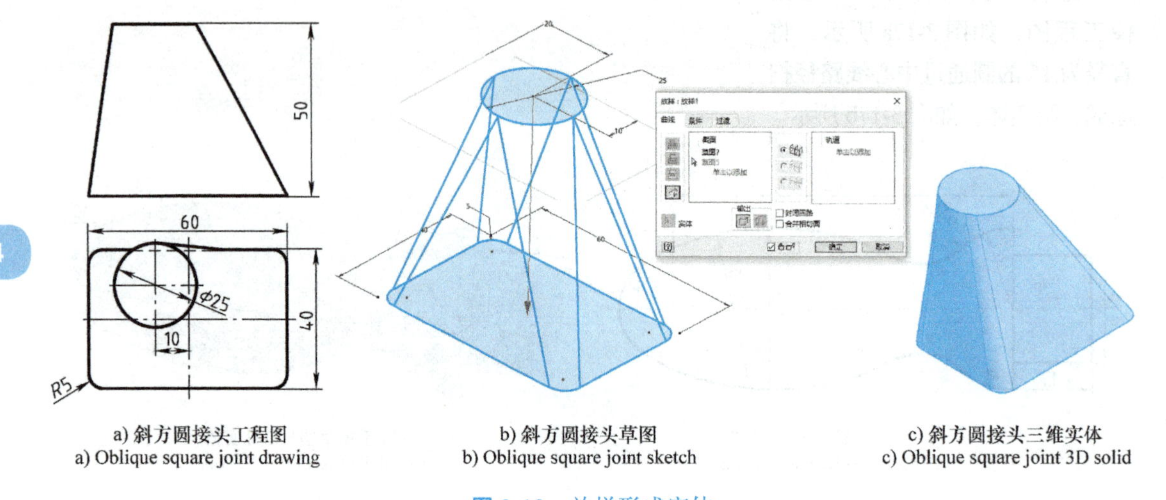

a) 斜方圆接头工程图
a) Oblique square joint drawing

b) 斜方圆接头草图
b) Oblique square joint sketch

c) 斜方圆接头三维实体
c) Oblique square joint 3D solid

图 3-18　放样形成实体

Fig.3-18　Loft to build a solid

（5）螺旋扫掠　将二维封闭草图沿一条螺旋路径扫描成实体特征，这是一种类似弹簧的螺旋造型工具。

绘制螺旋弹簧：已知圆柱螺旋压缩弹簧外径 D=45mm，簧丝直径 d=5mm，节距 t=10mm，有效圈数 n=8，支承圈数 n_1=2.5，右旋，画出其三维模型，如图 3-19b 所示。

（5）Coil　scan a 2D closed sketch along a spiral path into a solid feature, which is a spring spiral-shaped tool.

Drawing spring: the outer diameter of cylindrical spiral compression spring is D=45mm, spring wire diameter d=5mm, pitch t=10mm, effective number of coils n=8, number of support coils n_1=2.5, right-handed. Build 3D model that is shown in Fig.3-19b.

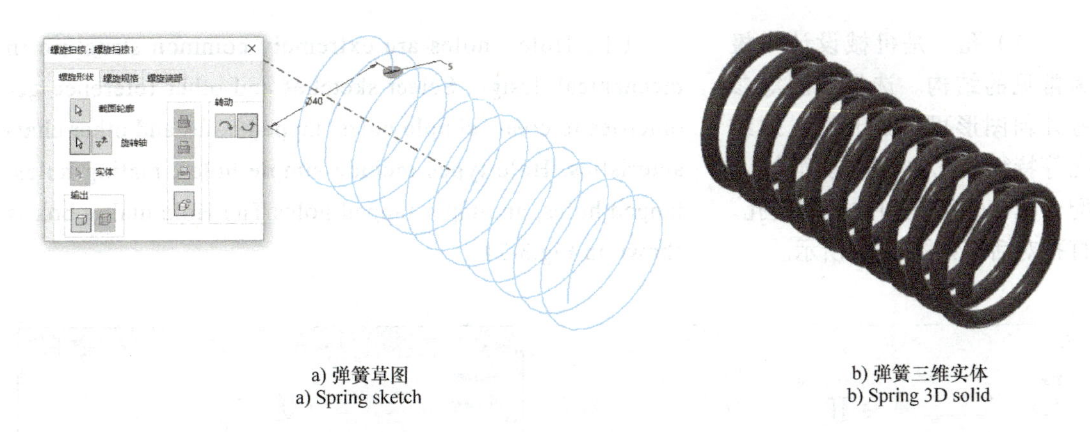

a) 弹簧草图
a) Spring sketch

b) 弹簧三维实体
b) Spring 3D solid

图 3-19　螺旋扫掠形成实体

Fig.3-19　Coil to form a solid

（6）加强筋　将二维草图按给定的厚度向实体方向延伸形成加强筋（肋）。

绘制加强筋：在 L 型板的对称中心面上绘制加强筋草图，即斜线，其末端距两个端面均为 8mm，如图 3-20a 所示，可通过“加强筋”命令直接形成加强筋，如图 3-20b 所示。

（6）Rib　according to the given thickness from 2D sketch, extend the material to form rib.

Drawing a rib: draw a sketch of the rib on the symmetrical center plane of the L-shaped plate, a oblique line which the end is 8 mm from both end faces, as shown in Fig.3-20a. Fig.3-20b illustrate that through the rib could be formed directly by the the “Rib” command.

2. 放置特征

常用的放置特征有孔、圆角、倒角、螺纹、抽壳、分割、矩形阵列、环形阵列和镜像。

2. Placed features

Placed features include hole, fillet, chamfer, thread, shell, rectangular pattern, circular pattern, mirror feature etc.

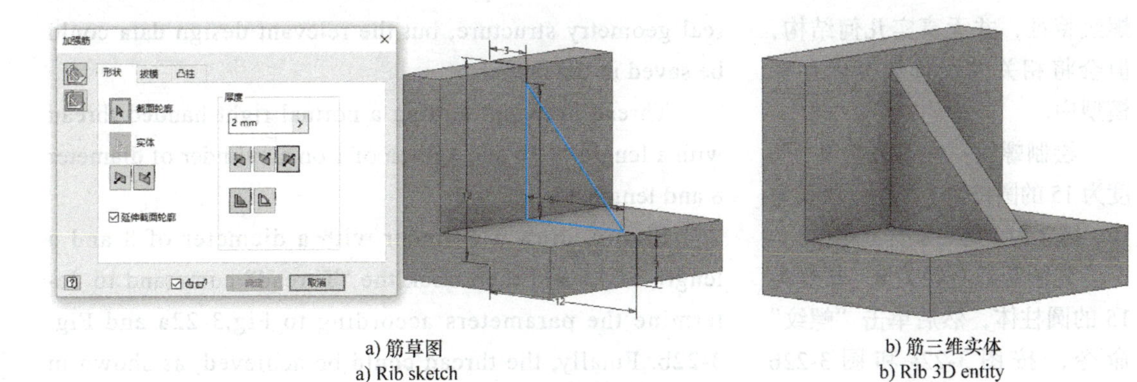

a) 筋草图
a) Rib sketch

b) 筋三维实体
b) Rib 3D entity

图 3-20　加强筋实体

Fig.3-20　Reinforced rib

（1）孔　是机械设计中极为常见的结构。选择草图等参考几何图形可创建光孔、螺纹孔等特征，孔的类型有简单孔、配合孔、螺纹孔和锥管螺纹孔。打孔对话框如图 3-21 所示。

（1）Hole　holes are extremely common structure in mechanical design. Select sketches and other reference geometries to create simple holes, tapped holes and other characteristics. Hole types include simple holes, mating holes, tapped holes, and taper tapped holes.The Hole dialog box is shown in Fig.3-21.

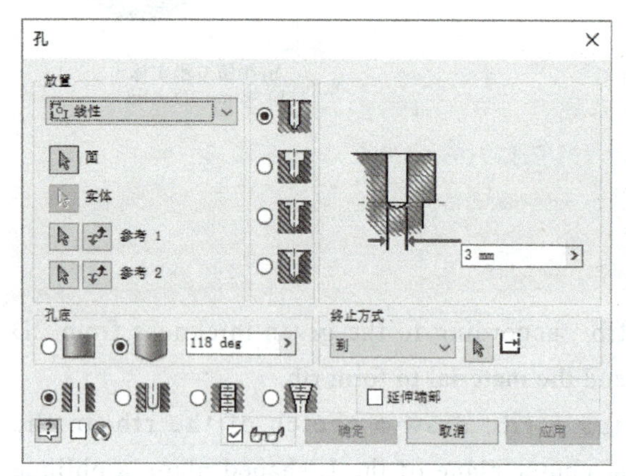

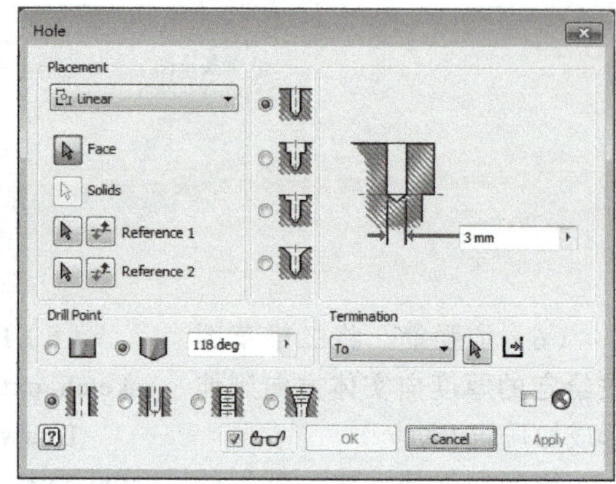

图 3-21　孔对话框

Fig.3-21　Hole dialog box

（2）圆角　为零件上一条或多条边添加内、外圆角的特征。

（2）Fillet　add inner and outer fillet features to one or more edges of the part.

（3）倒角　为零件上一条或多条边添加倒角特征。

（3）Chamfer　add chamfer features to one or more edges of the part.

（4）螺纹　在圆柱或圆锥面上生成螺纹效果的图像。

（4）Thread　an image that produces a thread effect on a cylindrical or conical surface.

使用表面贴图的形式表示螺纹特征，并无真实几何结构，但会将相关的设计数据保存在模型中。

Use surface maps to represent thread features without real geometry structure, but the relevant design data could be saved in the model.

绘制螺纹：在直径为 8、长度为 15 的圆柱体上车削一段长为 10、螺距为 1 的右旋普通螺纹。

Thread drawing: cutting a normal right-handed thread with a length of 10 and a pitch of 1 on a cylinder of diameter 8 and length 15.

先绘制好直径为 8、长度为 15 的圆柱体，然后单击“螺纹”命令，按图 3-22a 和图 3-22b 确定参数，最终生成螺纹，如图 3-22c 所示。

Firstly, draw a cylinder with a diameter of 8 and a length of 15 and then click the “Thread” command to determine the parameters according to Fig.3-22a and Fig. 3-22b. Finally, the thread could be achieved, as shown in Fig.3-22c.

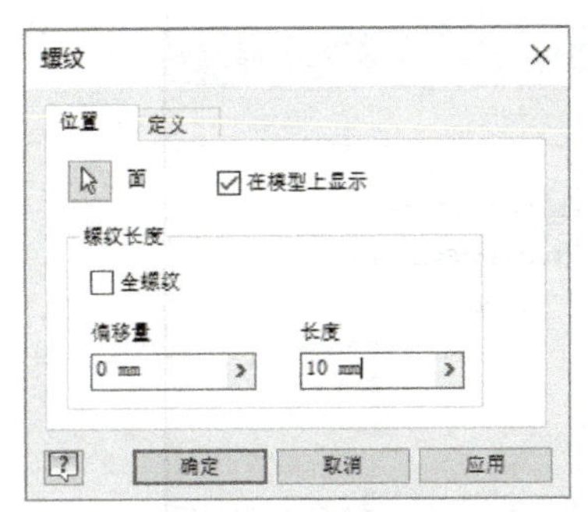

a) 确定螺纹位置
a) Thread position determination

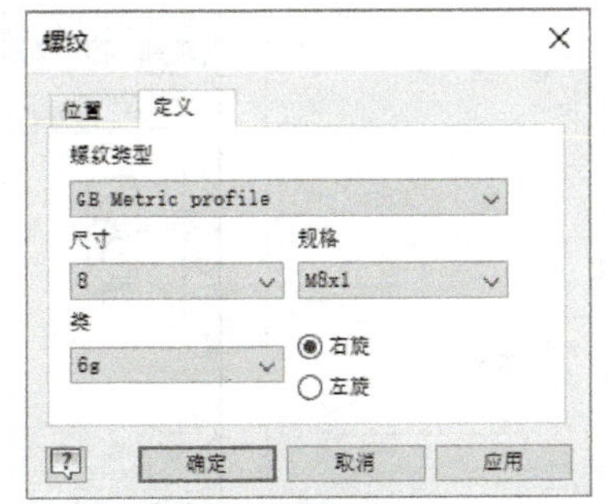

b) 确定螺纹参数
b) Thread parameter determination

c) 螺纹三维实体
c) Thread 3D solid

图 3-22　普通螺纹绘制
Fig.3-22　General thread drawing

（5）抽壳　从零件内部去除材料，创建一个具有指定厚度的空腔零件。

绘制肥皂盒盖：肥皂盒盖工程图如图 3-23a 所示，先生成实心的三维实体，然后通过“抽壳”命令形成肥皂盒盖，如图 3-23b 所示，其抽壳对话框如图 3-23c 所示。

（5）Shell　remove material from a part interior and then create a hollow cavity with walls with the specified thickness.

The soap box cover drawing: the soap box cover engineering drawing is shown in Fig.3-23a. Firstly, the solid 3D entity is generated, and then the soap box cover is generated by the “sucking shell” command, as shown in Fig.3-23b. The Shell dialog box is shown in Fig.3-23c.

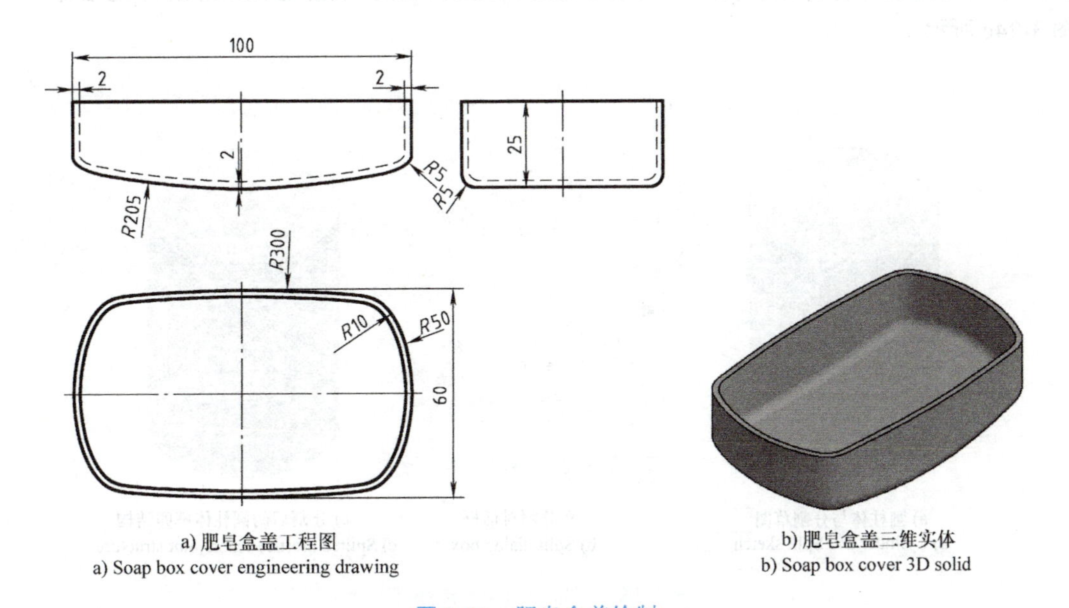

a) 肥皂盒盖工程图
a) Soap box cover engineering drawing

b) 肥皂盒盖三维实体
b) Soap box cover 3D solid

图 3-23　肥皂盒盖绘制
Fig.3-23　Drawing of the soap box cover

c) 抽壳对话框

c) Shell dialog box

图 3-23　肥皂盒盖绘制（续）

Fig.3-23　Drawing of the soap box cover

（6）分割　利用分割工具（线、面、曲面）分割零件、曲面。

绘制圆柱体的榫卯结构：圆柱体直径 15，高度 20，在圆柱体中部进行分割，其分割草图如图 3-24a 所示。

分割对话框如图 3-24b 所示，形成的圆柱体榫卯结构如图 3-24c 所示。

（6）Split　use “Split” command to split part or curve with lines, planes and curves.

Drawing the mortise-tenon joint structure of cylinder: the diameter and height of the cylinder is 15 and 20 separately. The split is in middle of cylinder and the split sketch is shown in Fig.3-24a.

The Split dialog box is shown in Fig.3-24b, and the mortise-tenon joint is generated, as shown in Fig.3-24c.

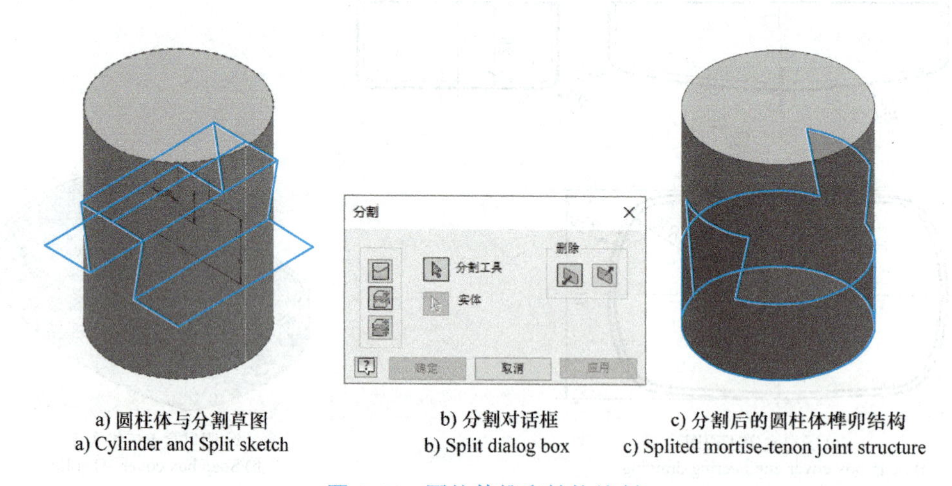

a) 圆柱体与分割草图　　b) 分割对话框　　c) 分割后的圆柱体榫卯结构

a) Cylinder and Split sketch　　b) Split dialog box　　c) Splited mortise-tenon joint structure

图 3-24　圆柱体榫卯结构绘制

Fig.3-24　Drawing the mortise-tenon joint structure of cylinder

（7）矩形阵列　按照矩形规律，对已有特征进行复制。阵列时需要指定行和列的方向以及行和列上特征的数量和间距。

（7）Rectangular pattern　copying the existing features according the rectangular rules. During array, the direction of the rows and columns need to be specified and the number and spacing of features on the rows and columns.

绘制座盖：座盖立体图如图 3-25a 所示，绘制座盖底板 4 个圆孔的矩形阵列对话框如图 3-25b 所示。

Drawing the cover: the 3D view of the cover is shown in Fig.3-25a. Fig.3-25b shows the rectangular pattern dialog box for drawing the four circular holes on the base of the cover.

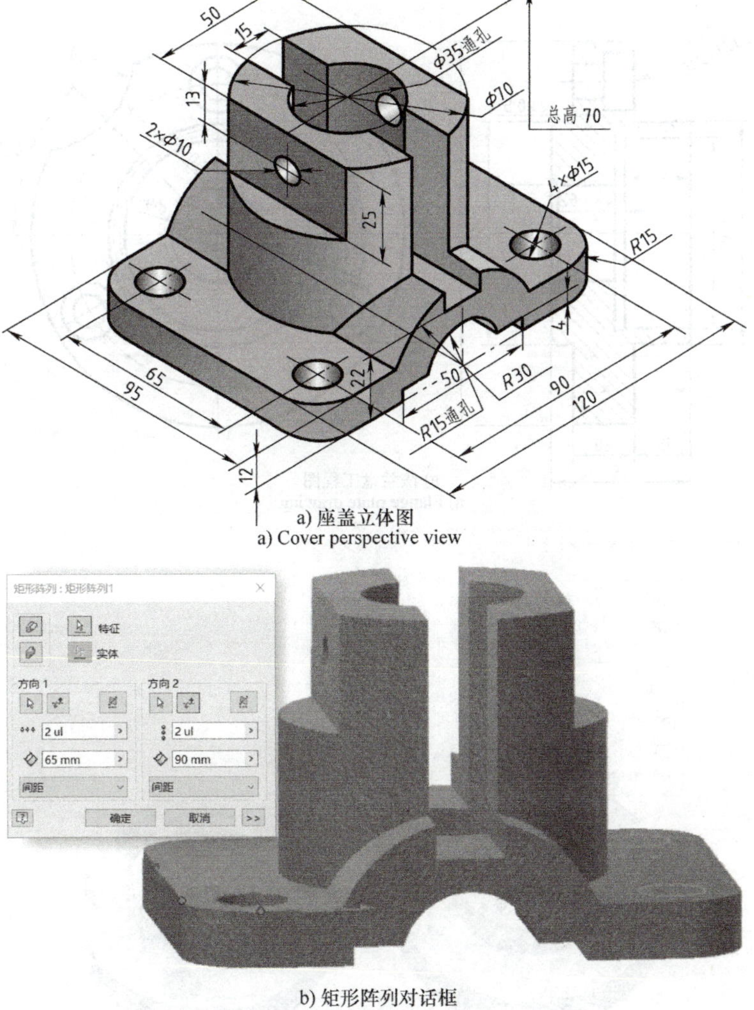

a) 座盖立体图
a) Cover perspective view

b) 矩形阵列对话框
b) Rectangular pattern dialog box

图 3-25　座盖绘制

Fig.3-25　Cover drawing

（8）环形阵列　按照环形规律，对已有特征进行复制。

绘制法兰盘：法兰盘工程图如图 3-26a 所示，生成基本回转形体后再打出一个沉头孔，调出“环形阵列”并按图 3-26b 所示进行设置和选择，最后形成法兰盘三维实体，如图 3-26c 所示。

（8）Circular pattern　duplicates one or more features or bodies and arranges the other occurrences by a specific count and spacing in an arc or circle.

Drawing the flange plate: the flange plate drawing is shown in Fig.3-26a. After the basic revolving shape is generated, a counterbore is made by “Hole” command and use “Circular pattern” to patten the other holes according to Fig.3-26b. Finally, the 3D view of the flange plate is shown in Fig.3-26c.

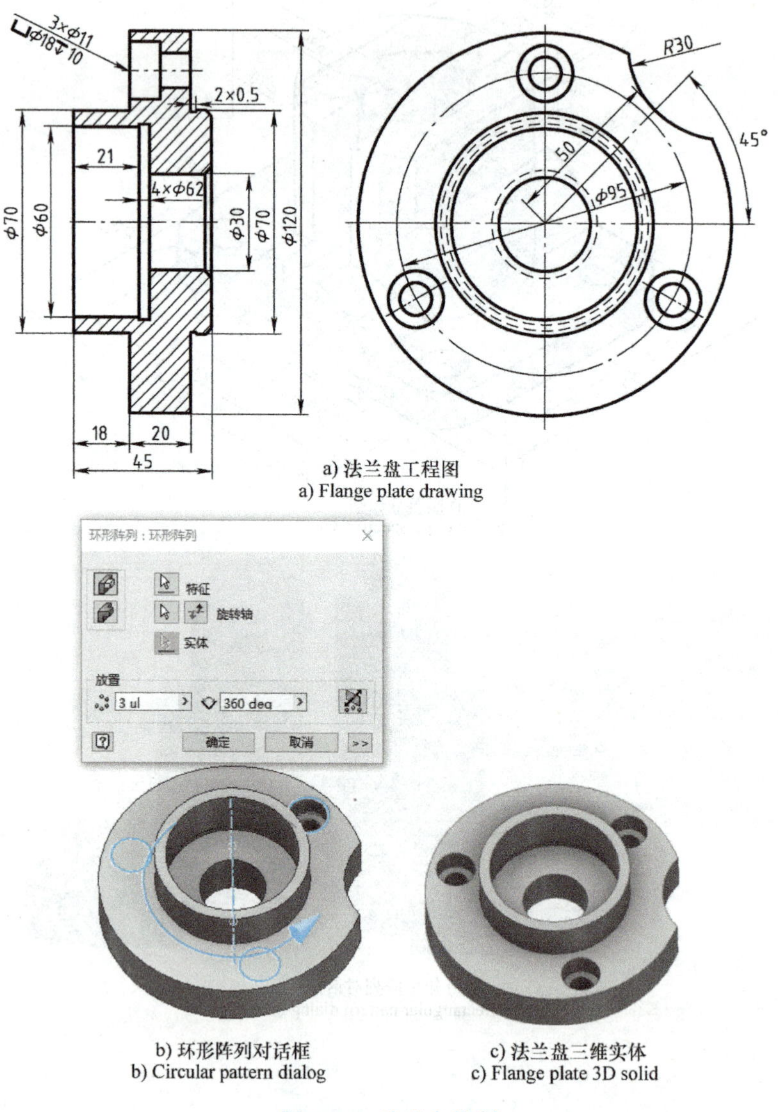

a) 法兰盘工程图
a) Flange plate drawing

b) 环形阵列对话框
b) Circular pattern dialog

c) 法兰盘三维实体
c) Flange plate 3D solid

图 3-26　法兰盘绘制

Fig. 3-26　Flange drawing

（9）镜像　将原有特征对称放置在镜像平面的另一侧，镜像对话框如图 3-27 所示。

(9) Mirror　the original features should be placed symmetrically on the other side of the mirror plane. A mirrored feature is a reverse copy of the selected feature on the other side of the mirror plane. The Mirror dialog box is shown in Fig.3-27.

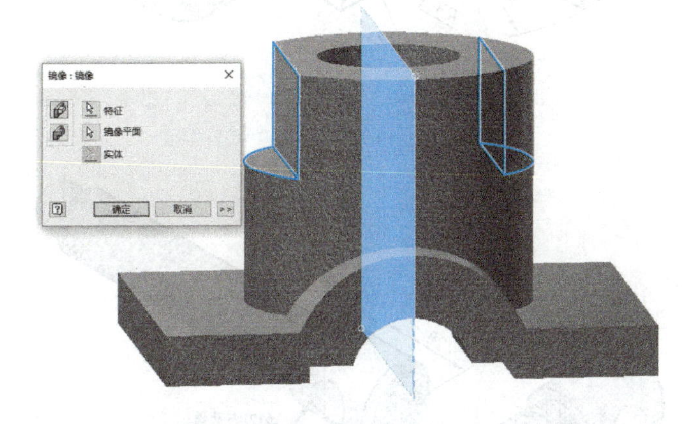

图 3-27　镜像对话框

Fig.3-27　Mirror dialog box

[本章习题]

[Chapter exercises]

根据压紧垫（图 3-28）、挡圈（图 3-29）、压紧螺杆（图 3-30）和横梁（图 3-31）的立体图构建物体的三维模型并生成其三视图。

Create a 3D model and its three-view according to the pictorial view of clamping pad (Fig.3-28), retaining ring (Fig.3-29), clamping worm(Fig.3-30) and crossbar (Fig.3-31).

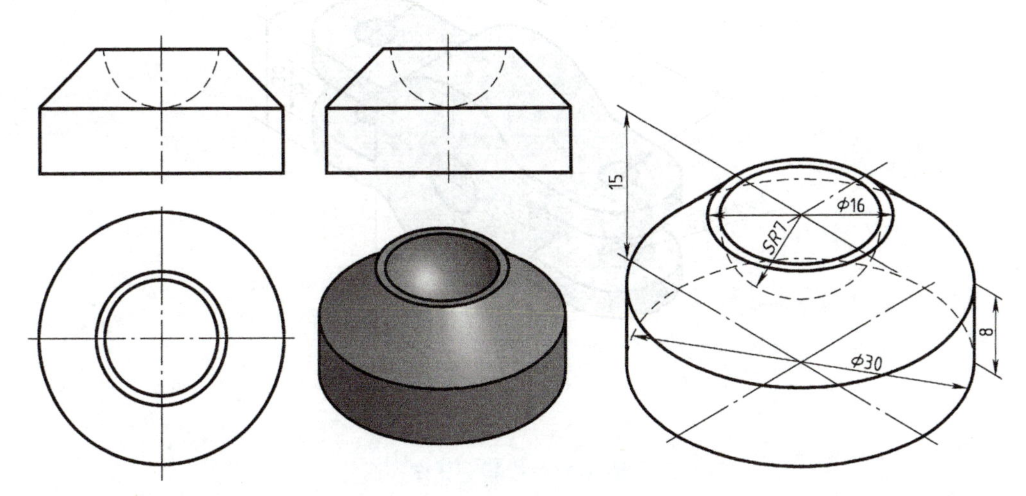

图 3-28　压紧垫三视图与立体图

Fig.3-28　Three view and pictorial view of clamping pad

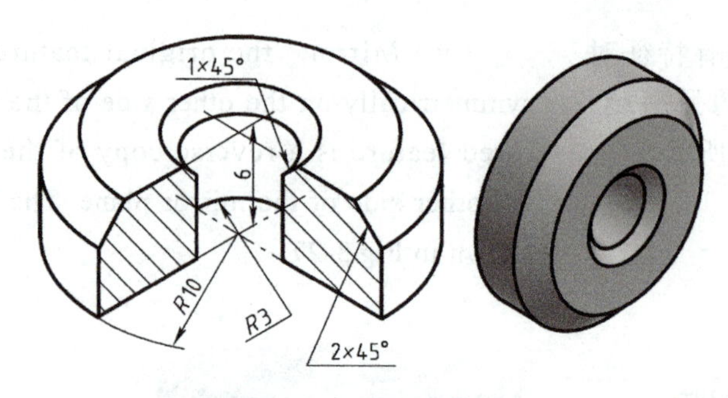

图 3-29　挡圈立体图

Fig.3-29　3D view of retaining ring

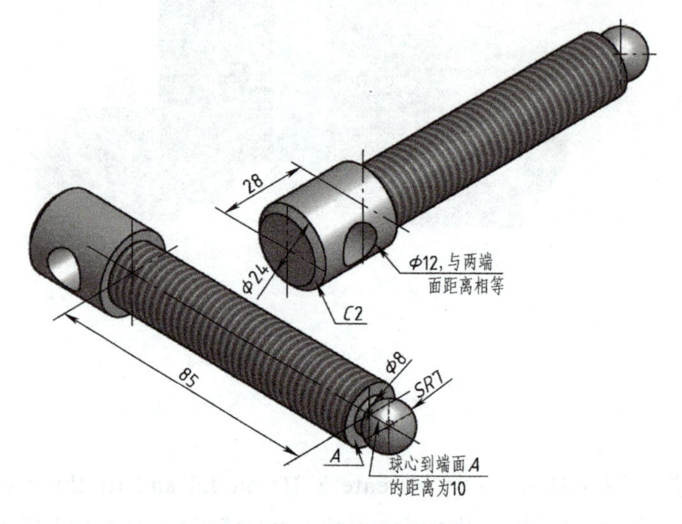

图 3-30　压紧螺杆立体图

Fig.3-30　3D view of the clamping worm

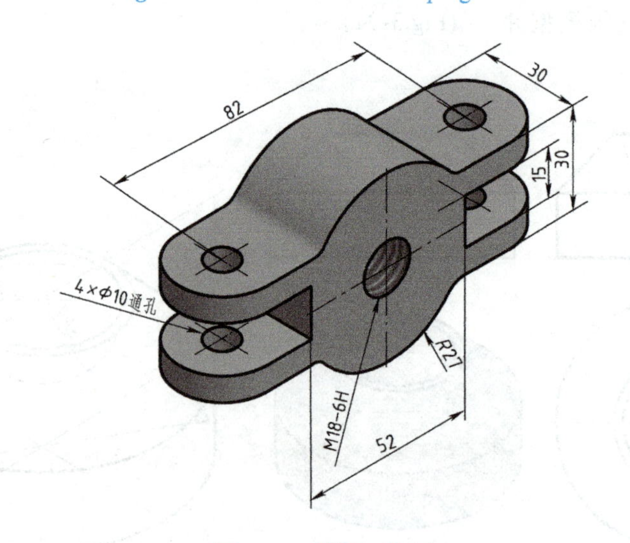

图 3-31　横梁立体图

Fig.3-31　3D view of the crossbar

第 4 章　机械制图标准

Chapter 4　Standards of Mechanical Drawing

4.1　机械制图国家标准的一般规定

4.1　General Provisions of National Standards of Mechanical Drawing

4.1.1　图纸幅面和格式（GB/T 14689—2008）

4.1.1　Drawing Sheet Size and Layout (GB/T 14689—2008)

1. 图纸幅面和尺寸

图纸幅面分为基本幅面和加长幅面，绘制图样时应优先选用表 4-1 中的基本幅面尺寸。

1. Size and layout of drawing sheet

Drawing sheets can be divided into standard sheet and extended sheet. Normally, standard sheet shown in the Table 4-1 should be chosen preferentially.

表 4-1　基本幅面尺寸（单位：mm）

Table 4-1　Standard sheet size (Unit: mm)

幅面代号 Sheet Code	A0	A1	A2	A3	A4
$B \times L$	841 × 1189	594 × 841	420 × 594	297 × 420	210 × 297
a	25				
c	10			5	
e	20		10		

注：B、L、a、c、e 尺寸含义见图 4-1。

Note: the meaning of dimension B, L, a, c, e are shown in Fig.4-1.

2. 图框格式

每张图纸在绘图前都必须先画出图框。图框有两种格式，一种是不留装订边的，另一种是留有装订边的。其周边尺寸 a、c、e 按表 4-1 中的规定选取。

2. Drawing frame format

Each drawing must be framed before drawing. There are two kinds of drawing frames, one has the binding edge, while the other dosen' t. The peripheral sizes of a, c and e are selected according to Table 4-1.

3. 标题栏（GB/T 10609.1—2008）

标题栏的位置一般应在图纸的右下角，标题栏的文字方向为看图方向，其格式按图 4-2 所示绘制，教学中采用的标题栏格式如图 4-3 所示。

3. Title block (GB/T 10609.1—2008)

Generally speaking, the title block should be in the lower right corner of the drawing. The reading direction for the text in the title block should be the same as the reading direction for the drawing. Fig.4-2 illustrates the format of a typical title block. For teaching and learning purposes, the format of title blocks used in this course should be simplified and it is recommended to use the format in Fig.4-3.

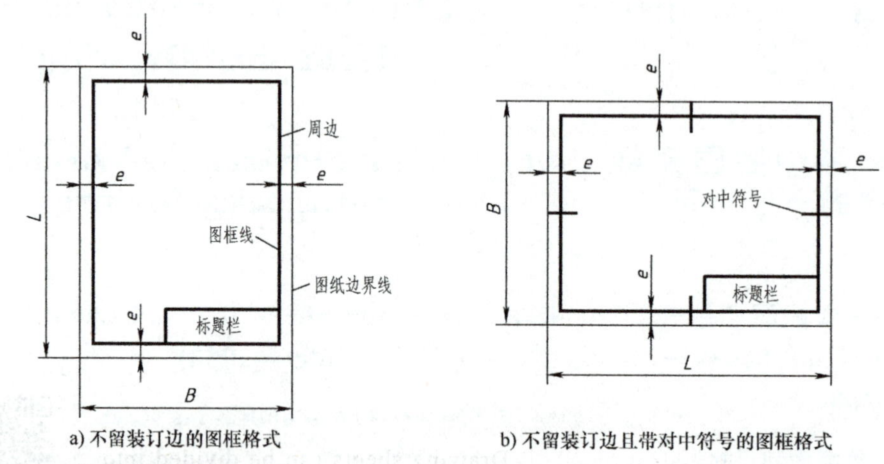

a) 不留装订边的图框格式
a) Drawing frame without binding edge

b) 不留装订边且带对中符号的图框格式
b) Drawing frame without binding edge but with centering marks

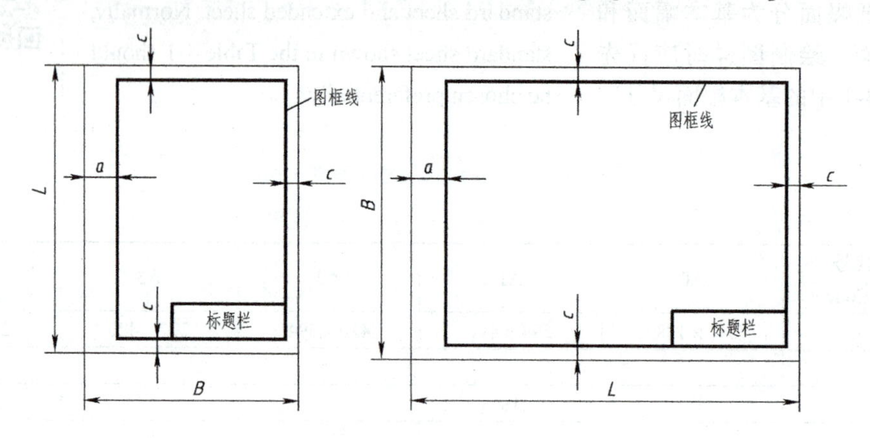

c) 留装订边的图框格式
c) Drawing frame with binding edge

图 4-1　图框格式
Fig.4-1　Drawing frame format

4.1.2　比例（GB/T 14690—1993）

图中图形与其实物相应要素的线性尺寸之比称为比例。绘图时应从表 4-2 中选取比例。比例一般注写在标题栏的“比例”栏内，必要时可在视图名称的下方或右侧标注，如 $\frac{I}{2:1}$、$\frac{B-B}{2.5:1}$ 等。

4.1.2　Scales (GB/T 14690—1993)

The scale of a drawing is the ratio between the size and dimensions of the drawn object to that of the actual object. The appropriate scale should be selected from Table 4-2. Scale should be noted in the “scale” block of the title block. If necessary, it can also be marked at the bottom or right side of the object, such as $\frac{I}{2:1}$ and $\frac{B-B}{2.5:1}$.

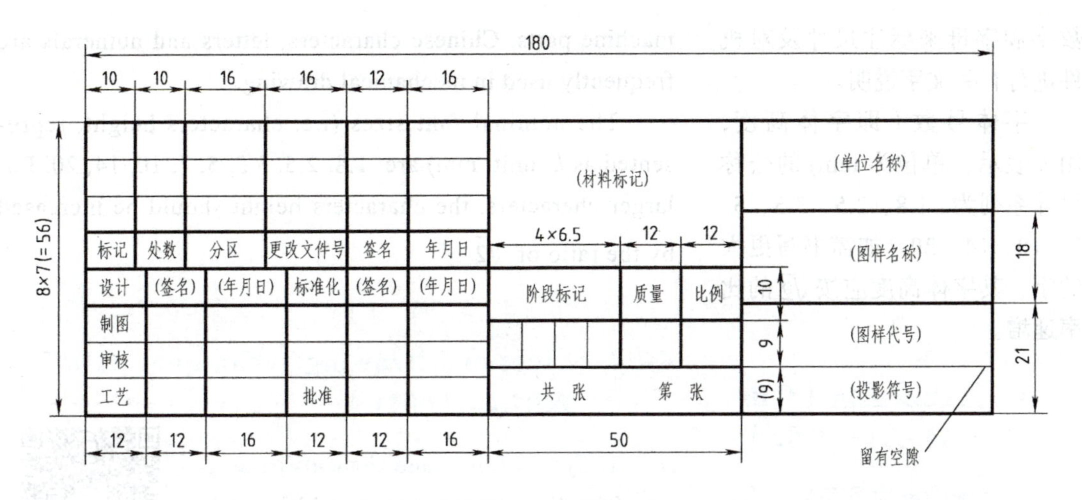

图 4-2　标题栏格式

Fig.4-2　Format of title block

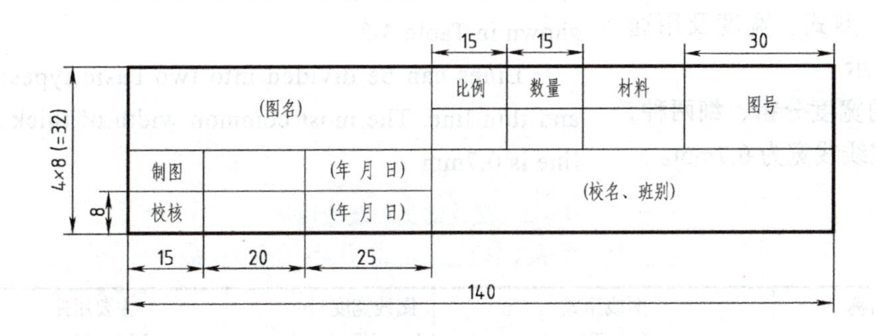

图 4-3　制图作业中推荐的标题栏格式

Fig.4-3　Recommended format of title block for exercise drawing

表 4-2　绘图比例

Table 4-2　Drawing scales

种类 Type	比例 Scales
原值比例 Full-Size	1:1
放大比例 Enlargement Scales	2:1　2.5:1　4:1　5:1 $2\times10^n:1$　$2.5\times10^n:1$　$4\times10^n:1$　$5\times10^n:1$　$1\times10^n:1$
缩小比例 Down Scales	1:1.5　1:2　1:2.5　1:3　1:4　1:5　1:6　1:10 $1:1.5\times10^n$　$1:2\times10^n$　$1:2.5\times10^n$　$1:3\times10^n$　$1:4\times10^n$　$1:5\times10^n$　$1:6\times10^n$　$1:1\times10^n$

注：1. n 为正整数。

2. 粗体字为优先选用的比例。

Notes: 1. n is a positive integer.

2. The preferred drawing scales are presented in bold font.

4.1.3　字体（GB/T 14691—1993）

4.1.3　Font(GB/T 14691—1993)

在图样中经常需要用汉字、

For the description of dimension and

数字和字母来标注尺寸及对机件进行有关文字说明。

字体号数（即字体高度，用 h 表示，单位为 mm）的公称尺寸系列为：1.8、2.5、3.5、5、7、10、14、20。如需书写更大的字，其字体高度应按 $\sqrt{2}$ 的比率递增。

4.1.4 图线及画法（GB/T 4457.4—2002）

1. 图线的型式及用途

国标中规定的机械图样的图线名称、型式、宽度及用途如表 4-3 所示。

图线的宽度分粗、细两种，常用的粗实线线宽为 0.7mm。

machine parts, Chinese characters, letters and numerals are frequently used in mechanical drawing.

The nominal font sizes (i.e. characters height, represented as h, unit: mm) are: 1.8, 2.5, 3.5, 5, 7, 10, 14, 20. For larger characters, the characters height should be increased by the ratio of $\sqrt{2}$.

4.1.4 Lines and Drawing Techniques (GB/T 4457.4—2002)

1. Types of lines and their utilization

The line names, types, widths and utilization of mechanical drawings are specified in the national standard are shown in Table 4-3.

Lines can be divided into two basic types: thick line and thin line. The most common width of thick continuous line is 0.7mm.

表 4-3 图线的类型及应用

Table 4-3 Types of Lines and Their Utilization

图线名称 Line Name	图线型式 Line Type	图线宽度 Line Width	主要用途 Major Usage
粗实线 Thick continuous line		d	可见轮廓线、移出断面轮廓线 Visible contour line, contour line of removed section.
细实线 Thin continuous line		$d/2$	尺寸线、尺寸界线、剖面线、辅助线、重合断面的轮廓线、引出线 Dimensions lines, extension line, section line, auxiliary line, outgoing line.
波浪线 Wave continuous line		$d/2$	断裂处的边界线、视图和剖视的分界线 Border line of the broken views, boundary line of the view and the section.
双折线 Continuous thin straight line with intermittent zigzags	(7.5d) 14d 30°	$d/2$	断裂处的边界线 Border line of the broken views.
细虚线 Thin dashed line	12d 3d	$d/2$	不可见的轮廓线 Invisible contour line.
细点画线 Thin dash-dotted line	6d 24d	$d/2$	轴线、对称中心线、齿轮的分度圆及分度线 Axis lines, symmetrical center line, pitch circle and graduation line.

（续）

图线名称 Line Name	图线型式 Line Type	图线宽度 Line Width	主要用途 Major Usage
粗点画线 Thick dash-dotted line		d	有特殊要求的线或表面的表示线 Indication line of lines and surfaces with special requirements.
双点画线 Thin double dot-dash line	9d 24d	$d/2$	相邻辅助零件的轮廓线、极限位置的轮廓线、假想投影轮廓线 Contour line of adjacent auxiliary parts, limiting position and imaginary projection.

2. 图线的应用

图线的应用如图 4-4 所示。

2. Application of lines

The application of lines is shown in Fig.4-4.

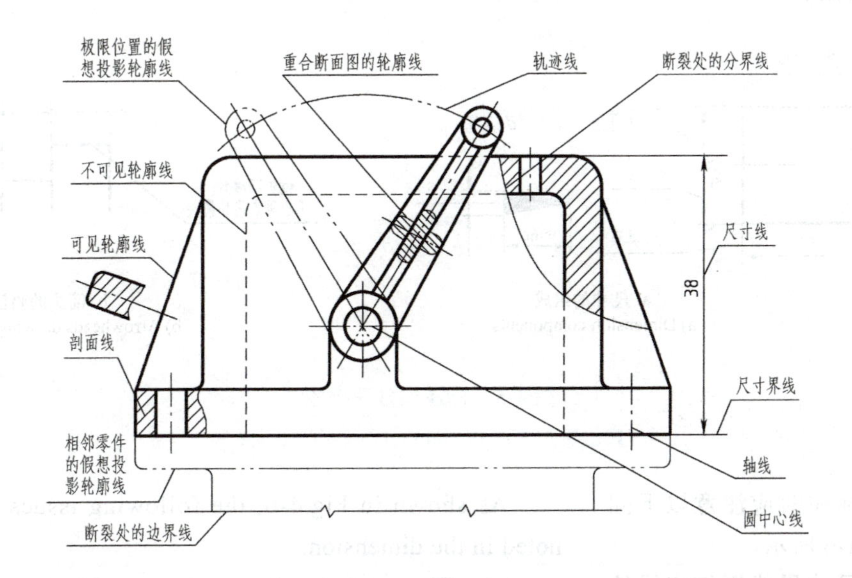

图 4-4　图线的应用

Fig.4-4　Application of lines

4.1.5　尺寸注法（GB/T 4458.4—2003）

机件结构形状的大小和相对位置需用尺寸表示。

1. 基本规则

（1）机件的真实大小应以图样中所标注的尺寸为依据，与图形的比例和绘图的准确度无关。

（2）图样中的尺寸，以毫

4.1.5　Dimensioning (GB/T 4458.4—2003)

The size and relative position of the structural shape of machine parts should be determined by dimensions.

1. Basic rules

（1）The actual size of machine parts should be based on the dimension marked in the drawing, which is irrelevant to the scale and accuracy of the drawing.

（2）When the drawing dimension unit is millimeter, the code and name of the measurement unit should not be

米为单位时，不需标注计量单位的代号或名称。

marked.

（3）机件的每一尺寸，一般只标注一次，并应标注在反映该结构最清晰的图形上。

（3）Each dimension of the machine part need to be marked generally only once and should be marked on the drawing that reflects the clearly structure.

2. 尺寸的组成

2. Dimension composition

每个完整的尺寸，一般包括尺寸界线、尺寸线和尺寸数字三个基本要素，尺寸线终端一般采用箭头的形式，位置不够时也可以用小黑点代替箭头，如图 4-5 所示。

Each complete dimension includes three basic elements, such as extension lines, dimension lines and numerals. The end of the dimension line is usually in the form of an arrowhead. It could be replaced by small black point if the position is not enough, as shown in Fig.4-5.

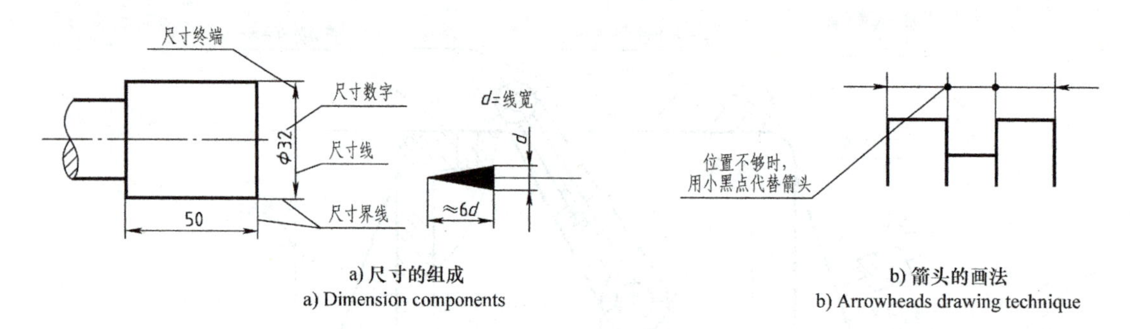

a) 尺寸的组成
a) Dimension components

b) 箭头的画法
b) Arrowheads drawing technique

图 4-5　标注尺寸三要素

Fig.4-5　Three elements of dimension

尺寸标注时应注意以下问题，如图 4-6 所示。

As shown in Fig.4-6, the following issues should be noted in the dimension.

（1）尺寸界线用细实线绘制，由图形的轮廓线、轴线或对称线引出，也可直接利用轮廓线、轴线或对称中心线等作为尺寸界线。尺寸界线应超出尺寸线约 2 ~ 3 mm。尺寸界线一般应与尺寸线垂直，必要时允许倾斜。

（1）The extension line is drawn with thin continuous line and derived from the contour line, axis or symmetric line of the drawing which could also act as the extension line directly. The extension line should exceed the dimension line with 2 to 3 millimeters. The extension line should be perpendicular to the dimension line usually and is allowed to tilt it if necessary.

（2）尺寸线用细实线绘制。标注线性尺寸时，尺寸线必须与所标注的线段平行。尺寸线一般不用其他图线代替，也不能与其他图线重合或在其延长线上，应尽量避免与其他尺寸线或尺寸界线相交。

（2）The dimension line should be drawn with a thin continuous line. When marking a linear dimension, the dimension line must be parallel to the marked line segment. In general, the dimension line could not be replaced by other line, or coincide with other line or on its extension line. Moreover, the dimension line should avoid intersection with other dimension line or extension line.

（3）尺寸数字不允许被任何图线通过，当无法避免时，必须将图线断开。

（3）The dimension numerals could not be allowed to overlap or be intersected by any graphic lines. If the intersection is unavoidable, the graphic line must be disconnected.

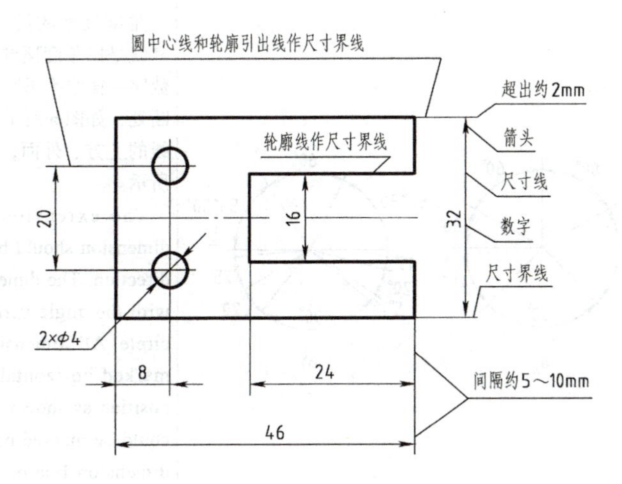

图 4-6　标注尺寸的注意事项

Fig.4-6　Notes for dimensioning

3. 常用尺寸的注法

常用尺寸的注法，如表 4-4 所示。

3. Commonly-used dimension examples

Table 4-4 provides some dimension examples that are frequently used.

表 4-4　尺寸注法

Table 4-4　Dimension Examples

尺寸标注的类型 Types of Dimensions	图例 Examples	说明 Description
线性尺寸 Linear dimension	30° 20 20 20 20 20 20 20 20 20 20 a) 16 16 16 b)	线性尺寸的尺寸数字注写如图 a 所示，水平尺寸数字字头朝上，垂直尺寸数字字头朝左，倾斜尺寸数字字头保持朝上的趋势，并尽量避免在图示 30° 范围内标注尺寸，当无法避免时可采用图 b 形式标注。 The marking of the dimension numeral of the linear dimension is shown in Fig.a. The numeral of the horizontal dimension is also horizontal. The numeral of the vertical dimension heads to the left. The numeral of the tilted dimension keeps an upward trend and should not be written within 30 ° as shown in the graph. When it is unavoidable, it should be written as Fig.b.

（续）

尺寸标注的类型 Types of Dimensions	图例 Examples	说明 Description
角度 Angle	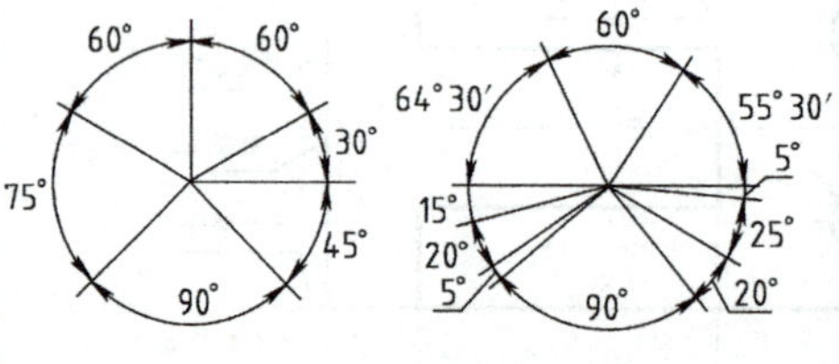	角度尺寸的尺寸界线沿径向引出，尺寸线是以角度顶点为圆心的圆弧线，尺寸数字一律水平书写，一般注写在尺寸线中断处，如图 a 所示。必要时可注写在尺寸线的上方、外面，也可引出标注，如图 b 所示。 The extension line of the angular dimension should be extended in the radial direction. The dimension line is an arc line with the angle vertex as the center of the circle. All dimension numerals should be marked horizontally in the disconnected position as shown in Fig.a. If necessary, it could be marked on the top, outside of the dimension line or outside of the graph as shown in Fig.b.
圆和圆弧 Circle and arc	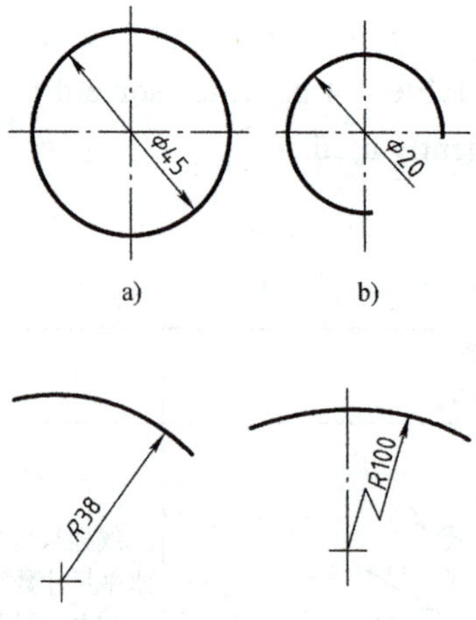	在直径、半径的尺寸数字前应分别加注符号“ϕ”“R”，如图 a、b、c 所示，大圆弧的标注如图 d 所示。 The symbol of “ϕ” and “R” should be written in front of the dimension numerals of the diameter and radius respectively as shown in Fig.a, b, c. The dimension method of the large arc is shown in Fig.d.
小尺寸 Babysize dimension	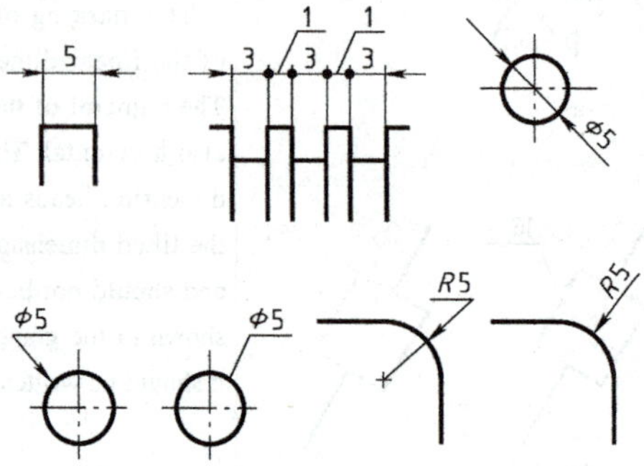	当没有足够的地方画箭头或注写尺寸数字时，可按如图的形式标注。 When there is not enough space to draw arrowheads or dimension numerals, the dimensioning method shown in the graph should be followed.

4. 尺寸标注中常用的符号

（1）尺寸数字前面的符号　尺寸标注中常用的符号含义如表 4-5 所示，常用符号的标注示例如图 4-7 所示。

4. Dimensioning symbols and abbreviations

（1）Symbols in front of dimension numerals. The meaning of commonly-used dimension symbols is shown in Table 4-5. The examples of commonly-used dimension symbols are shown in Fig.4-7.

表 4-5　尺寸标注中常用的符号

Table 4-5　Commonly-used dimensioning symbols

符号 Symbol	含义 Meaning	符号 Symbol	含义 Meaning
ϕ	直径 Diameter	⌵	埋头孔 Countersink
R	半径 Radius	⌴	沉孔或锪平 Counterbore/Spotface
S	球 Sphere	↧	深度 Depth
EQS	均布 Equivalents	□	正方形 Square
C	45° 倒角 Chamfer	∠	斜度 Obliquity
t	厚度 Thickness	▷	锥度 Taper
⌒	弧长 Arc length	○_	展开长 Expand length

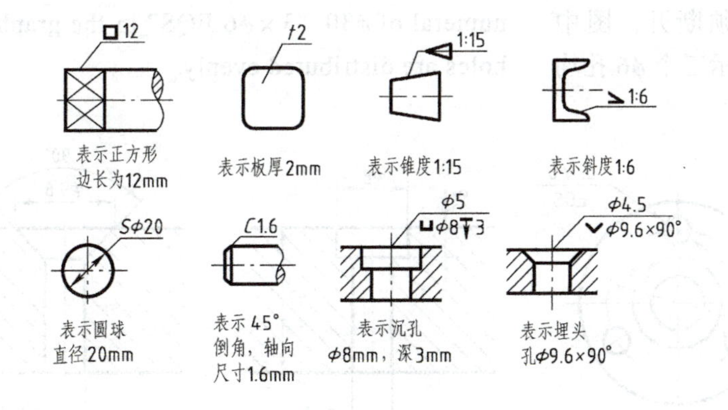

图 4-7　常用符号的标注示例

Fig.4-7　Examples of commonly-used dimensioning symbols

（2）对称机件的尺寸标注　对称机件的尺寸标注方法如图 4-8 所示。图 4-8a 中 80、92 两个尺寸线的一端无法注全时，它们的尺寸线要超过对称中心线一段，图中 4×ϕ6 表示有 4 个 ϕ6 的孔。图 4-8b 中分布在对称中心线两侧的相同结构，可仅标注其中一侧的结构尺寸。

（2）Dimensioning method of symmetrical machine parts. The dimensioning methods of symmetrical machine parts are shown in Fig.4-8. In Fig.4-8a, when one end of dimension line 80 and 92 cannot be fully dimensioned, they should surpass the symmetrical center line by a segment. The 4×ϕ6 in the graph means there are four ϕ6 holes. In Fig.4-8b, the two sides of the symmetrical center line have the same structure, so only one side needs to be marked.

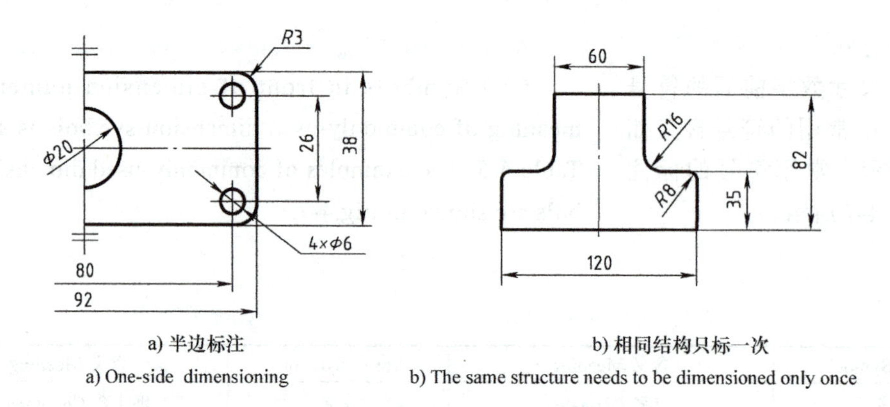

a) 半边标注

a) One-side dimensioning

b) 相同结构只标一次

b) The same structure needs to be dimensioned only once

图 4-8 对称机件的尺寸注方法

Fig.4-8 Dimensioning method of symmetrical machine parts

（3）图线通过尺寸数字要断开 尺寸数字不能被图样上的任何图线所通过，当不可避免时，必须将图线断开，如图 4-9 所示图线通过尺寸数字 ϕ30 时，图线必须断开，图中“3 × ϕ6 EQS”表示三个 ϕ6 孔均匀分布。

(3) Graphic lines must be disconnected, which have dimension numerals. The dimension numeral could not be allowed to be crossed by any lines in the graph. When it is unavoidable, the line must be disconnected. As shown in Fig.4-9, the line is disconnected as it passes the dimension numeral of ϕ30. “3 × ϕ6 EQS” in the graph means three ϕ6 holes are distributed evenly.

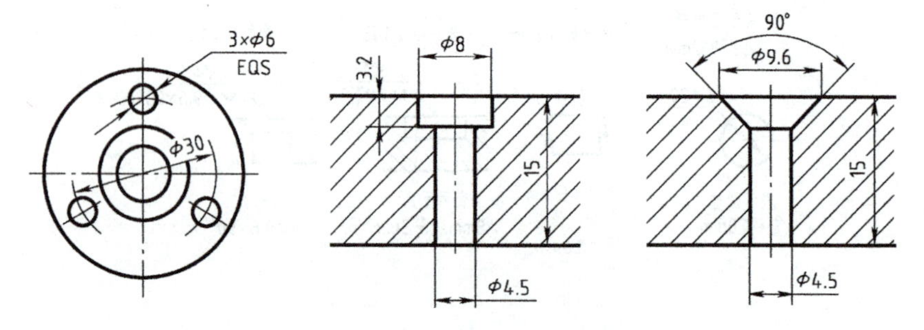

图 4-9 图线通过尺寸数字时要断开

Fig.4-9 Graphic lines should be disconnected near dimension numerals

4.2 机件表达方法

4.2 Representation of Machine Parts

在实际生产中，对机械图样的要求，一是要看图方便，二是在完整清晰地表达各部分形状的前提下力求制图简便。本节主要介绍国家标准《机械制图 图样画法视图》中“图样画法”

There are two requirements for mechanical drawings in the production process. One is for reading conveniently, another is for handy drawing while guaranteeing the complete and clear expression of the shapes of different parts. This section mainly introduces the view (GB/T 4458.1—2002), section view, cross section view (GB/T 4458.6—2002), local

规定的视图（GB/T 4458.1—2002）、剖视、断面（GB/T 4458.6—2002）、局部放大、简化画法（GB/T 16675.1—2012）和其他规定画法等。

enlarged view, simplified drawing method (GB/T 16675.1—2012) and other drawing methods specified in the "Drawing Methods" of the national standard, Mechanical Drawing, Drawing Method,View.

4.2.1　视图

视图主要是用于表达机件外部结构和形状的图形。

1. 基本视图和向视图

（1）基本概念　GB/T 4458.1—2002 规定：用正六面体的六个面作为基本投影面，把放在六个面中间的机件分别向这六个面投影，所得到的视图称为基本视图，如图 4-10 所示。在 inventor 中由基础视图和投影视图生成。

4.2.1　View

The view is mainly used to express the external structure and shape of the object.

1. Basic view and direction view

（1）Basic concepts. GB/T 4458.1—2002 stipulates that the six planes of a regular hexahedron are called basic projection planes. The machine part is placed in a regular hexahedron and its projection graphs on the six projection planes are called Basic views, as shown in Fig.4-10. It is generated with basic view and projection view of Inventor.

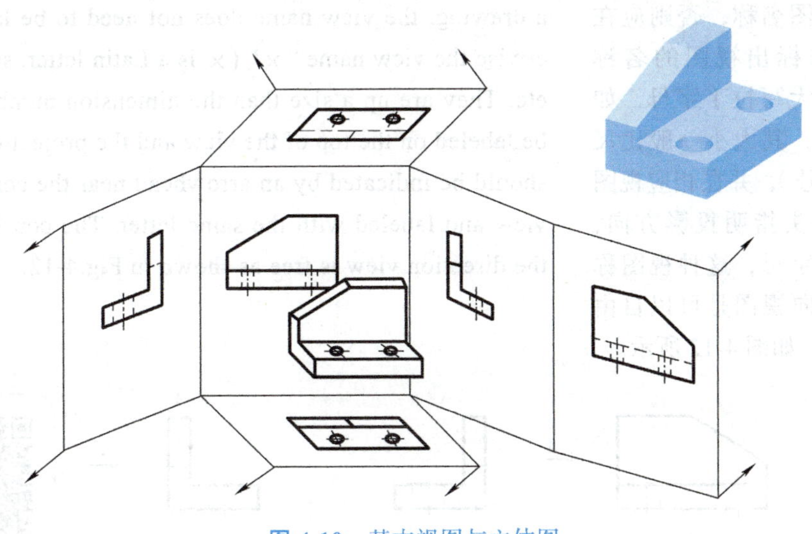

图 4-10　基本视图与立体图

Fig.4-10　The basic views and stereogram

基本视图是除了原来的主视图、俯视图和左视图外，还有从右向左、从下向上、从后向前投影得到的右视图、仰视图和后视图。为了把这六个基

Basic views not only include front view, top view and left view, but also include right view, upward view and rear view which are got from right to left, from bottom to top and from back to front respectively. In order to expand these six basic views onto a plane, it is required to expand as shown

本视图展开在一个平面上，规定按图 4-11 所示展开，从而得到六个基本视图的配置关系。

in Fig.4-11 to obtain the configuration relation of these six basic views.

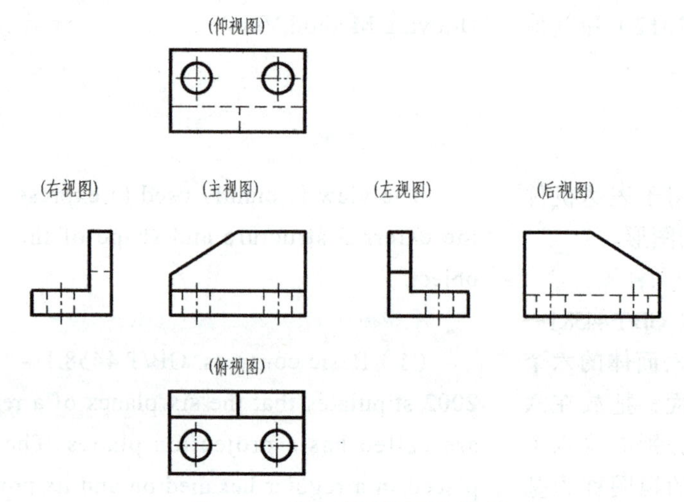

图 4-11 基本视图的配置关系

Fig.4-11 Configuration relationship of basic views

按图 4-11 配置视图时，一律不标注视图名称，否则应在视图的上方标出视图的名称“×”（× 为大写拉丁字母，如 *A*、*B*、……，其大小一般比尺寸数字大一号），并在相应视图的附近用箭头指明投影方向，注上相同的字母，这种视图称为向视图。向视图是可以自由配置的视图，如图 4-12 所示。

When the view is configurated according to Fig.4-11 in a drawing, the view name does not need to be labeled, otherwise the view name “×” (× is a Latin letter, such as *A*, *B*, etc. They are up a size than the dimension numbers) should be labeled on the top of the view and the projection direction should be indicated by an arrowhead near the corresponding view and labeled with the same letter. The configuration of the direction view is free as shown in Fig.4-12.

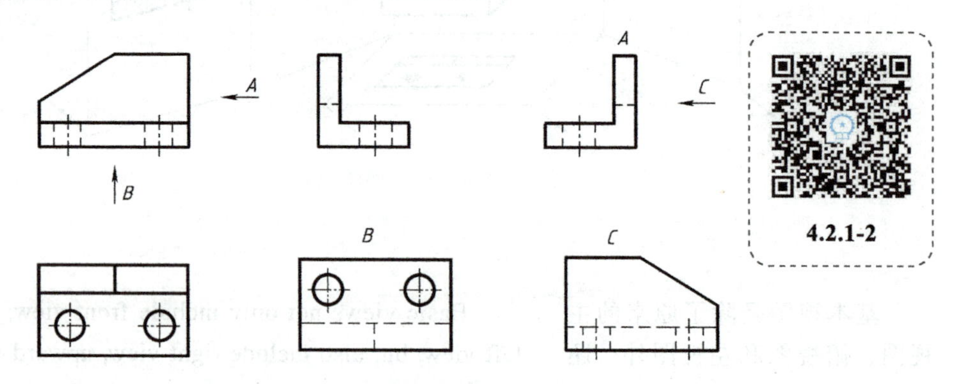

图 4-12 基本视图的其他配置方式

Fig.4-12 Other configuration of basic views

（2）Inventor 的基础视图和投影视图　Inventor 可创建的视图种类主要有基础视图、投影视图、斜视图、剖视图、局部视图、局部剖视图、断裂视图以及断面图等。工程图中第一个视图一般是自动投影零部件模型而生成的，也可以由部件的设计视图和表达视图创建工程视图。Inventor 的“放置视图”面板如图 4-13 所示。

（2）Basic views and projection views of Inventor. The views Inventor could create mainly include basic views, projection views, oblique views, section views, local views, local section views, broken views and cross-section views, etc. The first view of an engineering drawing is usually generated from the automatic projection of the part model, or from its design view and expression view. The panel of “Place View” of Inventor is shown in Fig.4-13.

图 4-13　Inventor 的“放置视图”面板

Fig.4-13　“Place View” panel in Inventor

1）进入工程图工作环境。单击“工具栏”中的“新建”按钮，双击“工程图”命令 Standard.idw 或 Standard.dwg，即可进入工程图环境，如图 4-14 所示。默认为 A2 图纸，可更改（在浏览器中“图纸”上单击右键→“编辑图纸”）。编辑图纸对话框如图 4-15 所示。

1）Enter the working environment of engineering drawing. Click the “new” button in the “toolbar” and double-click the “engineering drawing” command Standard.idw or Standard.dwg to enter the engineering drawing environment, as shown in Fig.4-14. The default A2 drawing width can be changed: click the right button on the “drawing” in the browser and select “edit drawing”. The dialog box of “edit drawing” is shown in Fig.4-15.

图 4-14　工程图环境

Fig.4-14　Engineering drawing environment

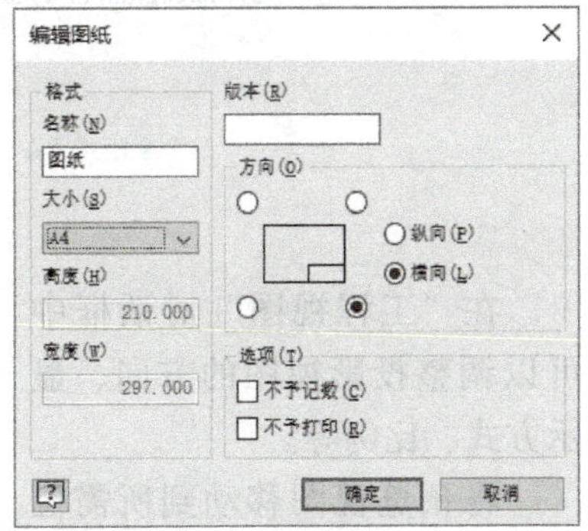

图 4-15　编辑图纸

Fig.4-15　Edit drawing

2）基础视图。在新的工程图中，首先需要独立创建一个基础视图，因为基础视图（如主视图）是生成其他从属视图（如俯视图、左视图等）的父视图。根据零部件表达的需要，在同一张工程图中可以添加多个基础视图。

在放置视图面板中，单击“基础视图”图标按钮，打开“工程视图”对话框，单击“文件”按钮，在打开的对话框中选择根据锤头立体图创建好的“锤头.ipt”文件，如图4-16所示。

2) Basic view. A basic view needs to be created in the new engineering drawing independently, as the basic view (such as the front view) is the parent view of other subordinate views (such as the top view, left view, etc.). Multiple basic views can be added to the same engineering drawing according to the expression need of the part.

Click the “basic view” icon in the “place view” panel to open the dialog box of “engineering view”, select the file of “hammer.ipt” created according to the stereogram of a hammer under “file”, as shown in Fig.4-16.

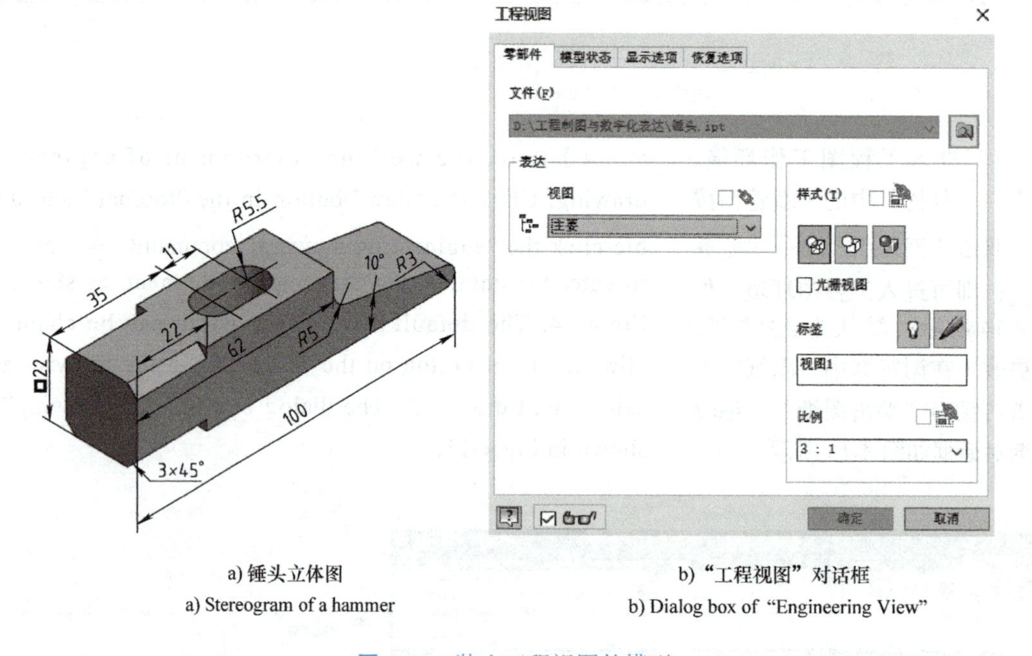

a) 锤头立体图

a) Stereogram of a hammer

b)“工程视图”对话框

b) Dialog box of “Engineering View”

图 4-16 装入工程视图的模型

Fig.4-16 Models included in the engineering view

在“工程视图”对话框中可以调整投影视图的方向、显示方式、比例等。

将预览视图移动到所需位置，单击对话框中的“确定”按钮，创建的基础视图如图4-17所示。

In the dialog box of “Engineering View”, the direction, display mode and scale of the projection view can be adjusted.

Move the preview view to the desired location, click the “OK” button in the dialog box and create the basic view as shown in Fig.4-17.

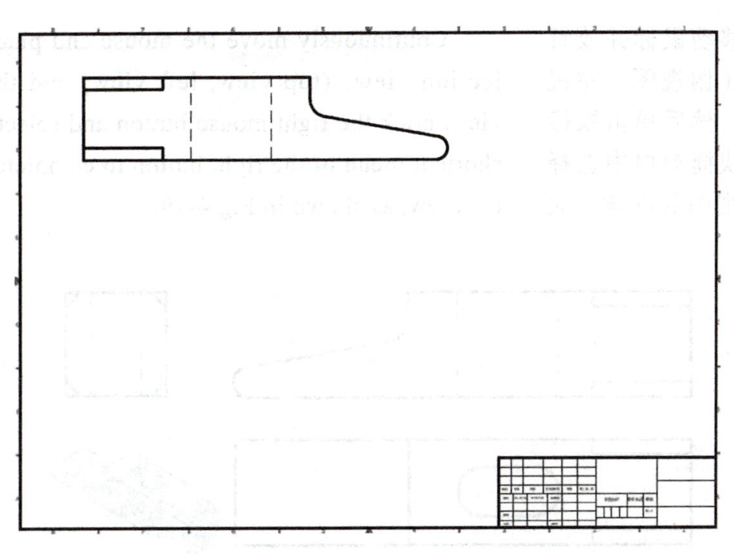

图 4-17　基础视图

Fig.4-17　Basic view

3）投影视图。用投影视图工具可以创建以基础视图为基础的其他基本视图或等轴测视图（简称轴测图，也可称为立体图）。投影视图的特点是默认与父视图对齐，并且继承父视图的比例和显示方式，若移动父视图，投影视图仍保持与它的正交对齐关系；若改变父视图的比例，正交投影视图的比例也随之改变。

在图 4-17 的基础上，单击放置视图面板中的“投影视图”图标，选取基础视图，移动鼠标进行投影，移动的同时可以预览投影视图，移动到适当位置单击放置视图，如图 4-18 所示。

3）Projection View. Using the projection view tool, other basic views or axonometric axis views (or stereogram) could also be created based on the basic view. As to the features of the projection view, it is alignment with the parent view by default and inherits the scale and display mode of the parent view. If the parent view is moved, the projection view remains its orthogonal alignment with it. If the scale of the parent view is changed, the scale of the orthogonal projection view is also changed.

On the basis of Fig.4-17, click the icon of “Projection View” in the “place view”, then select the “basic view”, and move the mouse for projection.While previewing the projection view, move it to the appropriate place then click the projection view to place it, as shown in Fig.4-18.

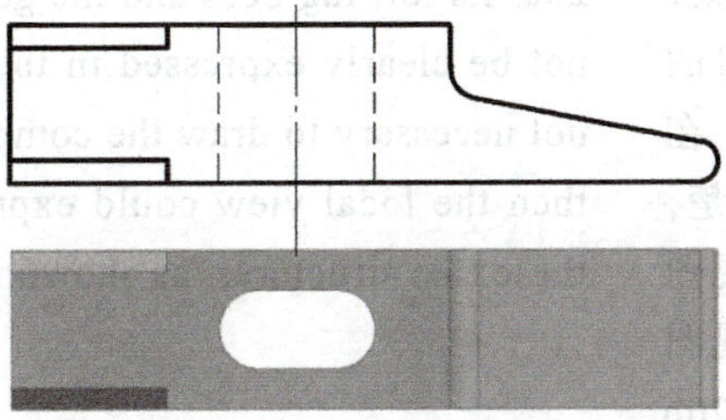

图 4-18　放置投影视图

Fig.4-18　Place projection view

可以连续移动鼠标并放置多个投影视图（俯视图、左视图）和轴测图，然后单击鼠标右键，在右键快捷菜单中选择“创建”，完成视图的创建，如图 4-19 所示。

Continuously move the mouse and place multiple projection views (top view, left view) and the axonometric view, click the right mouse button and select “create” in the shortcut menu of the right button to complete the creation of the view, as shown in Fig.4-19.

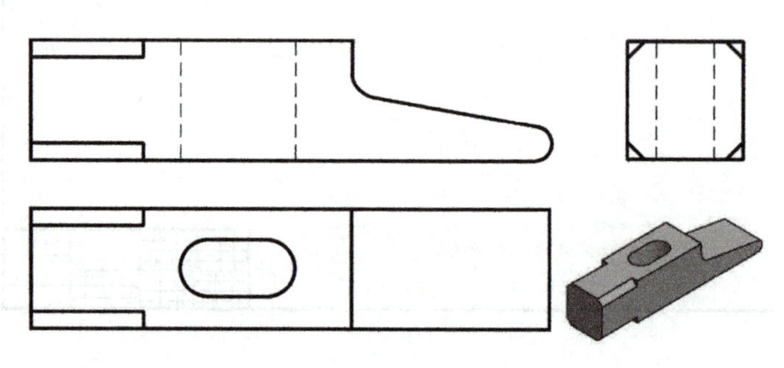

图 4-19 创建投影视图和轴测图

Fig.4-19 Create projection view and axonometric view

2. 局部视图

（1）基本概念 将机件的某一部分向基本投影面投射所得到的视图称为局部视图。

局部视图是一个不完整的基本视图，当机件上的某一局部形状没有表达清楚，而又没有必要用一个完整的基本视图表达时，可将这一部分单独向基本投影面投射，表达机件上局部结构的外形，避免因表达局部结构而重复画出别的视图上已经表达清楚的结构。利用局部视图可以减少基本视图的数量，如图 4-20a 所示支架，支架左侧凸台和右上角缺口的形状（以立体图左图为例），在主、俯视图上无法表达清楚，又没有必要画出完整的左视图和右视图，此时可用局部视图表示两处特征形状，如图 4-20b 所示。

2. Local View

（1）Basic concepts. The view obtained by projecting a part of the machine part onto the basic projection plane is called the local view.

The local view is an incomplete basic view. When a certain local structure of the machine part is not clearly expressed and it' s not necessary to express it with a complete basic view, this local structure could be projected separately onto the projection plane to express its shape, so as to avoid the drawing of the structure which has been clearly expressed in other views. The local view could reduce the number of basic views. Such as the holder shown in Fig.4-20a, Its left lug boss and the gap in the top right corner cannot be clearly expressed in the front and top view, and it's not necessary to draw the complete left view and right view, then the local view could express the feature and shape of these two structures, as shown in Fig.4-20b.

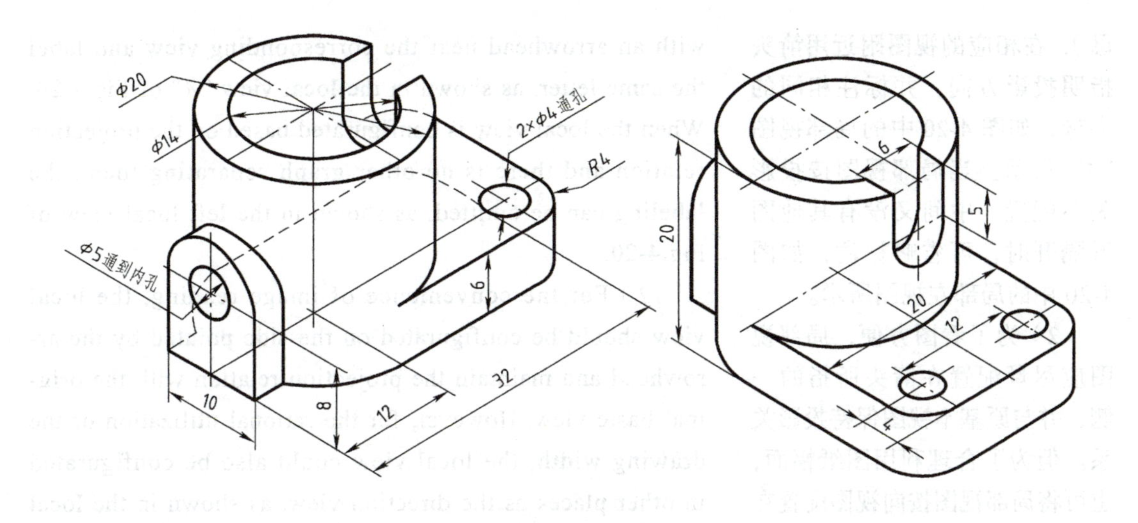

a) 支架立体图
a) Stereogram of the holder

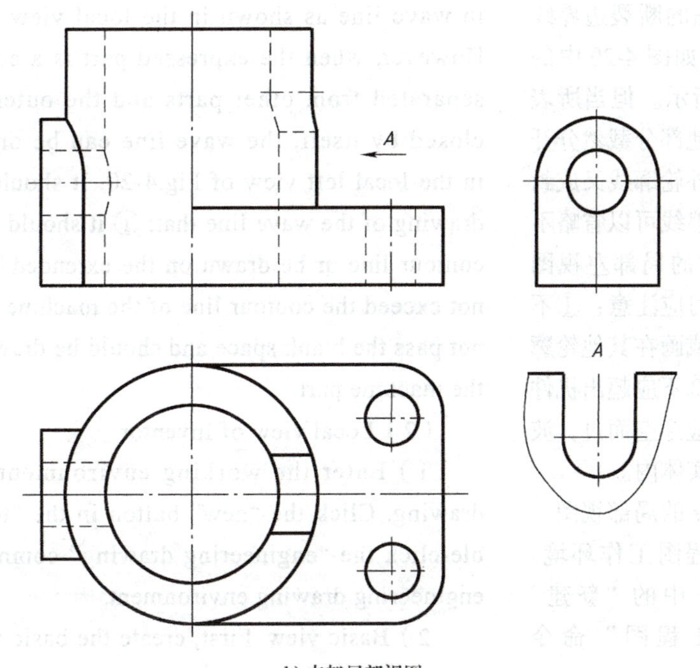

b) 支架局部视图
b) Local view of the holder

图 4-20　支架局部视图的配置与标注

Fig.4-20　Configuration and dimension of the local view of the holder

局部视图的配置与标注规定如下。

The configuration and dimension of the local view are specified as follows.

1）局部视图上方标出视图名称“×”（“×”为大写字

1）Write the view name “X” (“X” is a capital letter) on the top of the local view, indicate the projection direction

母），在相应的视图附近用箭头指明投影方向，并标注相同的字母，如图 4-20 中的局部视图“*A*”所示。当局部视图按投影关系配置，中间又没有其他图形隔开时，可省略标注，如图 4-20 中的局部左视图所示。

2）为了看图方便，局部视图应尽量配置在箭头所指的一侧，并与原基本视图保持投影关系。但为了合理利用图纸幅面，也可将局部视图按向视图配置在其他适当的位置，如图 4-20 中的局部视图“*A*”所示。

3）局部视图的断裂边界线用波浪线表示，如图 4-20 中的局部视图“*A*”所示。但当所表达的部分是与其他部分截然分开的完整结构，且外轮廓线又成封闭的情况时，波浪线可以省略不画，如图 4-20 中的局部左视图所示。画波浪线时应注意：①不应与轮廓线重合或画在其他轮廓线的延长线上；②不应超出机件的轮廓线；③不应穿空而过，波浪线应画在机件实体内。

（2）Inventor 的局部视图

1）进入工程图工作环境。单击“工具栏”中的“新建”按钮，双击“工程图”命令 Standard.idw，即可进入工程图环境。

2）基础视图。先创建基础视图，这里是主视图，再根据基础视图创建投影视图（俯视图、左视图和右视图），左视图和右视图要求隐藏虚线，如图 4-21 所示。

3）局部视图。将图 4-21 的左、右视图隐藏必要的图线，如图 4-22 所示。

with an arrowhead near the corresponding view and label the same letter, as shown in the local view “*A*” of Fig.4-20. When the local view is configurated based on the projection relation and there is no other graph separating them, the labeling can be omitted, as shown in the left local view of Fig.4-20.

2）For the convenience of image reading, the local view should be configurated on the side pointed by the arrowhead and maintain the projection relation with the original basic view. However, for the rational utilization of the drawing width, the local view could also be configurated in other places as the direction view, as shown in the local view “*A*” of Fig.4-20.

3）The section boundary of the local view is expressed in wave line as shown in the local view “*A*” of Fig.4-20. However, when the expressed part is a complete structure separated from other parts and the outer contour line is closed by itself, the wave line can be omitted, as shown in the local left view of Fig.4-20. It should be noted in the drawing of the wave line that: ① it should coincide with the contour line or be drawn on the extended line; ② it should not exceed the contour line of the machine part; ③ it should not pass the blank space and should be drawn in the entity of the machine part.

（2）Local view of Inventor

1）Enter the working environment of engineering drawing. Click the “new” button in the “toolbar” and double-click the “engineering drawing” command to enter the engineering drawing environment.

2）Basic view. First, create the basic view (front view) and create the projection view based on the basic view (top view, left view and right view). The dotted lines of the left and right views should be hidden as shown in Fig.4-21.

3）Local View. Hide the necessary lines in the left and right views of Fig.4-21, as shown in Fig.4-22.

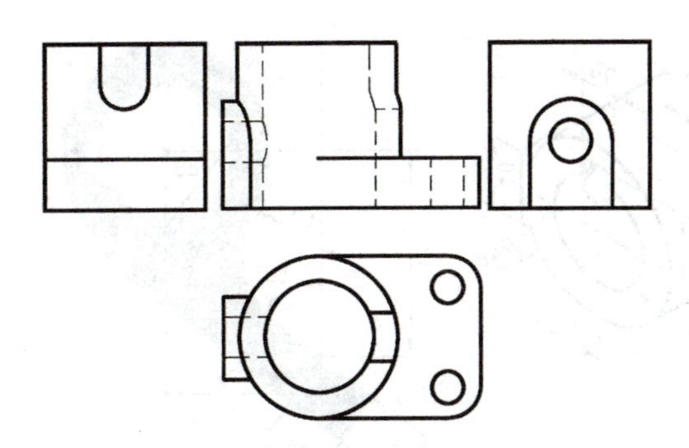

图 4-21　基础视图和投影视图

Fig.4-21　Basic view and projection view

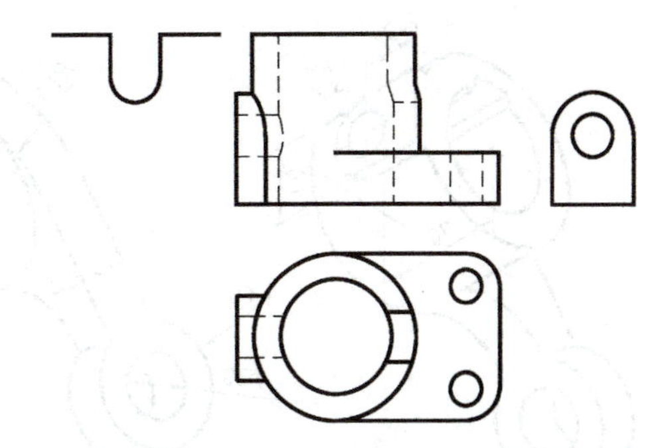

图 4-22　隐藏图线的基础视图和投影视图

Fig.4-22　Basic view and projection view with the necessary lines hidden

在图 4-22 的基础上添加必要的中心线，并断开右视图的对齐关系，将右视图拖放到图 4-20 的位置，修改默认的标注成如图 4-20 所示，最后画上波浪线即可。

Add the necessary center line on the basis of Fig.4-22, disconnect the alignment of right view, drag the right view to the position of Fig.4-20, modify the default labeling as shown in Fig.4-20, and finally draw the wave line.

3. 斜视图

（1）基本概念　机件向不平行于基本投影面的平面投射所得的视图，称为斜视图。

如图 4-23 所示的压紧杆，三视图如图 4-24 所示，其斜面圆孔在俯视图和左视图中投影出椭圆，给识图带来困难，也不便于标注尺寸。

当机件上某部分的倾斜结构不平行于任何基本投影面时，在基本视图中不能反映该部分的实际形状。这时，可增设一个新的辅助投影面，使其与机件的倾斜部分平行，且垂直于某一个基本投影面，如图 4-25 中的平面 *P*。然后将机件上的倾斜部分向新的辅助投影面投射，再将新投影面旋转到与其垂直的基本投影面重合的位置，即可得到反映该部分实际形状的视图。

3. Oblique View

（1）Basic concepts.　the view generated by the machine part projecting onto a plan that is not parallel to the basic projection plane is called the oblique view.

4.2.1-4

The three views of the compressor rod shown in Fig.4-23 is shown in Fig.4-24. The round hole in the slope would have ellipses in the top view and left view which brings difficulty in image reading and dimension.

When the oblique structure of the machine part is not parallel to any basic projection planes, the basic views cannot reflect the real shape of this structure. A new auxiliary projection plane should be added which would be parallel to the oblique structure of the machine part and perpendicular to a basic projection plane as the plane-*P* in Fig.4-25. The oblique structure would be projected to this auxiliary projection plane which would be rotated to the coincident place of the perpendicular basic projection plane, thus receiving the view which could reflect the real shape of this structure.

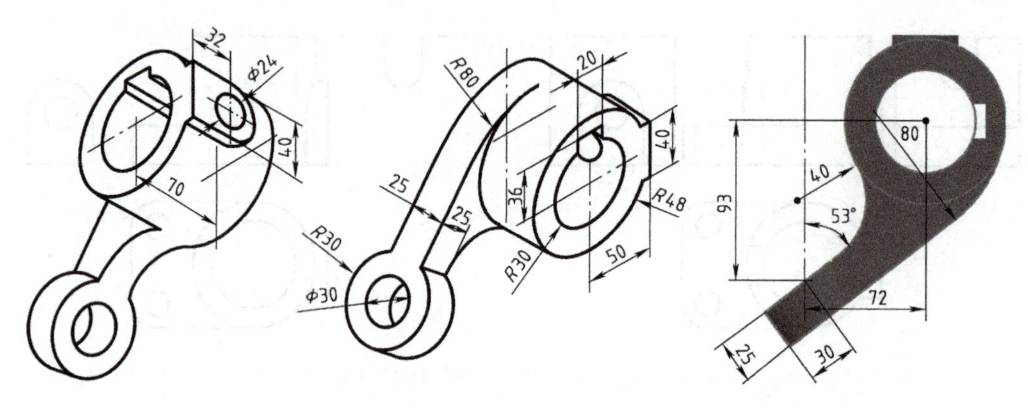

图 4-23 压紧杆立体图及尺寸

Fig.4-23 Stereogram and size of the compressor rod

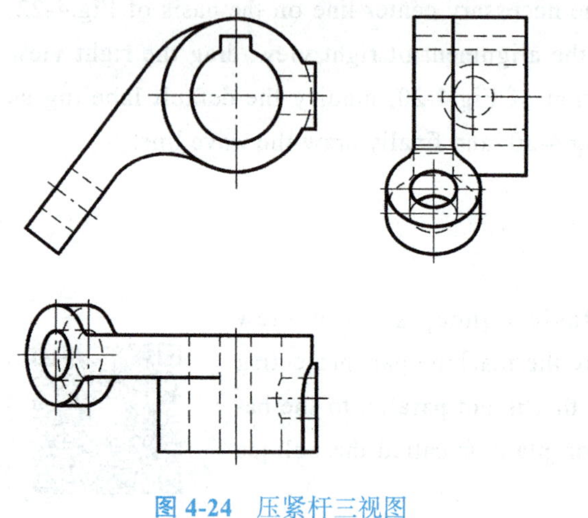

图 4-24 压紧杆三视图

Fig.4-24 Three views of the compressor rod

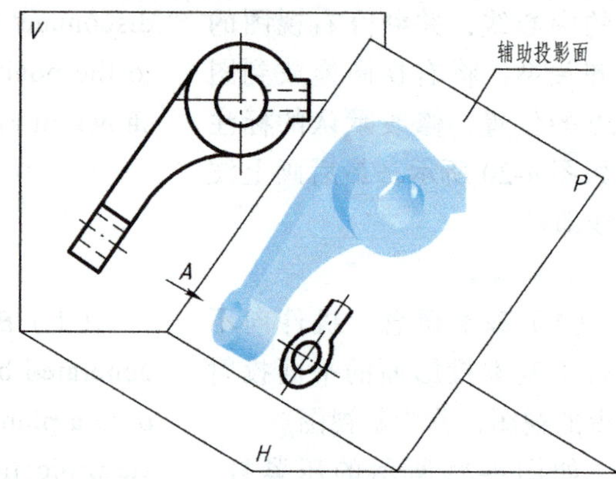

图 4-25 压紧杆斜视图的形成

Fig.4-25 The generation of the oblique view of the compressor rod

斜视图的配置与标注规定如下。

1）斜视图必须用带字母的箭头指明表达部位的投影方向，并在斜视图上方用相同的字母标注“×”（“×”为大写字母），如图4-26a所示的“*A*”。

2）斜视图一般配置在箭头所指方向的一侧，且按投影关系配置，如图 4-26a 中的斜视图“*A*”。有时为了合理地利用图纸幅面，也可将斜视图按向视图

The configuration and dimension of the oblique view are specified as follows.

1）The oblique view must use an arrowhead with a letter to indicate the projection direction and write “×” (“×” is a capital letter) using the same letter on the top of the oblique view, just like “*A*” in the Fig.4-26a.

2）The oblique view is usually configurated on the side pointed by the arrowhead following the projection relation as the oblique view “*A*” shown in Fig.4-26a. Sometimes, for the rational utilization of the drawing width, the oblique view could also be configurated in other places as the direction view.

配置在其他适当的位置。

3）斜视图一般只表达倾斜部分的局部形状，其余部分不必全部画出，可用波浪线断开，如图 4-26 所示的局部斜视图“*A*”。

4）斜视图允许旋转配置，如图 4-26b 所示，但计算机绘图时一般不做旋转，如图 4-26a 所示。

3）As the oblique view usually just describe the partial shape of the oblique structure, there is no need to draw all the other structures. The section could be expressed using the wave line, as shown in the local oblique view “*A*” in Fig.4-26.

4）If necessary, the oblique view could be rotated as shown in Fig.4-26b. However, the computer-aided drawing does not rotate generally as shown in Fig.4-26a.

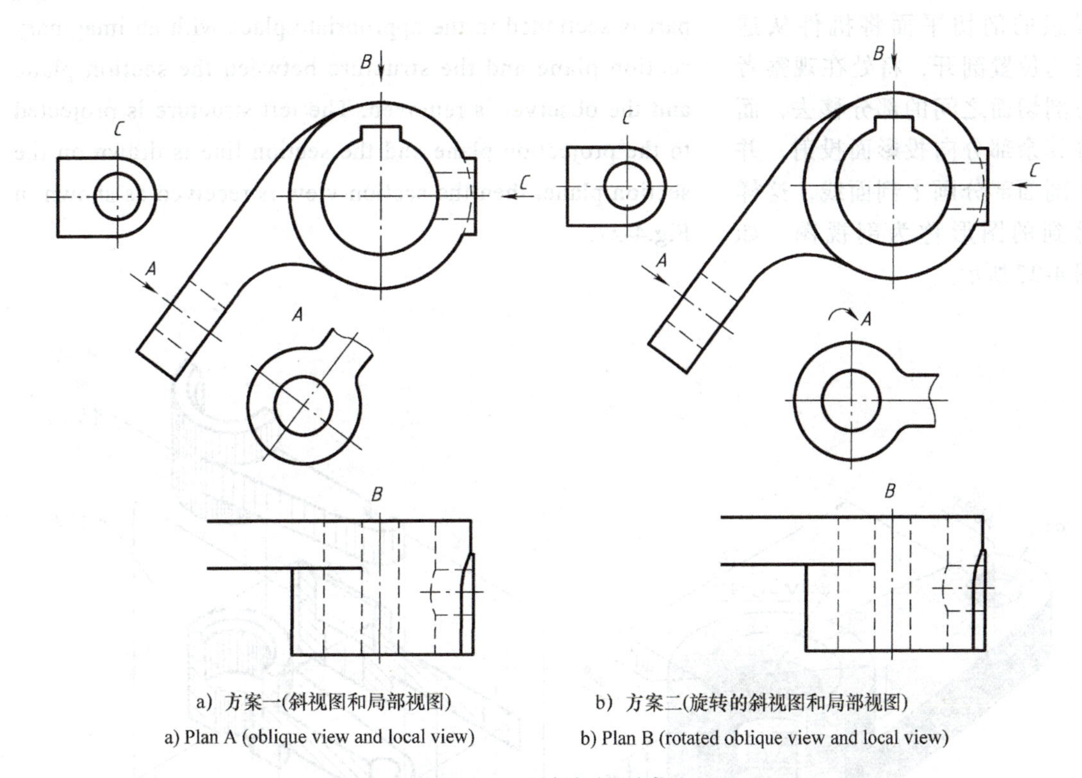

a）方案一(斜视图和局部视图)
a) Plan A (oblique view and local view)

b）方案二(旋转的斜视图和局部视图)
b) Plan B (rotated oblique view and local view)

图 4-26　压紧杆视图表达

Fig.4-26　Views of the compressor rod

（2）Inventor 的斜视图　斜视图的 Inventor 绘制方式和局部视图相同，完成的局部斜视图（方案一）如图 4-26a 所示。

（2）Oblique view of Inventor. The drawing method of the oblique view of Inventor is the same with that of the local view. The completed local oblique view (Plan A) is shown in Fig.4-26a.

4.2.2-1

4.2.2　剖视图

4.2.2　Section View

在用视图表达机件时，其内部结构都用虚线来表示，内

In the views of a machine part, its internal structure is expressed with dashed lines. With the

部结构形状越复杂，视图中的虚线就越多，这样会影响图面清晰，不便于看图和标注尺寸。为此，国家标准规定用“剖视”的方法来解决机件内部结构的表达问题。

more complicated internal structures, the more dashed lines are used in the views. It would affect the clarity of the drawing and make it harder to read and dimension. Therefore, the national standard specifies that the internal structure of the machine part should be expressed using the method of “section view”.

1. 剖视图的基本概念和作图方法

1. Basic concepts and drawing methods of the section view

（1）剖视图的基本概念　用假想的剖切平面将机件从适当的位置剖开，将处在观察者与剖切面之间的部分移去，而将其余部分向投影面投射，并在剖面部分画上剖面线，这样得到的图形称为剖视图，如图 4-27 所示。

(1) Basic concept of the section view.　The machine part is sectioned in the appropriate place with an imaginary section plane and the structure between the section plane and the observer is removed. The left structure is projected to the projection plane and the section line is drawn on the section plane, then the section view is received as shown in Fig.4-27.

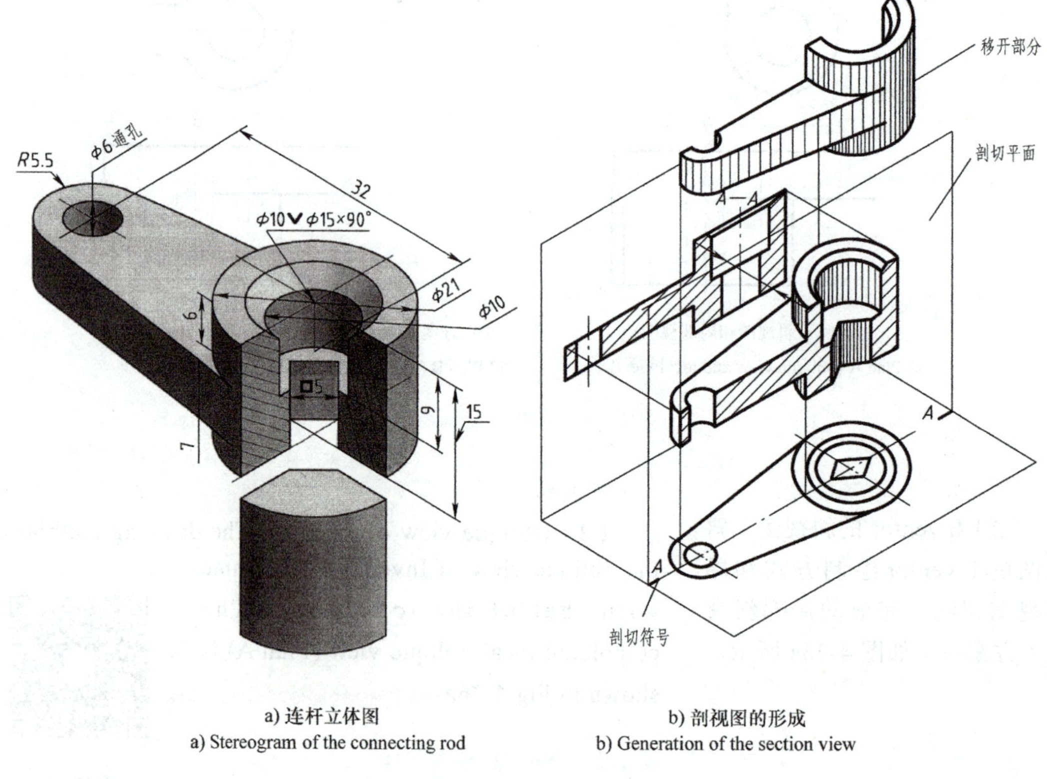

a) 连杆立体图
a) Stereogram of the connecting rod

b) 剖视图的形成
b) Generation of the section view

图 4-27　剖视图的形成

Fig.4-27　Generation of the section view

（2）剖视图的画法

1）确定剖切面的位置。剖切平面选择，一是要清楚地反映机件的内部形状，二是要便于看图。因此，剖切平面一般应通过机件的对称平面或轴线。剖切平面应平行于投影面，以反映剖面的实形，如图 4-28 所示。

2）画出留下部分的视图。剖切平面选定后，按选定投影方向，将相应留下的部分画出投影图，如图 4-28b 所示。

原来不可见的线条变为可见的线条，即虚线变为实线，并去掉剖切掉了的实线。

3）在剖面上画出剖面符号。在剖视图中，被剖切面剖切到的部分，称为剖面。为了在剖视图上区分剖面和其他表面，应在剖面上画出剖面符号（也称剖面线）。机件的材料不相同，采用的剖面符号也不相同，画图时应采用国家标准所规定的剖面符号，常见材料的剖面符号见表 4-6。

（2）Drawing methods of the section view

1）Determine the position of the section plane. The selection of the section plane should clearly reflect the internal structure of the machine part and be easy reading. Therefore, the section plane usually passes the symmetrical plane and axis. The section plane should be parallel to the projection plane to reflect the real shape of the section plane as shown in Fig.4-28.

2）Draw the left part of the view. After the section plane is selected, draw the projection drawing of the left structure following the selected projection direction, as shown in Fig.4-28b.

At this point, the original invisible lines become visible lines, that is, the dashed lines become continuous lines and the sectioned continuous lines are removed.

3）Draw the section plane symbol on the section plane. In the section view, where is sectioned by the section cutting plane is called the section plane. In order to distinguish the section planes from other planes in the section view, the section plane symbol should be drawn on the section plane (also known as the section line). The materials of machine parts are different, therefore the section plane symbols of them are also different. The section plane symbols specified in the national standard should be adopted in the drawing. The section plane symbols of regular materials are shown in Table 4-6.

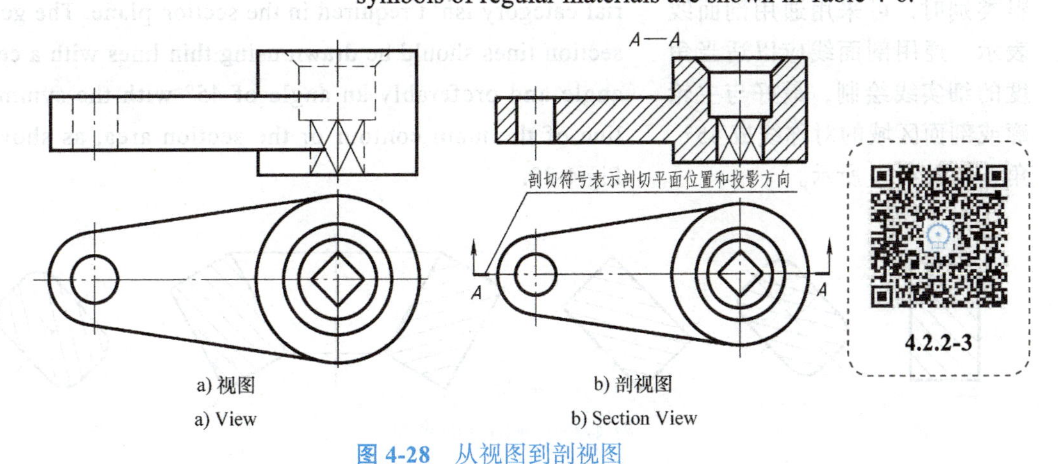

a) 视图
a) View

b) 剖视图
b) Section View

图 4-28　从视图到剖视图

Fig.4-28　From view to the section view

表 4-6 剖面符号（GB/T 4457.5—2013）

Table 4-6 Section plane symbols (GB/T 4457.5—2013)

材料 Material	符号	材料 Material	符号
金属材料 Metallic materials （已有规定剖面符号者除外）（Those with specified symbols not included）		木质胶合板 Wooden plywood （不分层数）（Different layers）	
非金属材料 Non-metallic materials （已有规定剖面符号者除外）（Those with specified symbols not included）		基础周围的泥土 Soil around the foundation	
转子、电枢、变压器和电抗器等的叠钢片 Steel laminates of rotors, armatures, transformers and electric reactors, etc.		混凝土 Concrete	
线圈绕组元件 Coil winding element		钢筋混凝土 Reinforced concrete	
型砂、填砂、粉末冶金、砂轮、陶瓷刀片、硬质合金、刀片等 Molding sand, sand filling, powder metallurgy, grinding wheel, ceramic blade, cemented carbide, blades, etc.		砖 Brick	
玻璃及供观察用的其他透明材料 Glass and other transparent materials for observation		格网 Grids 筛网、过滤网等 Screen, strainer, etc.	
木材 Wood — 纵剖面 Longitudinal section		液体 Liquid	
木材 Wood — 横剖面 Transverse section			

不需在剖面区域中表示材料类别时，可采用通用剖面线表示。通用剖面线应以适当角度的细实线绘制，最好与主轮廓或剖面区域的对称线成 45°角，如图 4-29 所示。

The general section line can be adopted when the material category isn' t required in the section plane. The general section lines should be drawn using thin lines with a certain angle and preferably an angle of 45° with the symmetric line of the main contour or the section area, as shown in Fig.4-29.

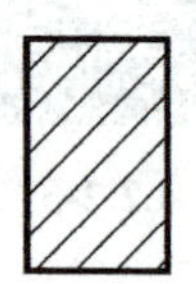
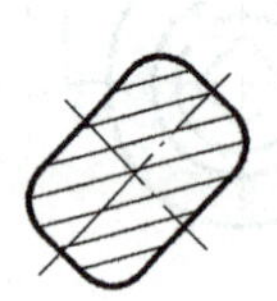
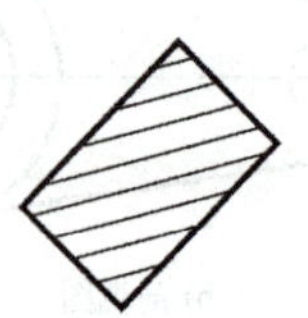
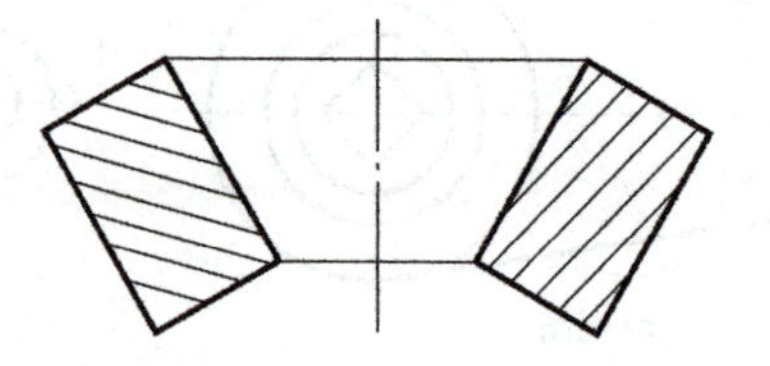
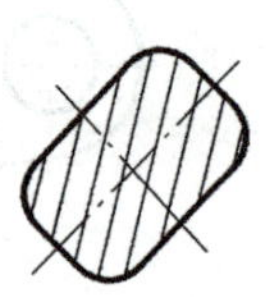

图 4-29 通用剖面线的画法

Fig.4-29 Drawing methods of general section lines

读图时，根据画剖面线部分是机件实体，未画剖面线部分是机件空心部分或剖面之后的部分，就容易想象出机件内部形状和远近层次。

When reading the drawing, it's easy to imagine the internal shape and different levels of the machine part according to that the structure with section lines is real part and the structure without section lines is the hollow Part or that behind the section plane.

（3）剖视图的标注　剖视图一般应进行标注，以指明剖切位置，指示视图间的投影关系。标注的内容如下。

（3）Labeling of the section view.　The labeling of the section view is to clearly indicate the section position and the projection relations among different views. The labeling contents are as follows.

1）剖切符号。指示剖切面起、讫和转折位置（用粗短画表示）及投影方向（用箭头表示）的符号，如图 4-28 所示。注有字母“*A*”的两段粗实线及两端箭头，即为剖切符号。

1）Section symbol. The symbols indicating the starting point, ending point, turning position (using short and thick lines) and projection direction as shown in Fig.4-28. The section symbol is two thick heavy lines with both arrowheads marked with “*A*”.

2）剖视图的名称。在剖切符号处应用相同的大写字母标出，并在相应的剖视图上方标注相同的字母“X—X”，如图 4-28 中的“*A*—*A*”，以便对照看图。一个机件同时有几个剖视图时，名称应用不同字母按顺序书写，不得重复。

2）Name of the section view. For comparative reading, the same capital letter “X—X” should be written in the place of the section symbol and on top of the corresponding section view, as shown in Fig.4-28. If a machine part has several section views, the names should be written using different letters without repeat.

若遇下列情况，剖视图的标注可省略或简化。

In the following cases, the labels of the section view could be omitted or simplified.

① 当单一剖切平面通过机件的对称平面或基本对称平面，且剖视图按投影关系配置，中间又没有其他图形隔开时，可省略标注，如图 4-30、图 4-31 中的主视图。

① When the single section plane passes the symmetrical plane or the basic symmetrical plane, and the section view is configurated based on the projection relation and there is no other graph separating them, the labeling can be omitted, as shown in the front views of Fig.4-30 and Fig.4-31.

② 当剖视图按投影关系配置，中间又没有其他图形隔开时，可省略箭头，如图 4-31 中的左视图。

② When the section view is configurated based on the projection relation and there is no other graph separating them, the arrowhead can be omitted, as shown in the left view of Fig.4-31.

（4）画金属材料的剖面符号时，应遵守下列规定

（4）The drawing of section symbols of metallic materials must comply with the following requirements

1）同一机件的零件图中，剖视图、断面图的剖面符号，

1）In the parts drawing of the same machine part, the section plane symbols of the section view and cross-section view must be drawn as the thin line with the same gap, direction and an angle of 45° with the horizontal direction (could tilt to the left or the right), as shown in Fig.4-30.

应画成间隔相等、方向相同且为与水平方向成45°（向左、向右倾斜均可）的细实线，如图4-30所示。

2）当图形的主要轮廓线与水平线成45°时，该图形的剖面线应画成与水平成30°或60°的平行线，其倾斜方向仍与其他图形的剖面线一致，如图4-29所示。

（5）关于剖视图上虚线问题

1）对已表达清楚的结构，其虚线应省略不画。

2）对于一些定位的虚线，若没有其他视图表达，则不能省略，如图4-30所示，画出虚线表示支座底板的厚度，则可以省去左视图。

原则：在剖视图上尽可能不画虚线，应根据机件的各种表达方式，将不可见虚线变为可见实线。

2）When the angle between the major contour line and the horizontal line is 45°, the section line of the graph should have an angle of 30° or 60° with the horizontal line and the inclined direction should be consistent with the section line of another graph, as shown in Fig.4-29.

（5）Issues about dashed lines in the section view

1）The dashed lines of the structure that has been clearly expressed should be omitted.

2）For some positioning dashed lines, if there is no other view expression, they cannot be omitted. As shown in Fig.4-30, the dashed lines indicate the thickness of pedestal chassis, then the left view can be omitted.

Principle: dashed lines can't be omitted in the section view as far as possible. Based on the expression form of the machine part, the invisible dashed lines could become visible continuous lines.

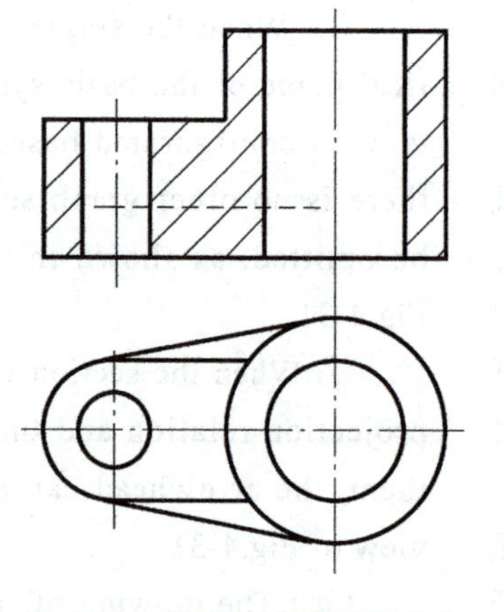

图4-30　必要的虚线不能省略

Fig.4-30　The necessary dashed lines should not be omitted

2. 剖视图的种类

剖视图按剖切机件的范围分为全剖视图、半剖视图和局部剖视图三大类，现分述如下。

2. Types of the Section View

According to the scope of the sectioned machine part, the section view is divided into three categories: full section view, half section view and local section view, which are described as follows.

（1）全剖视图

1）基本概念。用剖切平面完全地剖开机件所得到的剖视图称作全剖视图（简称全剖视），如图 4-28、图 4-30~ 图 4-32 所示。

全剖视图主要用于外形简单、内形复杂的不对称机件。有些外形简单的对称机件，为了将内形显示完整，便于标注尺寸，也常采用全剖视图，如图 4-30、图 4-33 所示。全剖视图采用剖视图的标注方法。

2）Inventor 的全剖视图。以轴架为例，其立体图如图 4-33a 所示，完成其三维模型的创建后，在工程图环境下生成轴架的基础视图（俯视图），然后在工程图视图面板单击，选择基础视图，在现有视图上绘制剖切线，如图 4-33b 所示，在右键菜单中单击“继续”，然后用鼠标单击放置剖视图，在标注面板下标注中心线，完成的全剖视图如图 4-33c 所示。

（1）Full section view

1）Basic concepts. The section view which is completely sectioned by the full section plane, as shown in Fig.s 4-28, 4-30, 4-31, 4-32.

4.2.2-5

The full section view is mainly used for asymmetrical machine parts with simple external shape and complicated inner structure. Some symmetrical machine parts also use the full section view to display the complete inner structure and make it easier for dimensioning, as shown in Fig.4-30 and Fig.4-33. The full section view adopts the dimensioning method of the section view.

2）The full section view of Inventor. Taking the shaft bracket as an example, its stereogram is shown in Fig.4-33a. After the 3D model is created, generate the basic view (top view) of the shaft bracket in the engineering drawing environment, click in the engineering drawing view panel, select the basic view, draw the section line in the present view as shown in Fig.4-33b, click “continue” in the pop-up menu, click and place the section view using the mouse and label the center line in the labeling panel. The complete full section view is shown in Fig.4-33c.

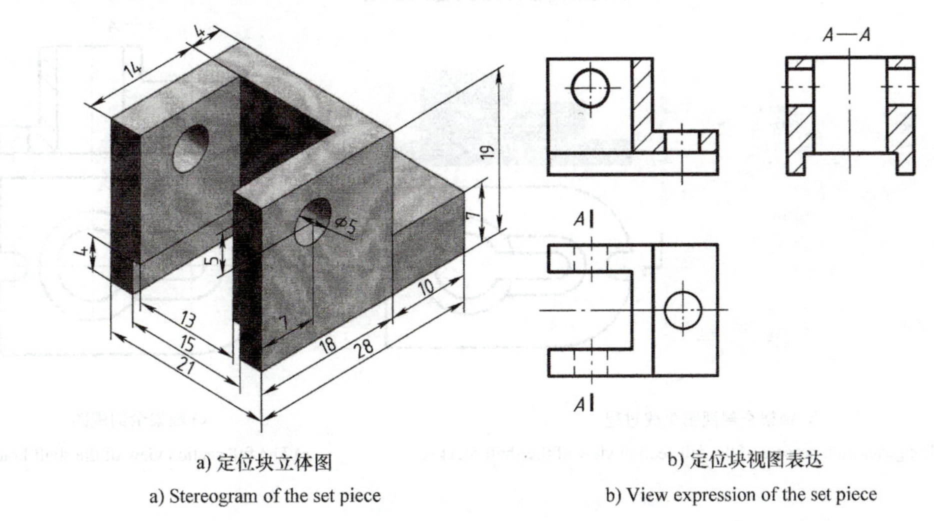

a) 定位块立体图

a) Stereogram of the set piece

b) 定位块视图表达

b) View expression of the set piece

图 4-31　全剖视图（一）

Fig.4-31　Full section view (1)

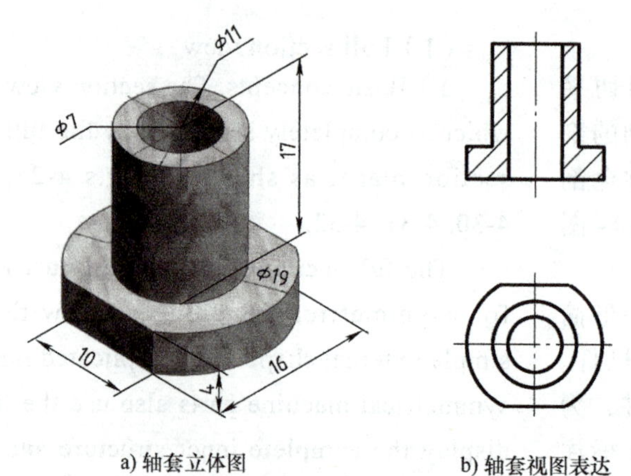

a) 轴套立体图
a) Stereogram of the shaft sleeve

b) 轴套视图表达
b) View expression of the shaft sleeve

图 4-32　全剖视图（二）

Fig.4-32　Full section view (2)

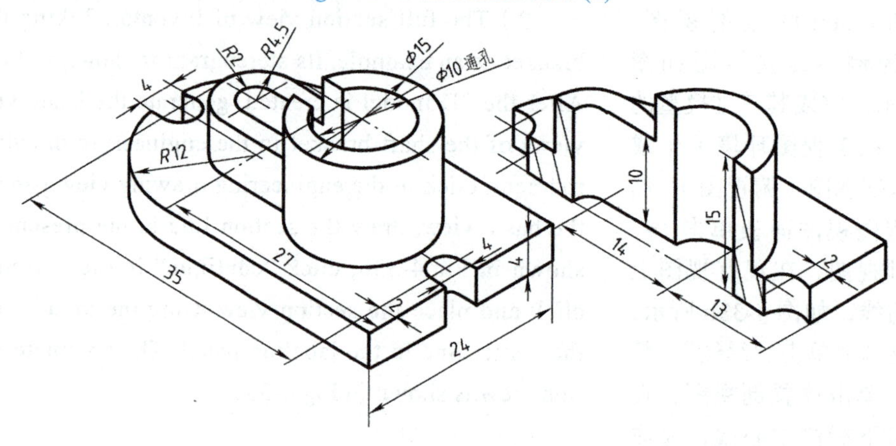

a) 轴架立体图
a) Stereogram of the shaft bracket

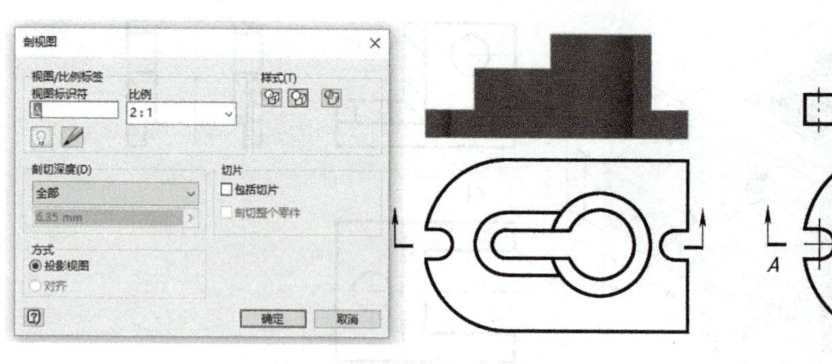

b) 轴架全剖视图生成过程
b) The generation process of the full section view of the shaft bracket

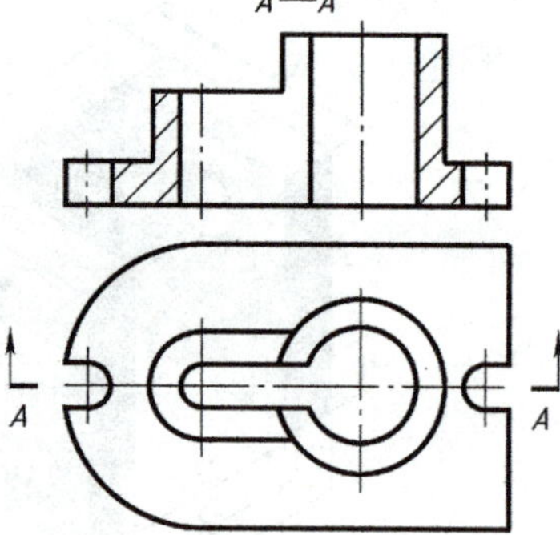

c) 轴架全剖视图
c) The full section view of the shaft bracket

图 4-33　轴架立体图与剖视图

Fig.4-33　Stereogram and section view of the shaft bracket

（2）半剖视图

1）基本概念。如图 4-34a 所示阀体 02，当机件具有对称平面时，在垂直于对称平面的投影面上投影所得到的图形，以中心线为界，一半画成剖视，另一半画成视图，这样得到的剖视图称为半剖视图（简称半剖视），如图 4-34d 所示，其主视图和俯视图的剖切情况如图 4-34b、c 所示。

半剖视图主要用于内外形状都需要表达的对称机件。对于形状接近于对称，且不对称部分已另有图形表达清楚的机件，也可以画成半剖视，如图 4-35 所示阀盖。

4.2.2-6

（2）Half section view

1）Basic concepts. Take a look at the valve body shown in Fig.4-34a, when the machine part has a symmetrical plane, the graph received by projecting on the projection plane which is perpendicular to the symmetrical plane may be divided into two halves with the center line as the boundary. One half could be drawn as the section view, the other half could be drawn as the view. This kind of section view is called the half section view, as shown in Fig.4-34d. Its section in the front view and top view is shown in Fig.4-34b and c.

The half section view is mainly used for symmetrical machine parts whose inner and outer shape should both be described. For the machine part whose shape is nearly symmetrical and the asymmetrical structure has been clearly expressed by other graphs, it could also be drawn as the half section view, such as the valve cover shown in Fig.4-35.

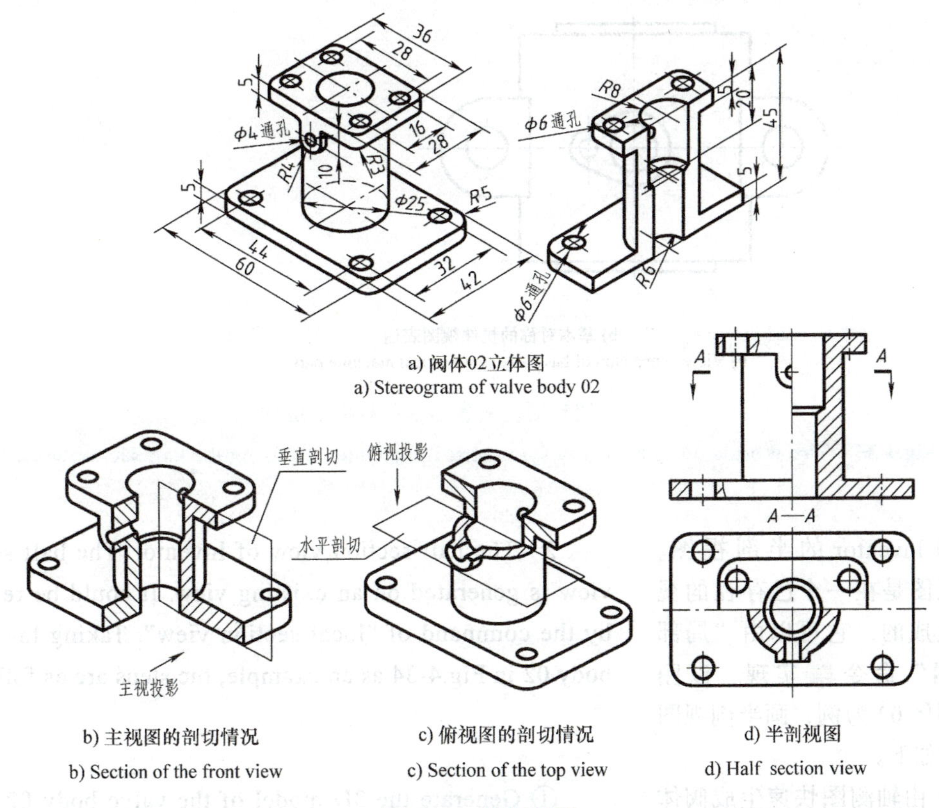

a) 阀体02立体图
a) Stereogram of valve body 02

b) 主视图的剖切情况
b) Section of the front view

c) 俯视图的剖切情况
c) Section of the top view

d) 半剖视图
d) Half section view

图 4-34　半剖视图
Fig.4-34　Half section view

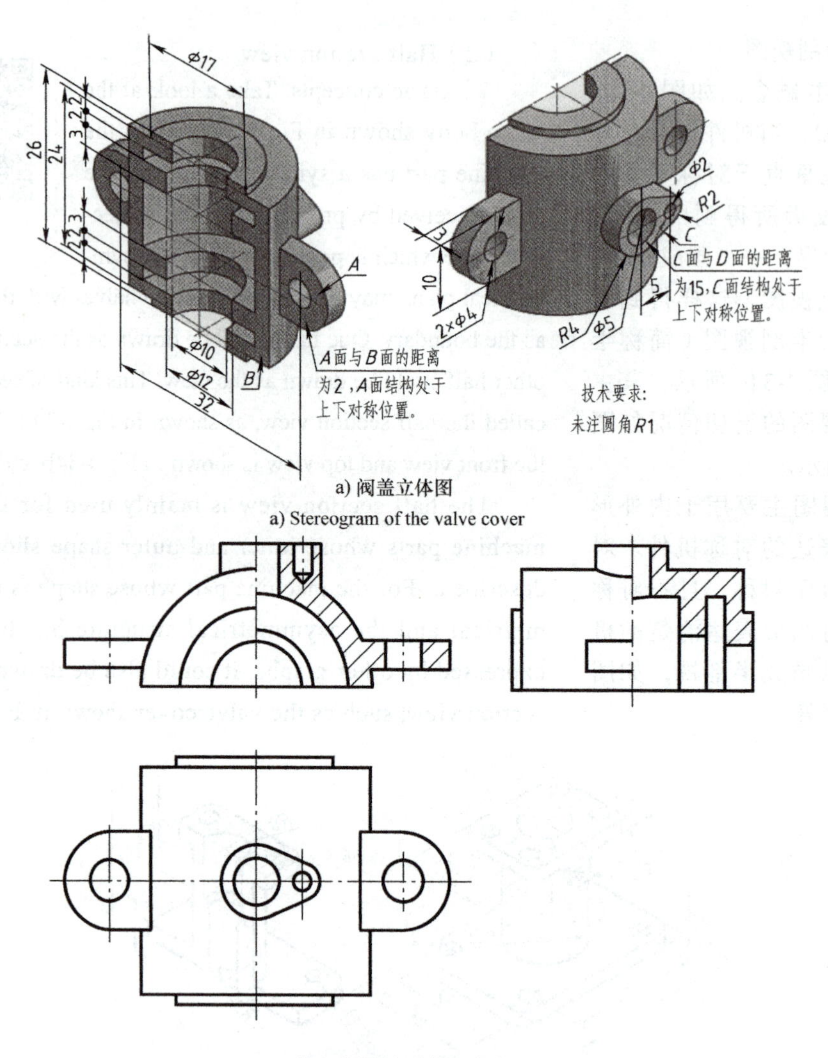

a) 阀盖立体图

a) Stereogram of the valve cover

b) 基本对称的机件视图表达

b) View expression of basically symmetrical machine part

图 4-35 用半剖视图表示基本对称的机件

Fig.4-35 The expression of basically symmetrical machine part using the half section view

2）Inventor 的半剖视图。半剖视图是在一个已存在的视图上生成的，它可以由“局部剖视图”命令实现。以图 4-34 阀体 02 为例，画半剖视图的步骤如下。

① 由轴测图快速生成阀体 02 的三维模型。

② 生成主视图和俯视图。

2）The half section view of Inventor. The half section view is generated on an existing view. It could be realized by the command of “local section view”. Taking the valve body 02 in Fig.4-34 as an example, the steps are as follows.

① Generate the 3D model of the valve body 02 using the axonometric drawing.

② Generate the front view and the top view. Generate

由“基础视图”生成俯视图，由“投影视图”生成主视图。

the top view using the “basic view”; generate the front view using the “projection view”.

③ 绘制剖切草图。单击已有主视图，将其激活。再单击功能块上的“草图”按钮，进入草图工作环境，按需绘制草图，该草图与视图相关联。

③ Draw the section sketch. Click and activate the existing front view. Click the “sketch” button in the functional block to enter the sketch working environment. Draw the sketch as needed. The sketch is associated with the view.

④将主视图改成半剖视图。单击“返回”按钮，退出草图环境，单击“局部剖视图”按钮，按要求完成半剖主视图（重复② - ③完成半剖俯视图）。

④ Change the front view to the half section view. Click the “back” button, exit the sketch environment, click the “local section view” button, complete the half section front view as needed (repeat ② - ③ to complete the half section top view).

画半剖视图时应注意以下几点。

Attention should be paid in the drawing of the half section view.

① 半个视图和半个剖视图的分界线是对称中心线，不能画成实线或波浪线。

① The boundary between the half view and half section view is the symmetrical center line which should not be drawn into continuous line or wave line.

② 在表示外形的半个视图中，一般不画虚线。

② Generally, the dotted line should not be drawn in the half view describing the shape.

③ 半剖视图的标注与全剖视图的标注完全相同。

③ The labeling of the half section view is the same with that of the full section view.

（3）局部剖视图

（3）Local section view

1）基本概念。用剖切平面局部地剖开机件所得到的剖视图，称为局部剖视图（简称局部剖），如图 4-36a 所示阀体 03。

1）Basic Concepts. The section view which is locally sectioned by the section plane is called the local section view, such as the valve body 03 shown in Fig.s 4-36a.

a) 阀体03立体图
a) Stereogram of valve body 03

图 4-36　阀体 03 局部剖视图
Fig.4-36　Local section view of valve body 03

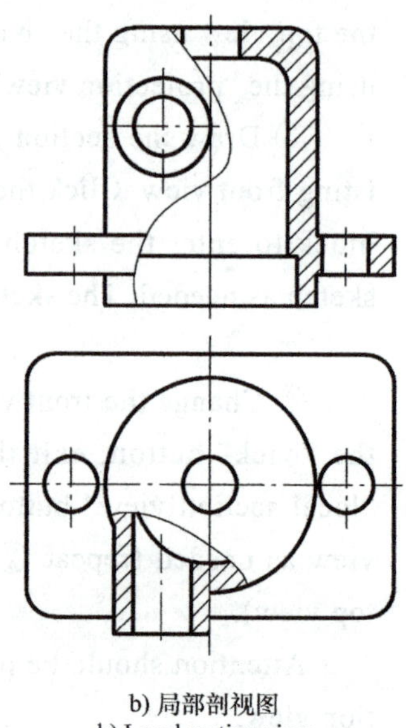

b) 局部剖视图
b) Local section view

图 4-36　阀体 03 局部剖视图（续）
Fig.4-36　Local section view of valve body 03

局部剖视图的应用比较广泛而且灵活，常用于图形不对称，不能采用半视图同时表达内外结构时，或者表达底板、凸缘上的小孔时，如图 4-36b 所示。

画局部剖视图应注意以下几点。

① 局部剖视和视图之间用波浪线分界。波浪线表示机件上断裂的痕迹，它不应与图样上的其他图线重合，更不要超出机件的实体部分，如图 4-37 所示。

② 当被剖切结构为回转体时，允许将该结构的中心线作为局部剖视和视图的分界线，如图 4-38 的主视图所示。

③ 有些机件虽然对称，但轮廓与对称中心线重合，不宜采用半剖视，而以采用局部剖视为宜，如图 4-39 所示的主视图。

The application of the local section view is very extensive and flexible. It is mainly applied when the graph is asymmetrical and the internal and external structure cannot be simultaneously expressed, or to express the holes in the baseboard and flange.

The following points should be noted in the drawing of the local section view.

① The boundary between the local section view and view is expressed in wave lines. The wave line represents the breakage of the machine part and shouldn't coincide with other lines in the graph or exceed the material part of the machine part, as shown in Fig.4-37.

② When the sectioned structure is an axisymmetric body, the center line of the structure is allowed to be the boundary between the local section view and the view, such as the front view in Fig.4-38.

③ Although some machine parts are symmetrical, their contour lines coincide with the symmetrical center line. So, the local section view should be adopted instead of the half section view, such as the front view in Fig.4-39.

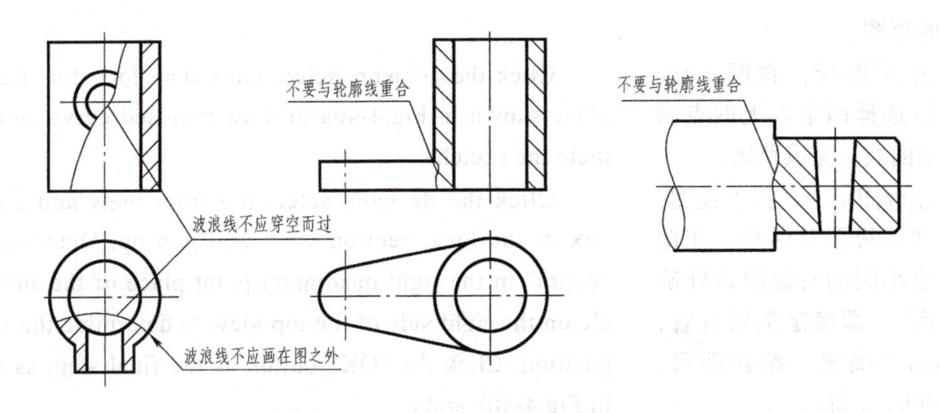

图 4-37　波浪线的错误画法

Fig.4-37　False drawing method of the wave line

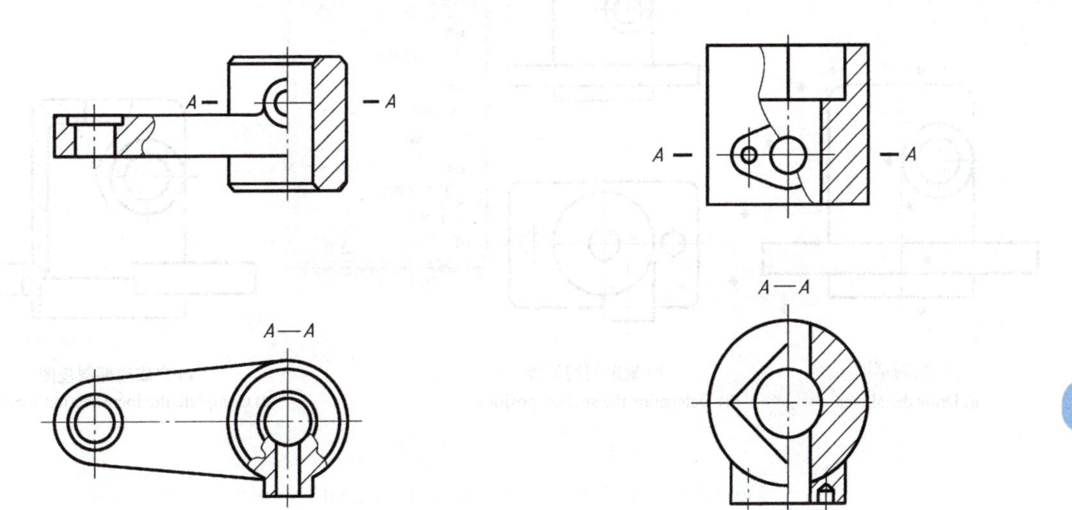

图 4-38　中心线作为局部剖视图和视图的分界线

Fig.4-38　The center line is the boundary between the local section view and the view

图 4-39　轮廓线与中心线重合的采用局部剖视

Fig.4-39　The local section view is adopted when the contour line coincides with the center line

④ 局部剖视一般应作标注，标注方法如图 4-38、图 4-39 中的 “*A—A*” 局部剖视所示。对于剖切位置明显的局部剖视，可不标注，如图 4-36 所示。

④ The local section view should be labeled generally and the labeling method is shown by the local section view “*A—A*” in Fig.4-38 and Fig.4-39. The local section view with clear section place doesn't have to be labeled, as shown in Fig.4-36.

2）Inventor 的局部剖视图。如图 4-40 所示，单击俯视图的红色点线边框，移动鼠标，红色点线边框变为绿色点线边框时，单击图标，创建与俯视

2）Local section view of Inventor. As shown in Fig.4-40, click the red dotted line frame of the top view. Move the mouse, when the red dotted line becomes the green, click the icon to create the sketch associated with the top view.

图关联的草图。

单击图标，在图4-40a所示部位选择四个以上的点画一个封闭曲线，完成草图。

单击图标，选择主视图，弹出局部剖视图对话框，在俯视图右边小圆的右象限点处确定“深度”，即确定剖切位置，最后单击“确定”按钮即可，如图4-40b、c所示。

Click the icon, select more than four dots from the place shown in Fig.4-40a to draw a closed curve and complete the sketch.

Click the icon, select the front view and a dialog box of the local section view will pop up. Determine the “depth” in the right quadrantal point place of the small circle on the right side of the top view to determine the section position. Click the “OK” button as the final step, as shown in Fig.4-40b and c.

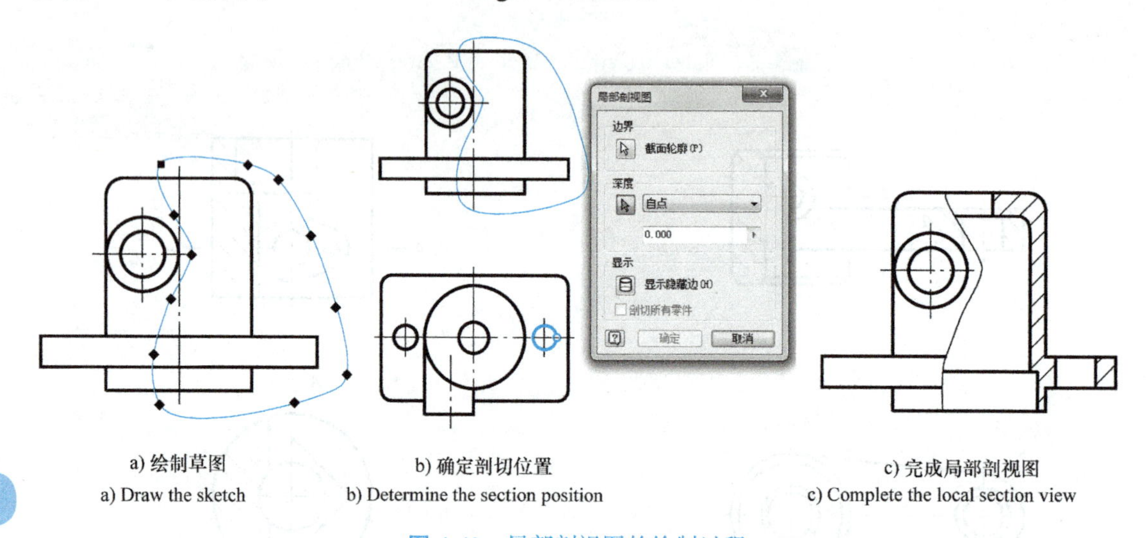

a) 绘制草图
a) Draw the sketch

b) 确定剖切位置
b) Determine the section position

c) 完成局部剖视图
c) Complete the local section view

图 4-40 局部剖视图的绘制过程
Fig.4-40 The drawing process of local section view

3. 剖切面的种类及剖切方法

剖切面可以是平面或柱面，一般是用平面剖切机件。剖切面可分为单一剖切面、几个相交的剖切面、几个相互平行的剖切平面三种。这三种剖切面剖开机件均适合绘制全剖视图、半剖视图和局部剖视图。画剖视图时，可根据机件的结构特点，选用适当的剖切面形式。

（1）单一剖切面 用一个剖切平面剖切机件来表达处于同一位置的内部结构，如图4-36、图4-38、图4-39等所示。

3. Types of section planes and section methods

The section plane could be a plane or a cylinder. The machine part is usually sectioned with a section plane. The section plane can be divided into three types: single section plane, several intersecting section planes and several parallel section planes. These three types are suitable for the drawing of full section view, half section view and local section view. The drawing of the section view should select appropriate type based on the structural feature of the machine part.

（1）Single section plane. A single section plane is used to section the machine part to describe the internal structure at the same position, as shown in Fig.s4-36, 4-38and 4-39. The “*A—A*” section view in Fig.4-41describes the elbow,

图 4-41 中 “*A—A*” 剖视图表达了弯管及顶部凸缘、凸台和通孔结构。

top flange, lug boss and through hole.

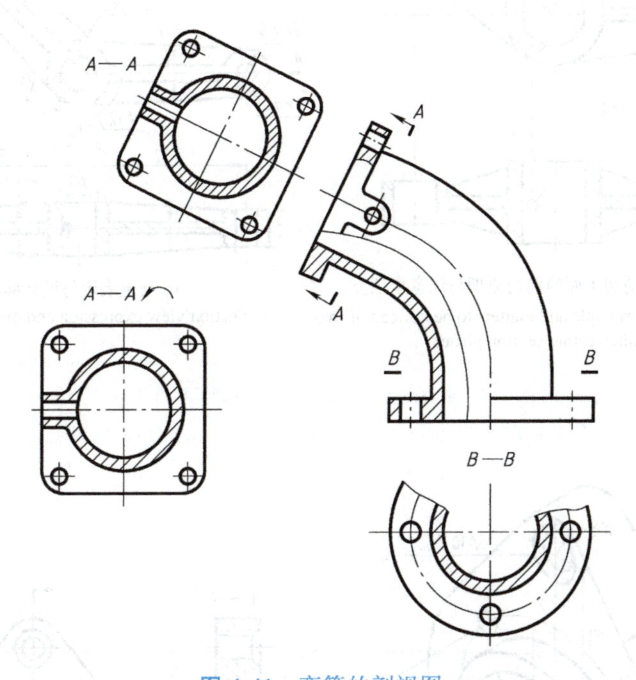

图 4-41　弯管的剖视图

Fig.4-41　The section view of the elbow

剖视图可按投影关系配置在与剖切符号相对应的位置。必要时，可以配置在其他适当的位置或将其旋转成水平画出，但必须标注旋转符号，如图 4-41 中的 “*A—A*” 所示。

（2）几个相交的剖切面　用几个相交的剖切平面（交线垂直于某一基本投影面）剖开机件，将被倾斜剖切平面剖开的结构要素及其有关部分旋转到与选定的投影面平行再进行投影。如图 4-42、图 4-43 所示。

此类剖视图常用于表达具有公共回转轴线的杆类机件（图 4-42）及盘盖类机件（图 4-43）。

The section view should be configurated on the corresponding position of section symbols according to the projection relation. If necessary, it could also be configurated on other position or rotated to the horizontal direction while labeling the rotating symbol, such as “*A—A*” in Fig.4-41.

(2) Several intersecting section planes.　The machine part is sectioned using several intersecting section planes (the intersecting line is perpendicular to a basic projection plane). The structural elements sectioned by the inclined section planes and relevant segments are rotated to the parallel direction with the selected projection plane for projection. As shown in Fig.4-42, Fig.4-43.

4.2.2-8

This type of section view is usually used to describe machine parts like rods (Fig.4-42), plates and covers (Fig.4-43) with common axes of rotation.

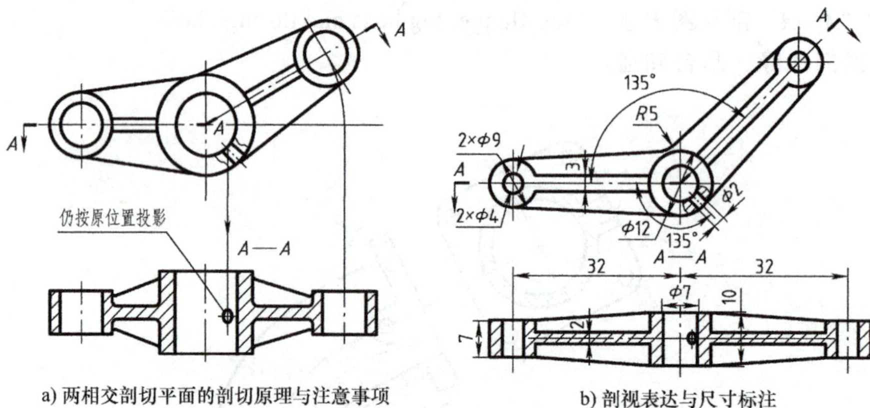

a) 两相交剖切平面的剖切原理与注意事项

a) The section principle and matters to be noticed of two intersecting section planes

b) 剖视表达与尺寸标注

b) Section view expression and dimensioning

图 4-42 连杆 01 的剖视图

Fig.4-42 The Section view of the connecting rod 01

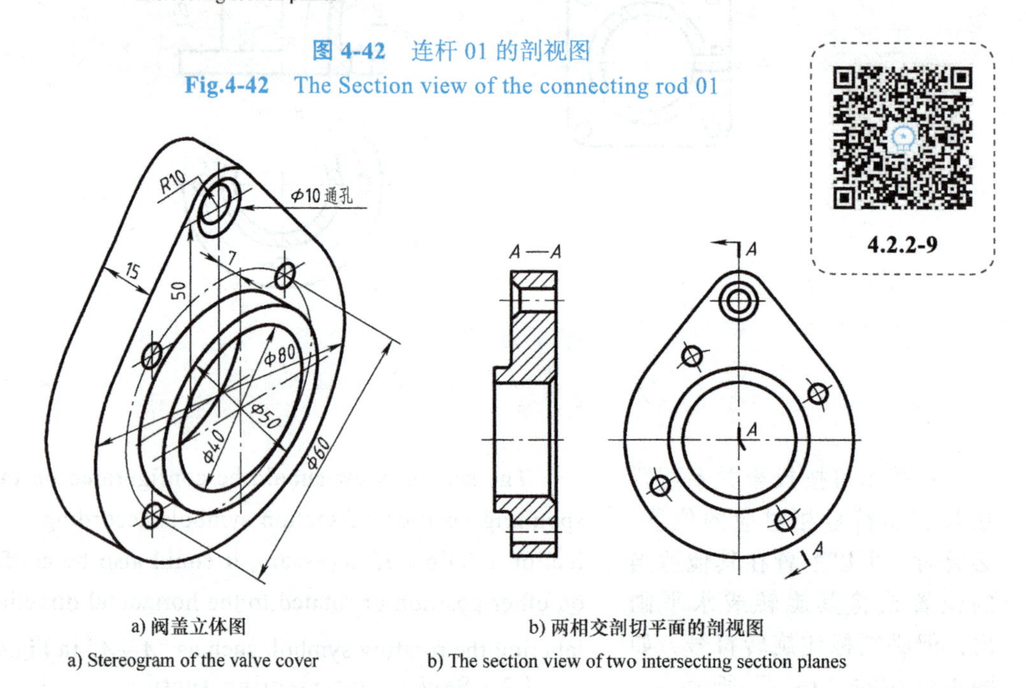

a) 阀盖立体图

a) Stereogram of the valve cover

b) 两相交剖切平面的剖视图

b) The section view of two intersecting section planes

图 4-43 阀盖的剖视图

Fig.4-43 The section view of the valve cover

画此类剖视图还应注意以下几点。

The following should be noted in the drawing of this type of section view.

1）剖切平面的交线应与机件上的公共回转轴线重合。

1）The intersecting line of the section plane should coincide with the public axis of rotation of the machine part.

2）剖切平面后的其他结构一般仍按原来位置投影，如图 4-42 中的油孔。

2）Other structure behind the section plane should be projected as their original position, such as the oil hole in Fig.4-42.

3）当剖切后产生不完整要素时，应将此部分按不剖绘制，

3）When incomplete elements are generated after section, this part should be drawn in the form without section,

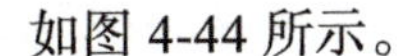

如图 4-44 所示。

as shown in Fig.4-44.

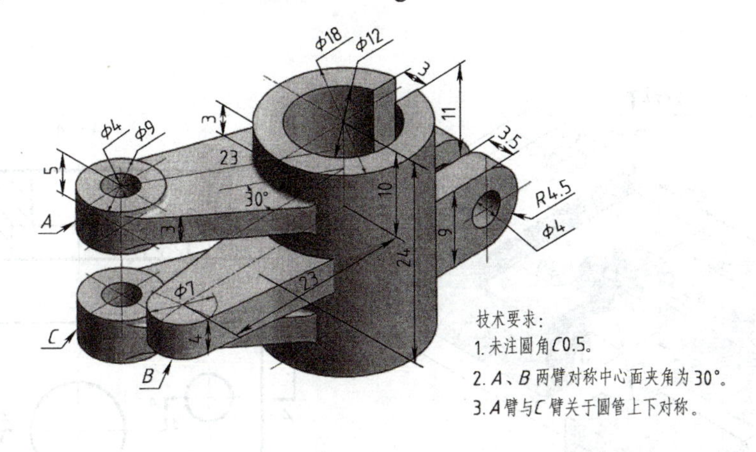

a) 连杆02立体图
a) Stereogram of connecting rod 02

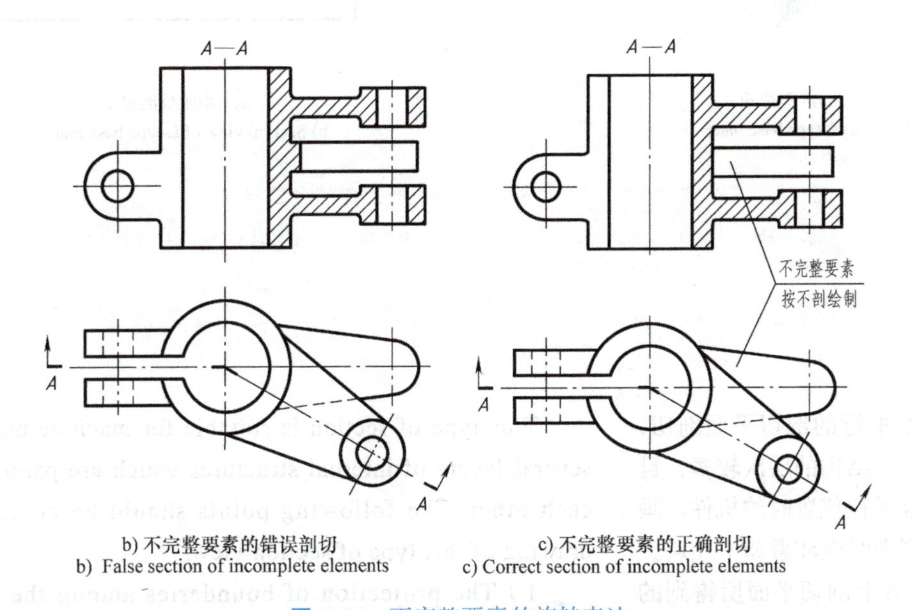

b) 不完整要素的错误剖切
b) False section of incomplete elements

c) 不完整要素的正确剖切
c) Correct section of incomplete elements

图 4-44　不完整要素的旋转表达

Fig.4-44　Rotated expression of incomplete elements

（3）几个平行的剖切平面　用几个平行的剖切平面剖开机件获得的剖视图如图 4-45 所示。假想用两个以上平行于基本投影面的剖切平面剖切机件，将每一个剖切平面所得到的结构向该基本投影面投影，画出其剖视图。

（3）Several parallel section planes. The section view generated by sectioning the machine part with several parallel section planes is shown in Fig.4-45. Suppose the machine part is sectioned by two section planes which are parallel to the basic projection plane, and the structure sectioned by them are projected to the basic projection plane, then the section view is drawn.

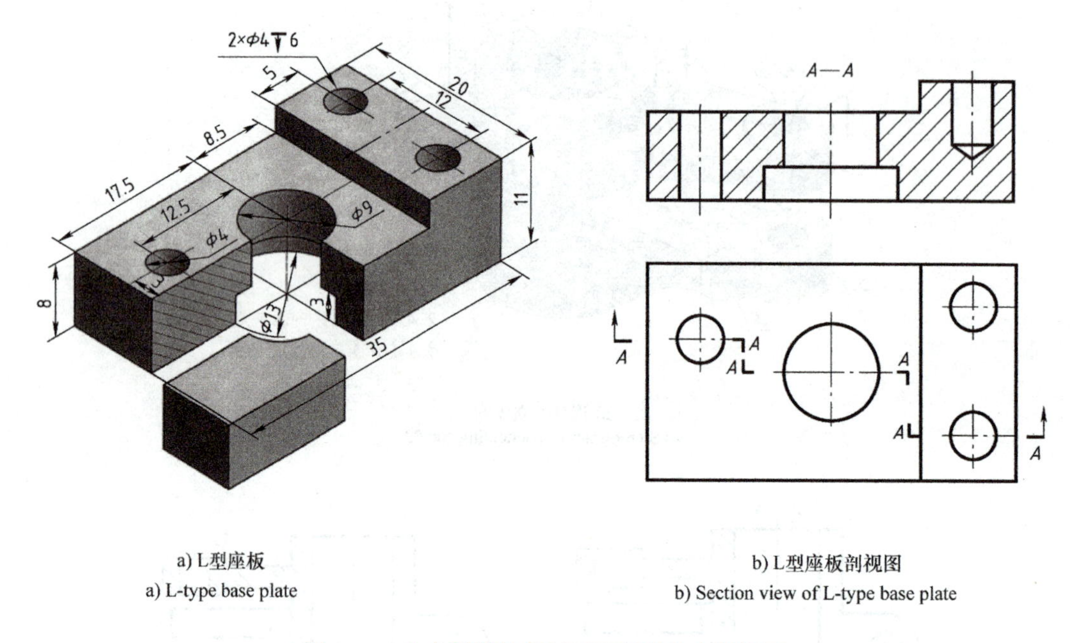

a) L型座板
a) L-type base plate

b) L型座板剖视图
b) Section view of L-type base plate

图 4-45 几个平行的剖切平面剖得的全剖视图
Fig.4-45 The full section view sectioned by several parallel section planes

几个平行的剖切平面剖切，适用于内部结构的层次较多，且位于相互平行位置时的机件。画此类剖视图时应注意如下几点。

1）各个剖切平面所得到的剖视之间不应画出分界线的投影，如图 4-45 所示。

2）在图形内不要出现不完整的要素，如图 4-47 所示。由于剖切平面只剖到左边半个孔，因此在剖视图上就出现了不完整孔的投影。只有当两个结构在图形上具有公共对称中心线时，可以各画出一半，这时应以对称中心线或轴线分界，如图 4-48 所示。

This type of section is suitable for machine parts with several layers of internal structures which are parallel with each other. The following points should be noted in the drawing of this type of section view.

1）The projection of boundaries among the section views sectioned by those section planes should not be drawn, as shown in Fig.4-46.

2）The graph should not have incomplete elements, as shown in Fig.4-47. As the section plane only sections the left half of the hole, the section view has a projection of the incomplete hole. Only when two structures have a public symmetrical center line could half of each of them be drawn. The boundary should be the symmetrical center line or the axis, as shown in Fig.4-48.

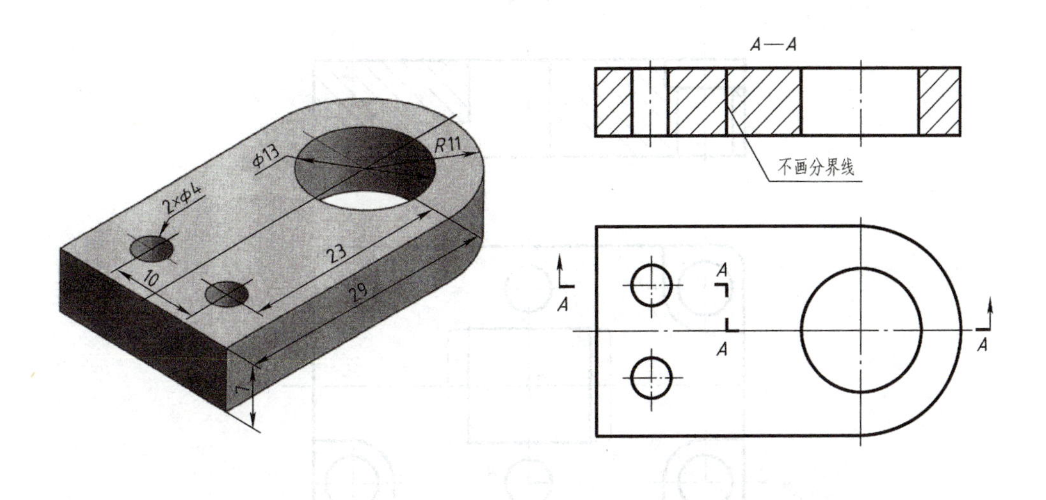

a) 座板立体图
a) Stereogram of base plate

b) 座板剖视图与注意事项
b) Section view and matters to be noticed of base plate

图 4-46　不画分界线

Fig.4-46　Don't draw the boundary line

a) 平行剖切注意事项
a) Note for parallel section

b) 剖视表达与尺寸
b) Expression and dimension of section view

图 4-47　不应出现不完整要素

Fig.4-47　The graph should not have incomplete elements

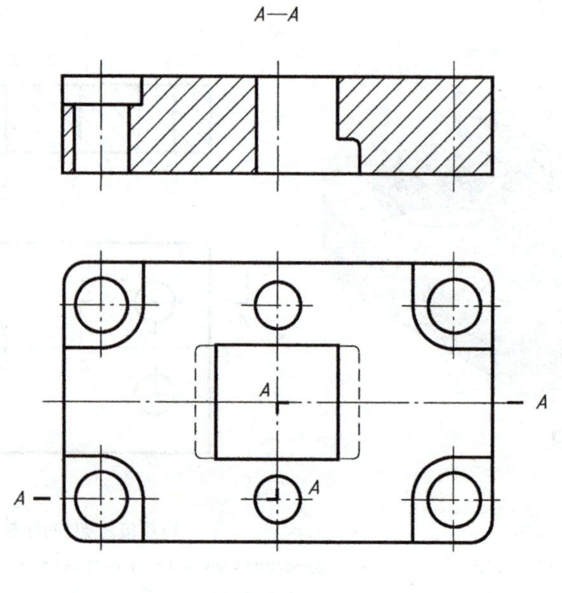

图 4-48 两个结构具有公共对称面的表达

Fig.4-48 The expression of two structures with the public symmetrical plane

4.2.3 断面图

1. 断面图的概念

假想用剖切面将机件的某处切断，仅画出断面的形状，并在断面上画出剖面符号的图形称为断面图，可简称为断面，如图 4-49 所示。

4.2.3 Cross-section View

1. Concept of the cross-section view

Suppose the machine part is sectioned by the section plane at a certain place, the graph describing the fracture surface with the section plane symbols is called the cross-section view, as shown in Fig.4-49.

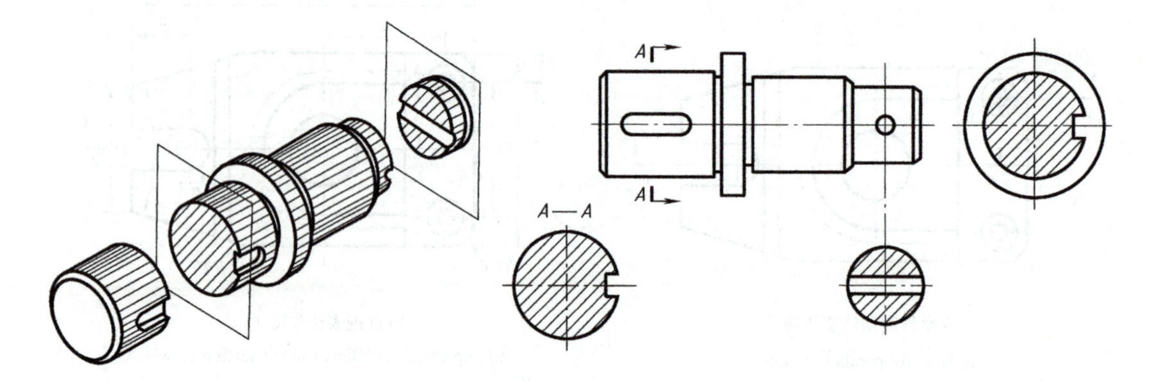

图 4-49 断面与剖视

Fig.4-49 Fracture surface and section view

断面图常用于表达机件上某些常见的结构，如：筋、轮辐、孔、槽等的断面形状。

断面图和剖视图的区别在于：断面图是机件上剖切处断面的形状，而剖视图则是剖切平面之后机件的全部投影。

2. 断面图的种类和画法

根据断面图的位置不同，断面图可分为移出断面图和重合断面图。

（1）移出断面图　画在视图轮廓线之外的断面图称为移出断面图，如图 4-50 所示。

画移出断面图时，应注意以下几点。

1）移出断面图的轮廓线用粗实线绘制。

2）移出断面图应尽量配置在剖切符号或剖切线延长线上，如图 4-50 右面的图形所示。剖切平面迹线是剖切平面与投影面的交线，用细点画线表示。必要时也可配置在其他适当的位置，如图 4-50 中的“*A—A*”“*B—B*”断面图所示。

The cross-section view is often used to describe the common structures on the machine part, such as the fracture surface structure of rib, spoke, hole and groove, etc.

The difference between the cross-section view and the section view is that the cross-section view is the shape of the fracture surface of the machine part, while the section view is the entire projection behind the section plane of the machine part.

2. Types and drawing methods of the cross-section view

According to the different position of the cross-section view, it could be divided into removed cross-section view and coincident cross-section view.

(1) Removed cross-section view.　The cross-section view drawn outside of the contour line is called the removed cross-section view, as shown in Fig.4-50.

The following should be noted in the drawing of the removed cross-section view.

1) The contour line of the removed cross-section view should be drawn in heavy line.

2) The removed cross-section view should be configurated on the extension line of the section symbol and section line, such as the right graph of Fig.4-53. The section plane trace is the intersecting line between the section plane and the projection plane. They should be drawn in thin dash-dotted line. If necessary, it could also be drawn in another appropriate place, such as the “*A—A*” and “*B—B*” cross-section view of Fig.4-50.

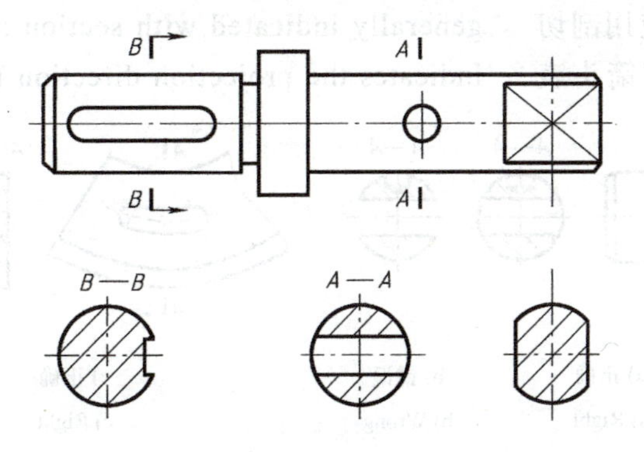

图 4-50　移出断面图

Fig.4-50　Removed cross-section view

3）由两个或多个相交的剖切平面剖切得出的移出断面图，中间一般应断开，如图 4-51 所示。

4）断面图形对称时可画在视图的中断处，如图 4-52 所示。

3) The removed cross-section view sectioned by two or more intersecting section planes should be disconnected in the middle, as shown in Fig.4-51.

4) When the cross-section view is symmetrical, it could be drawn at the interrupted position, as shown in Fig.4-52.

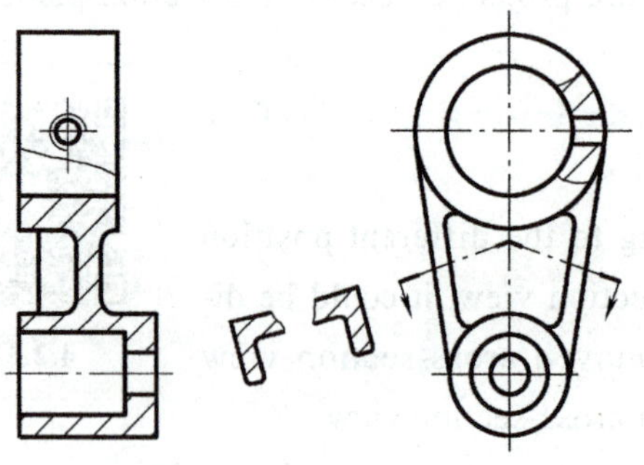

图 4-51 断开的移出断面图

Fig.4-51 Disconnected removed cross-section view

实长

图 4-52 对称断面的画法

Fig.4-52 Drawing method of symmetrical cross-section view

5）当剖切平面通过回转面形成的孔或凹坑的轴线时，这些结构按剖视绘制，如图 4-53a 所示，图 4-53b 为错误画法。

6）当剖切平面通过非圆孔，导致出现完全分离的两个断面时，这些结构应按剖视绘制，如图 4-53c 所示，图 4-53d 为错误画法。

7）移出断面的标注。

① 移出断面一般应用剖切符号表示剖切位置，用箭头表

5) When the section plane passes the holes or axes of pits of revolution surface, these structures should be drawn as the section view, as shown in Fig.4-53a. Graph b is incorrect.

6) When the section plane passes non-circular holes and results in two fully separated section planes, these structures should be drawn as the section view, as shown in Fig.4-53c. Graph b is incorrect.

7) Labeling of removed cross-section view.

① The section place of the removed cross-section is generally indicated with section symbols. The arrowhead indicates the projection direction labeled with letters. The

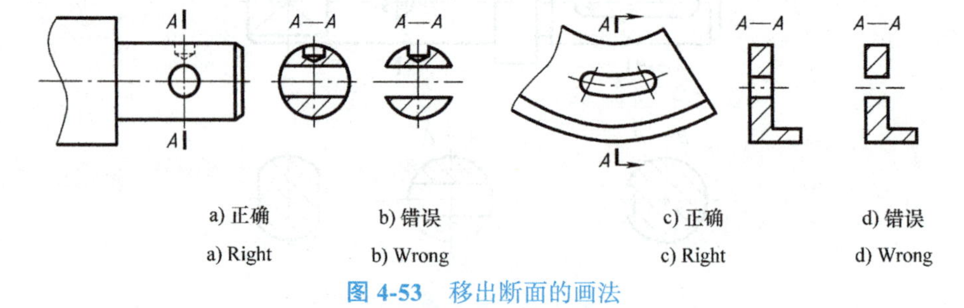

图 4-53 移出断面的画法

Fig.4-53 Drawing method of removed cross-section view

示投影方向，并注上字母，在断面图的上方应用同样字母标出相应的名称“×—×”，如图 4-50 中的“*B*—*B*”所示。

② 配置在剖切线延长线上的移出断面可省略字母，如图 4-50 中的右侧断面图。

③ 移出断面图形对称，即与投影方向无关时，可省略箭头，如图 4-50 中的“*A*—*A*”断面图所示。

④ 配置在剖切线延长线上而又对称的移出断面，和配置在视图中断处的移出断面图可以不标注，如图 4-51、图 4-52 所示。

（2）重合断面。画在视图轮廓线之内的断面称作重合断面图，如图 4-54 所示。

重合断面图的轮廓线规定用细实线绘制。当视图中的轮廓线与重合断面图重叠时，视图中的轮廓线仍应连续画出，不可间断。对称的重合断面不必标注，如图 4-54a 所示。不对称的重合断面可省略标注，如图 4-54b 所示。

corresponding names, “×—×” should be wrote on top of the cross-section view using the same letters, such as “*B*—*B*” in Fig.4-50.

② The removed cross-section view configurated on the extension line of the section line can omit the letter, such as the right cross-section view in Fig.4-50.

③ The arrowhead could be omitted when the removed cross-section view is symmetrical which means it's irrelevant to the projection direction, such as the “*A*—*A*” cross-section view in Fig.4-50.

④ The removed cross-section view which is symmetrical and configurated on the extension line of the section line, or configurated on the disconnected position of the view could omit the labeling, as shown in Fig.4-51and Fig.4-52.

（2）Coincident cross-section view. The cross-section view drawn inside of the contour line is called the coincident cross-section view, as shown in Fig.4-54.

The contour line of the coincident cross-section view is specified to be drawn in thin lines. When the contour line coincides with the coincident cross-section view, the contour line should be drawn continuously and cannot be disconnected. The symmetrical coincident cross-section view need not be labeled, as shown in Fig.4-54a. The labels of asymmetrical coincident cross-section view could be omitted, as shown in Fig.4-54b.

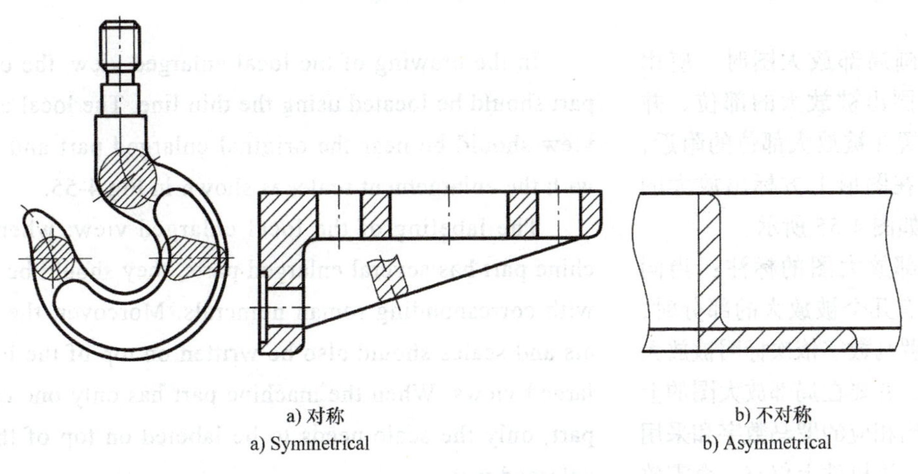

图 4-54 重合断面图
Fig.4-54 Coincident cross-section view

4.2.4 其他表达方法

1. 局部放大图

（1）概述　将机件的部分结构用大于原图形所采用的比例画出的图形，称为局部放大图。局部放大图可画成视图、剖视图或断面图，它与原图形的表达方式无关。当机件上某些细小结构在原图形中表达不清楚或不便于标注尺寸时，可采用局部放大图，如4-55、图4-56所示。

4.2.4 Other Representation

1. Local enlarged view

（1）Overview.　The graph of partial structure drawn on a scale larger than that of the original graph is called the local enlarged view. The local enlarged view can be drawn into view, section view or cross-section view. It's irrelevant with the representation of the original graph. The local enlarged view can be applied when some small structures of the machine part is difficult to be described clearly or dimensioned, as shown in Fig.4-55 and 4-56.

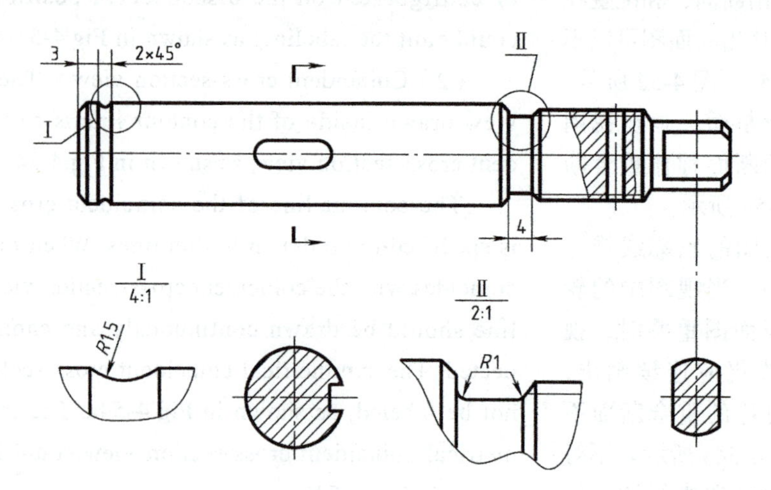

图 4-55　局部放大图

Fig.4-55　Local enlarged view

在画局部放大图时，应用细实线圈出被放大的部位，并尽量配置在被放大部位的附近，而且要在图形上方标出放大的比例，如图4-55所示。

局部放大图的标注：当同一机件有几个被放大的部分时，必须用罗马数字依次标明被放大的部位，并要在局部放大图的上方标注出相应的罗马数字和采用的比例。当机件上仅有一个需放大的部位时，在局部放大图的上

In the drawing of the local enlarged view, the enlarged part should be located using the thin line. The local enlarged view should be near the original enlarged part and labeled with the enlargement scale, as shown in Fig.4-55.

The labeling of the local enlarged view: when a machine part has several enlarged parts, they should be labeled with corresponding roman numerals. Moreover, the numerals and scales should also be written on top of the local enlarged views. When the machine part has only one enlarged part, only the scale needs to be labeled on top of the local enlarged view.

方只需标注采用的比例。

（2）Inventor 的局部放大图

1）依据图 4-56 绘制轴的三维模型。

2）基础视图、断面图与局部视图创建：进入工程图环境，先创建基础视图 - 主视图，再根据基础视图依次创建 *A—A*、*B—B* 断面图。

3）局部放大图创建：最后通过“局部视图”命令完成局部放大图的创建，如图 4-56 所示。

（2）Local enlarged view in Inventor

1）Draw the 3D model of the shaft.

2）Create the basic view, cross-section view and local view: enter the engineering drawing environment, create the basic view-front view at first, then create *A—A*, *B—B* cross-section views successively based on the basic view.

3）Create the local enlarged view: finally, complete the creation of the partial enlarged view through the command of “local view”, as shown in Fig.4-56.

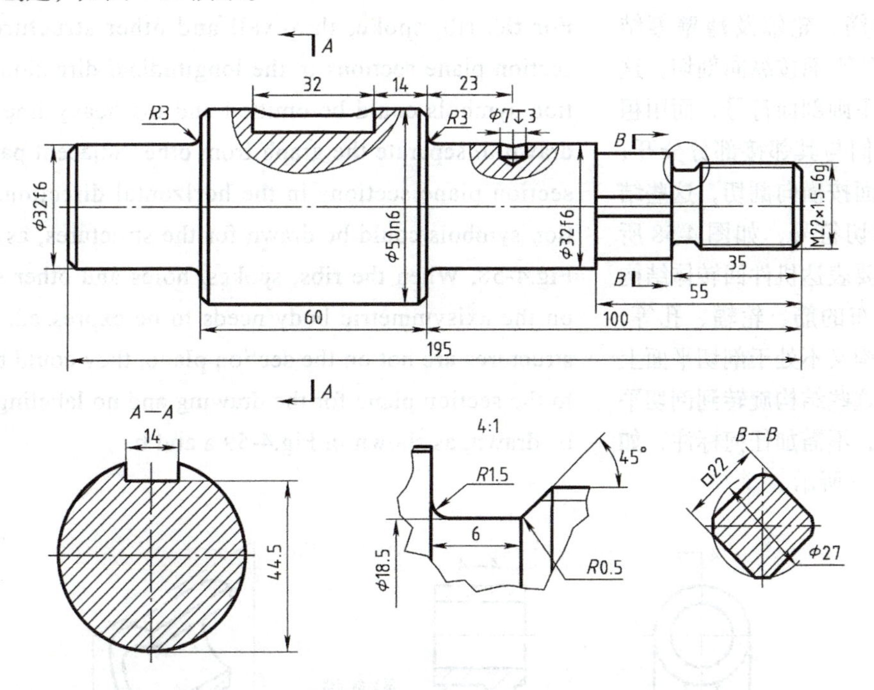

图 4-56　轴的视图表达与尺寸

Fig.4-56　View expression and dimension of the shaft

2. 简化画法和规定画法

（1）相同结构要素的省略画法　机件上相同的结构要素（如齿、孔、槽等），按一定规律分布时，可以只画出几个完整的要素，其余用细实线连接，或画出它们的中心位置，但图中必须注出该要素的总数，如图 4-57 所示。

2. Simplified drawing method and specified drawing method

（1）The simplified drawing method of same structural elements. When the same structural elements (such as teeth, holes, grooves, etc.) are distributed according to certain rules, only a few complete elements can be drawn. The rest are connected with thin lines, or their central locations are drawn, but the total number of those elements must be noted in the graph, as shown in Fig.4-57.

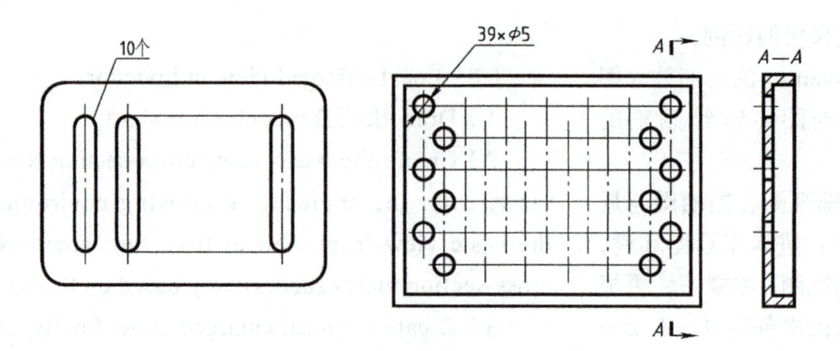

图 4-57 相同要素的省略画法

Fig.4-57 The simplified drawing method of same elements

（2）筋和轮辐的规定画法 对于机件的筋、轮辐及薄壁等结构，如剖切平面按纵向剖切，这些结构都不画剖面符号，而用粗实线将它们与其邻接部分分开；如剖切平面按横向剖切，这些结构画出剖切符号，如图 4-58 所示。当需要表达机件回转体结构上均匀分布的筋、轮辐、孔等，而这些结构又不处于剖切平面上时，可将这些结构旋转到剖切平面上画出，不需加任何标注，如图 4-59a、b 所示。

(2) Specified drawing method of the rib and spoke. For the rib, spoke, thin-wall and other structures, if the section plane sections in the longitudinal direction, the section symbols could be omitted and the heavy line could be drawn to separate the graph from other adjacent parts. If the section plane sections in the horizontal direction, the section symbols could be drawn for the structures, as shown in Fig.4-58. When the ribs, spokes, holes and other structures on the axisymmetric body needs to be expressed, but these structures are not on the section plane, they could be rotated to the section plane for the drawing and no labeling needs to be drawn, as shown in Fig.4-59 a and b.

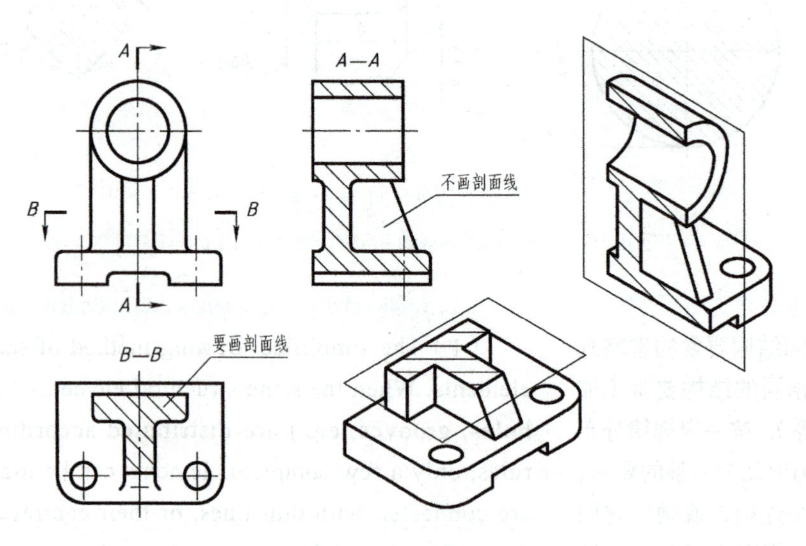

图 4-58 筋的画法

Fig.4-58 The drawing method of the rib

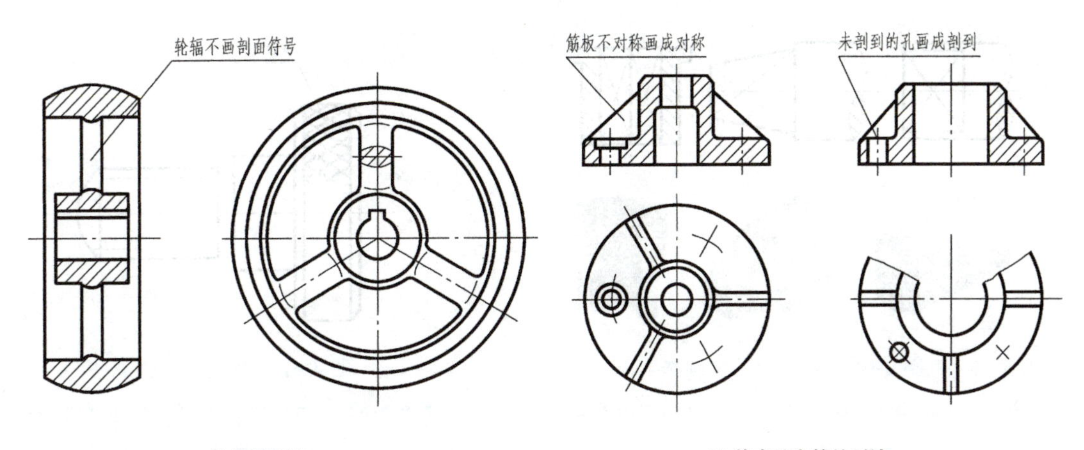

a) 轮辐的画法
a) The drawing method of the spoke

b) 均布孔和筋的画法
b) The drawing method of uniformly distributed holes and ribs

图 4-59　轮辐与均布孔、筋的规定画法

Fig.4-59　The specified drawing method of the spoke, uniformly distributed holes and ribs

（3）断开画法　对较长的机件沿长度方向的形状一致或按一定规律变化时，例如轴、杆、型材、连杆等，可以断开后缩短表示，但要标注实际尺寸。画图时，可按图 4-60 中所示方法表示。

（3）Disconnected drawing method. For those long machine parts whose shape in the long direction is the same or changes in the same pattern, such as shaft, rod, section bar and connecting rod, they could be disconnected to shorten the expression, but the real dimension should be labeled. The method shown in Fig.4-60 could be applied in the drawing.

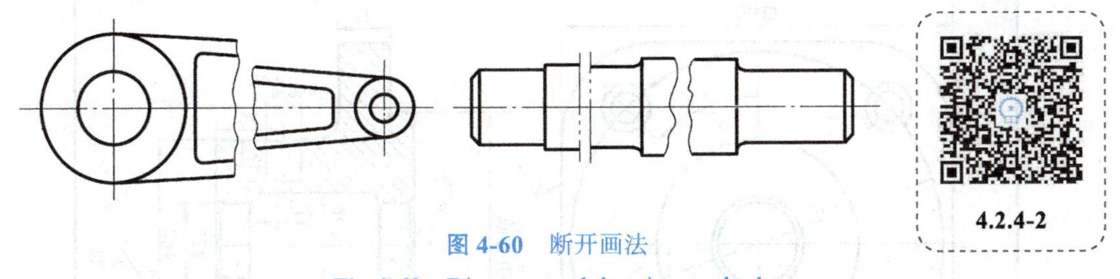

4.2.4-2

图 4-60　断开画法

Fig.4-60　Disconnected drawing method

（4）其他简化画法

1）当平面在图形中不能充分表达时，可用平面符号（相交的两条细实线）表示，如图 4-61 所示。

2）机件上的滚花部分，可在轮廓线附近用粗实线示意画出，如图 4-62 所示。

（4）Other simplified drawing methods

1）When the plane cannot be fully described in the graph, it could be described using the plane symbol (two intersecting thin lines), as shown in Fig.4-61.

2）The knurl on the machine part could be drawn using heavy lines near the contour line, as shown in Fig.4-62.

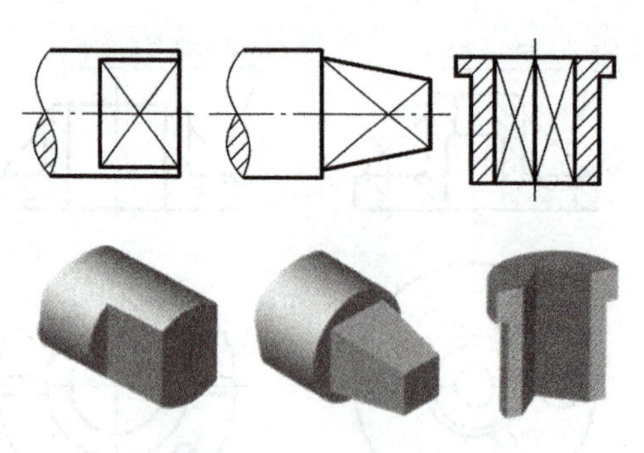

图 4-61　平面的简化画法

Fig.4-61　The simplified drawing method of the plane

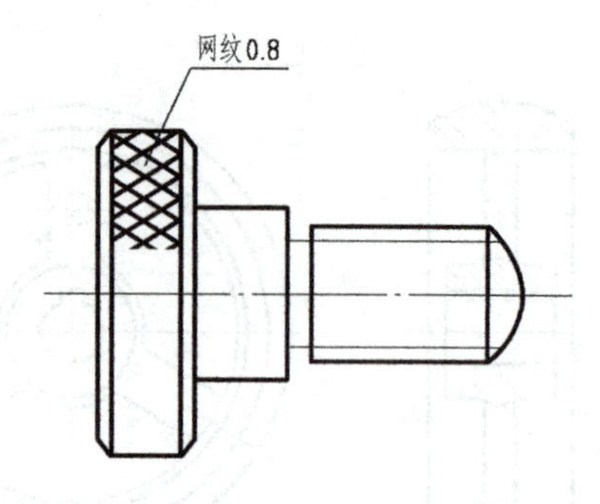

图 4-62　网纹的简化画法

Fig.4-62　The simplified drawing method of the overlapping curve

4.3　技术要求

机械图样上的技术要求是零件在设计、加工和使用中应达到的技术性能指标，主要包括表面结构、极限与配合、几何公差、热处理以及其他有关制造的要求等，如图 4-63 所示。

4.3　Technical Requirements

The technical requirements of mechanical drawings are that the technical performance indexes parts should meet in the design, processing and application, mainly including surface structure, limits and fit, geometric tolerance, heat treatment and other manufacturing requirements, as shown in Fig.4-63.

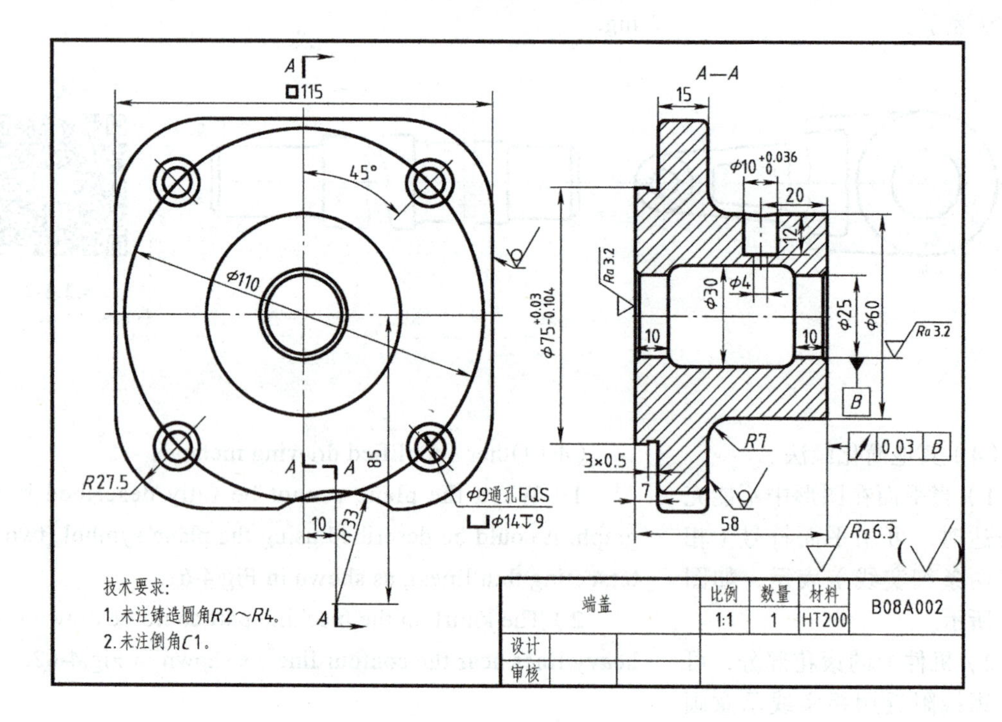

图 4-63　零件图上的技术要求

Fig.4-63　Technical requirements on the parts drawing

4.3.1　表面结构的表示法

为了保证零件的使用性能，在机械图样中需要对零件的表面结构给出要求。表面结构就是表面粗糙度、表面波纹度、表面缺陷和表面几何形状的总称。表面结构的各项要求在图样上的表示法在 GB/T 131—2006 中均有具体规定。本节只介绍常用的表面粗糙度表示法。

表面粗糙度的单位是 μm，其数值越小工件表面越光滑。国家标准规定用 *Ra* 和 *Rz* 两个高度参数来评定表面粗糙度。

轮廓的算术平均偏差 *Ra* 是指在一个取样长度内，纵坐标 $z(x)$ 绝对值的算术平均值。

轮廓的最大高度 *Rz* 是指在同一取样长度内，最大轮廓峰高和最大轮廓谷深之和。

优先选用轮廓算术平均偏差 *Ra*。

4.3.1　The Representation of the Surface Structure

In order to guarantee the performance of the parts, the surface structure of the parts should be specified in the mechanical drawing. Surface structure is a general term of surface roughness, surface waviness, surface defects and surface geometry shape. The representation of surface structure requirements on the drawing has been specified in GB/T 131—2006. This section only introduces the representation of commonly used surface roughness.

The unit of the surface roughness is. The smaller the value, the smoother the surface of the machine part. The national standard stipulates that the surface roughness is evaluated with two height parameters, *Ra* and *Rz*.

Ra is the arithmetic average deviation of the contour and refers to the arithmetic mean value of the absolute value of the vertical coordinates $z(x)$ within a sampling length.

Rz refers to the distance between the maximum peak and maximum valley of the contour within the same sampling length.

The arithmetic average deviation, *Ra*, is preferred in application.

1. 标注表面结构的图形符号

（1）图形符号及其含义　在图样中，可以用不同的图形符号来表示对零件表面结构的不同要求。标注表面结构的图形符号及其含义如表 4-7 所示。

1. Graphic symbols labeling the surface structure

（1）Graphic symbols and their meaning. In the drawing, the different requirements on the surface structure of parts could be represented with different graphic symbols. The graphic symbols and their meaning labeling the surface structure are shown in Table 4-7.

表 4-7　表面结构图形符号及其含义

Table 4-7　Graphic symbols of the surface structure and their meaning

符号名称 Symbol name	符号样式 Symbol pattern	含义及说明 Meaning and description
基本图形符号 Basic graphic symbol	√	未指定工艺方法的表面，基本图形符号仅用于简化代号标注，当通过一个注释解释时可单独使用，没有补充说明时不能单独使用 Surfaces without specified processing method. Basic graphic symbols are used only for simplified code labeling. It could be used alone when it can be explained with a note. It can't be used alone when there is no supplementary instruction.

（续）

符号名称 Symbol name	符号样式 Symbol pattern	含义及说明 Meaning and description
扩展图形符号 Extended graphic symbol		用去除材料的方法获得表面，如通过车、铣、刨、磨等机械加工的表面，仅当其含义是“被加工表面”时可单独使用 The surface is processed by removing the material, such as lathing, milling, planing, grinding and other machining methods. It can be used only when its meaning is “machined surface”.
		用不去除材料的方法获得表面，如铸、锻等，也可用于保持上道工序形成的表面，无论这种状况是通过去除材料或不去除材料形成的 The surface obtained without removing materials, such as casting and forging. This symbol could also be used for the surface formed by the former procedure, regardless of whether it's obtained with or without removing materials.
完整图形符号 Complete graphic symbol		在基本图形符号或扩展图形符号的长边上加一横线，用于标注表面结构特征的补充信息 A transverse line is added on the long edge of the basic graphic symbol or the extended graphic symbol to represent the supplementary information of structural features of the surface.
工件轮廓各表面图形符号 Graphic symbols of workpiece contour surfaces		当在某个视图上组成封闭轮廓的各表面有相同的表面结构要求时，应在完整图形符号上加一圆圈，标注在图样中工件的封闭轮廓线上 When different surfaces constituting a closed contour in a certain view have the same surface structure requirement, a circle should be added on the complete graphic symbol on the closed contour line of the workpiece in the drawing.

（2）表面结构要求在图形符号中的注写位置　在完整符号中，对表面结构的单一要求和补充要求应注写在图中所示的指定位置，如图 4-64 所示。

表面结构补充要求包括：表面结构参数代号；数值；传输带/取样长度。

(2) The labeling position of surface structure requirements in graphic symbols. In the complete symbol, the single requirement and supplementary requirement on the surface structure should be labeled in the designated position shown in the graph, as shown in Fig.4-64.

The supplementary requirements of the surface structure include: parameter code of the surface structure; values; transmission band/ sampling length.

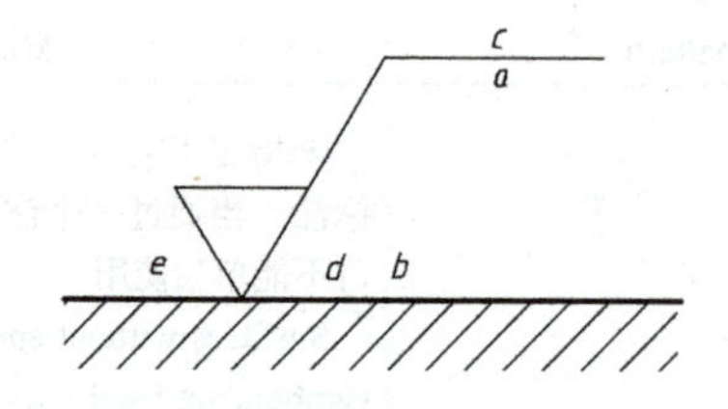

图 4-64　补充要求的注写位置

Fig.4-64　Labeling position of supplementary requirements

a 注写表面结构的单一要求，*a* 和 *b* 同时存在时，*a* 注写第一表面结构要求。*b* 注写第二表面结构要求。*c* 注写加工方法，如“车”“铣”“镀”等。*d* 注写表面纹理方向，如“=”“×”“M”等。*e* 注写加工余量。

a represents the single requirement of the surface structure. When *a* and *b* simultaneously exist, *a* represents the first surface structure requirement. *b* represents the second surface structure requirement. *c* represents the processing methods, such as lathing, milling and plating, etc. *d* represents the surface texture direction, such as “=”, “×” and “M”, etc. *e* represents the working allowance.

（3）表面结构要求在图样中的标注　表面结构要求在图样中的标注实例如表 4-8 所示。

（3）The labeling of surface structure requirements in the drawing. The labeling examples of surface structure requirements in the drawing are shown in Table 4-8.

表 4-8　表面结构要求在图样中的标注实例

Table 4-8　The labeling examples of surface structure requirements in the drawing

说明 Description	实例 Example
表面结构要求对每一表面一般只标注一次，并尽可能注在相应的尺寸及其公差的同一视图上 The surface structure requirements are labeled only once on each surface and should be labeled in the same view of the corresponding dimensions and their tolerances as far as possible. 表面结构的注写和读取方向与尺寸的注写和读取方向一致 The labeling and reading direction of the surface structure should be the same with that of the dimension.	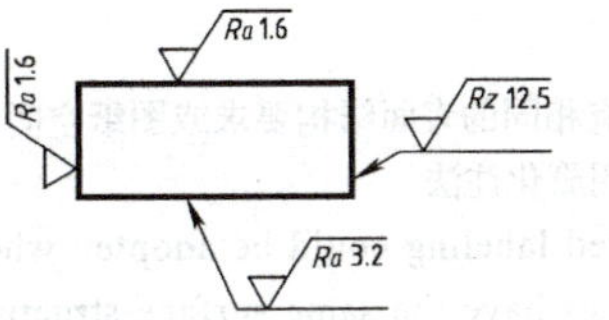 Ra 1.6　Ra 1.6　Rz 12.5　Ra 3.2
表面结构要求可标注在轮廓线或其延长线上，其符号应从材料外指向并接触表面。必要时表面结构符号也可用带箭头和黑点的指引线引出标注 The surface structure requirement should be labeled on the contour line or its extension line. Its symbol should point to and contact the surface from the outside of the material. If necessary, the surface structure symbol could also be labeled outside with the leading line having the arrowhead and black dots as the indication.	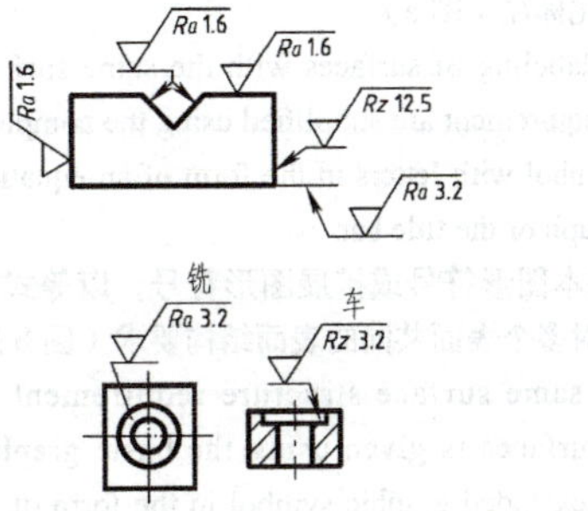 Ra 1.6　Ra 1.6　Ra 1.6　Rz 12.5　Ra 3.2 铣 Ra 3.2　车 Rz 3.2
在不致引起误解时，表面结构要求可以标注在给定的尺寸线上 Without causing misunderstanding, the surface structure requirements can be marked on a given dimension line.	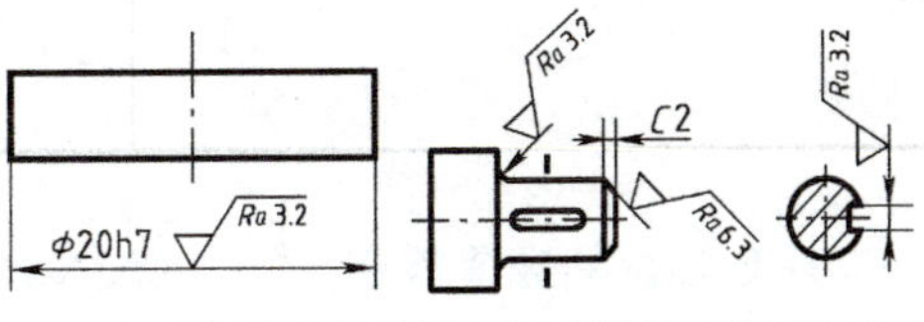 ϕ20h7　Ra 3.2　Ra 3.2　C2　Ra 6.3　Ra 3.2
表面结构要求也可以标注在几何公差框格的上方 The surface structure requirements can also be marked above the geometric tolerance bar.	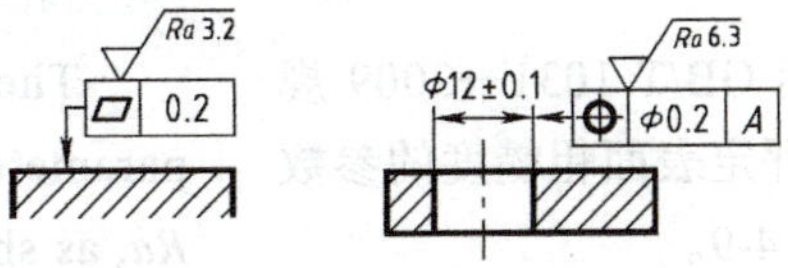 Ra 3.2　▱ 0.2　Ra 6.3　ϕ12±0.1　◎ ϕ0.2 A

（续）

说明 Description	实例 Example
如果在工件的多数表面有相同的表面结构要求，则其表面结构要求可统一标注在图样的标题栏附近，此时，表面结构要求的代号后面应有以下两种情况：①在圆括号内给出无任何其他标注的基本符号（图 a）；②在圆括号内给出不同的表面结构要求（图 b） If most surfaces of the workpiece have the same surface structure requirement, it could be labeled near the title bar together. Under this occasion, there are two situations behind the code of the surface structure requirement: ① the basic symbol without other labeling is given in the parentheses (Fig.a); ② different surface structure requirements are given in the parentheses (Fig.b).	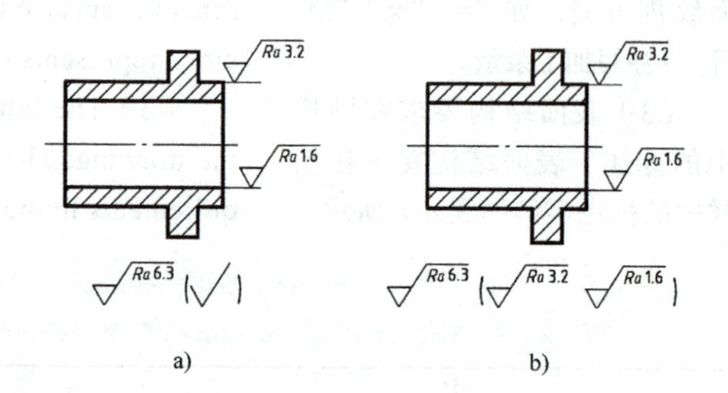
当多个表面有相同的表面结构要求或图纸空间有限时，可以采用简化注法。 The simplified labeling could be adopted when multiple surfaces have the same surface structure requirement or the drawing width is limited. ① 用带字母的完整图形符号，以等式的形式，在图形或标题栏附近，对有相同表面结构要求的表面进行简化标注（图 a）。 ① The labeling of surfaces with the same surface structure requirement are simplified using the complete graphic symbol with letters in the form of an equation near the graph or the title bar. ② 用基本图形符号或扩展图形符号，以等式的形式给出对多个表面共同的表面结构要求（图 b） ② The same surface structure requirement of multiple surfaces is given using the basic graphic symbol or extended graphic symbol in the form of an equation.	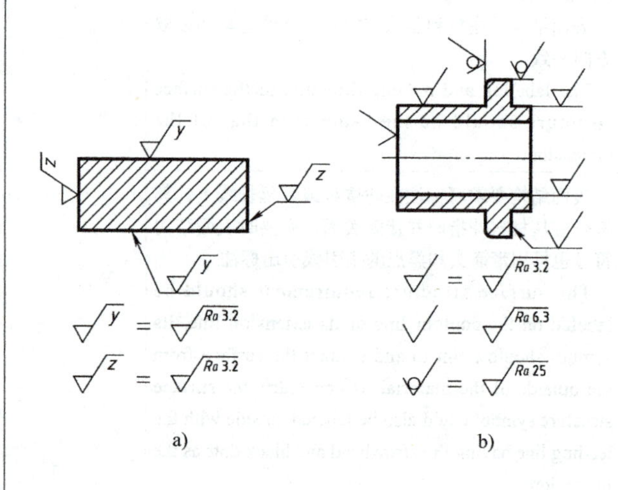

2. *Ra* 评定参数的数值规定与选用

国标 GB/T 1031—2009 规定了 *Ra* 评定表面粗糙度的参数值，见表 4-9。

2. Numerical value specification and selection of *Ra* evaluation parameters

The national standard GB/T 1031—2009 specifies the parameter values of the surface roughness evaluated using *Ra*, as shown in Table 4-9.

表 4-9 轮廓算术平均偏差 *Ra* 的数值 （单位：μm）

Table 4-9 The value of the arithmetic average deviation *Ra* （unit: μm）

基本系列 Basic series	补充系列 Supplementary series	基本系列 Basic series	补充系列 Supplementary series	基本系列 Basic series	补充系列 Supplementary series	基本系列 Basic series	补充系列 Supplementary series
	0.008	0.100			1.25		16.0
	0.010		0.125	1.60			20
0.012			0.160		2.0	25	
	0.016	0.20			2.5		32
	0.020		0.25	3.2			40
0.025			0.32		4.0	50	
	0.032	0.40			5.0		63
	0.040		0.50	6.3			80
0.050			0.63		8.0	100	
	0.063	0.80			10.0		
	0.080		1.00	12.5			

4.3.2 极限与配合

从一批规格相同的零（部）件中任取一件，不经修配，就能装到机器上去，并能保证使用要求，零件具有的这种性质称为互换性。现代化工业要求机器零（部）件具有互换性，这样，既能满足各生产部门广泛的协作要求，又能进行高效率的专业化生产。

1. 尺寸公差

制造零件时，为了使零件具有互换性，要求零件的尺寸在一个合理范围之内，由此就规定了极限尺寸。制成后的实际尺寸，应在规定的上极限尺寸和下极限尺寸范围内。允许尺寸的变动量称为尺寸公差，简称公差。

（1）公称尺寸与极限尺寸　公称尺寸是设计给定的尺寸，如 ϕ35 是根据计算和结构上的需要，所决定的尺寸。极限尺寸是允许尺寸变动的两个极限值，分别是上极限尺寸和下极限尺寸，它

4.3.2 Limits and Fits

The interchangeability of a part refers to that any piece from a batch with the same specification could be installed on the machine without repair and guarantee the operating requirements. The modern industry requires the interchangeability of machine parts, which can not only meet the extensive collaboration requirements of various production departments, but also carry out highly efficient specialized production.

1. Dimensional tolerance

4.3.2-1

In order to make a part interchangeable, the size of the part is required to be within a reasonable range, so the limit dimension is stipulated. The actual dimension of the finished product should be within the specified upper and lower limit dimensions. The amount of change of allowable dimension is called the dimensional tolerance.

(1) Nominal dimension and limit dimension. Nominal dimension is a given dimension, such as ϕ35 which is determined according to the computational and structural requirements. Limit dimensions are two limit values of allowable dimension changes including the upper limit dimension and lower limit dimension which are determined with the nominal dimension as the cardinal number. As shown in Fig.4-65,

是以公称尺寸为基数来确定的。如图 4-65 中孔的上极限尺寸为 35+0.025=35.025，下极限尺寸为 30−0=35，如图 4-65 所示。

the upper limit dimension of the hole is 35+0.025=35.025; the lower limit dimension is 30−0=35, as shown in Fig.4-65.

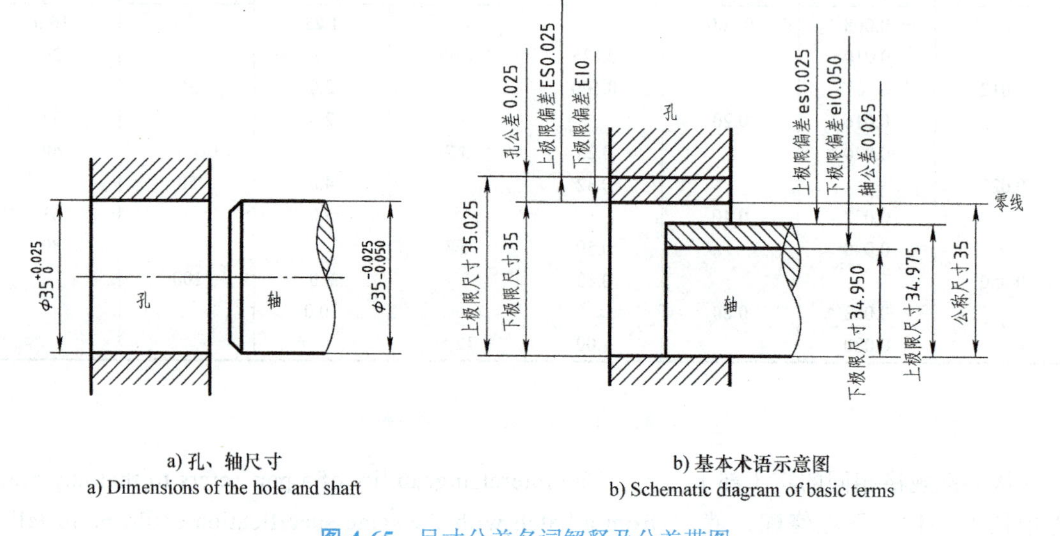

a) 孔、轴尺寸
a) Dimensions of the hole and shaft

b) 基本术语示意图
b) Schematic diagram of basic terms

图 4-65　尺寸公差名词解释及公差带图

Fig.4-65　Term explanation of dimensional tolerance and drawing of tolerance zone

（2）极限偏差与尺寸公差　偏差指某一实际尺寸减去其公称尺寸所得的代数差。极限偏差指上极限偏差和下极限偏差。上极限尺寸减去其公称尺寸所得的代数差就是上极限偏差；下极限尺寸减去其公称尺寸所得的代数差即为下极限偏差。

国标规定的偏差代号：孔的上、下极限偏差分别用 *ES* 和 *EI* 表示；轴的上、下极限偏差分别用 *es* 和 *ei* 表示，如图 4-65 所示。

孔的上极限偏差为 *ES*=35.025−35=+0.025。孔的下极限偏差为 *EI*=35−35=0。

尺寸公差（简称公差，用 *T* 表示）为允许尺寸的变动量。即上极限尺寸与下极限尺寸之

(2) Limit deviation and dimensional tolerance. Deviation is the algebraic difference obtained by the actual dimension subtracting the nominal dimension. Limit deviation refers to upper limit deviation and lower limit deviation. The algebraic difference obtained by the upper limit dimension subtracting the nominal dimension is the upper limit deviation. The algebraic difference obtained by the lower limit dimension subtracting the nominal dimension is the lower limit deviation.

Deviation codes stipulated in the national standard: The upper and lower limit deviations of the hole are respectively represented by *ES* and *EI*. The upper and lower limit deviations of the shaft are respectively represented by *es* and *ei*, as shown in Fig.4-65.

The upper limit deviation of the hole: *ES*=35.025−35=+0.025. The lower limit deviation of the hole: *EI*=35−35=0.

Dimensional tolerance (tolerance, *T*) is the amount of change of allowable dimension. It is the difference between

差，35.025−35=0.025；也等于上极限偏差与下极限偏差代数差的绝对值 |0.025−0|=0.025，如图 4-65 所示。

（3）公差带　零线是在公差带图中表示公称尺寸的一条直线，以其为基准确定公差和偏差，如图 4-65 所示。公差带是在公差带图中由代表上、下极限偏差的两条直线所限定的区域。图 4-65b 就是图 4-65a 的公差带图。

（4）标准公差与基本偏差　标准公差的数值与公称尺寸和公差等级有关。其中公差等级确定尺寸的精确程度，决定着加工的难易程度。标准公差分为 20 个等级，即 IT01、IT0、IT1 至 IT18。IT 表示标准公差，阿拉伯数字表示公差等级，它是反映尺寸精度的等级。IT01 公差数值最小，精度最高；IT18 公差数值最大，精度最低。

基本偏差是指在国家标准的极限与配合制中，决定公差带相对零线位置的那个极限偏差。它可以是上极限偏差或下极限偏差，一般是指靠近零线的那个偏差。

基本偏差系列图只表示公差带的位置，不表示公差的大小。因此，公差带一端是开口的，开口的一端由标准公差限定，国家标准对孔和轴分别规定了 28 个基本偏差，如图 4-66 所示。

孔和轴的公差带代号由基本偏差代号与公差等级代号组成。公差带代号标注含义如图 4-67 所示。

the upper limit dimension and the lower limit dimension. 35.025−35=0.025. It is also equal to the absolute value of the algebraic difference between the upper limit deviation and the lower limit deviation, |0.025−0|=0.025, as shown inn Fig.4-65.

（3）Tolerance zone. Zero line is a straight line representing the nominal dimension in the drawing of tolerance range, based on which the tolerance and deviation are determined, as shown in Fig.4-65. Tolerance zone is the area defined by two straight lines representing the upper and lower limit deviations in the drawing of tolerance zone. Fig.4-65b is the drawing of tolerance zone of Fig.4-65a.

（4）Standard tolerance and basic deviation.

The value of the standard tolerance is related to the nominal dimension and grade of tolerance. The grade of tolerance determines the accuracy of the dimension and the difficulty of processing. The standard tolerance is divided into 20 grades: IT01, IT0, IT1 to IT18. IT represents the standard tolerance; the number represents the grade of tolerance. They reflect the dimensional accuracy. IT01 has the smallest tolerance value and highest accuracy. IT18 has the largest tolerance value and the lowest accuracy.

Basic deviation refers to the limit deviation value determining the relative zero-line position of the tolerance zone within the limits and fits of the national standard. It could be the upper limit deviation or the lower limit deviation and usually the deviation near the zero line.

The basic deviation series diagram represents only the position of the tolerance zone, not the size of the tolerance. Therefore, the tolerance zone has one open end. It is determined by the standard tolerance. The national standard stipulates 28 basic deviation values for holes and shafts, as shown in Fig.4-66.

The tolerance zone codes of holes and shafts are mainly made up of codes of basic deviation and grade of tolerance. The meanings of tolerance zone codes are shown in Fig.4-67.

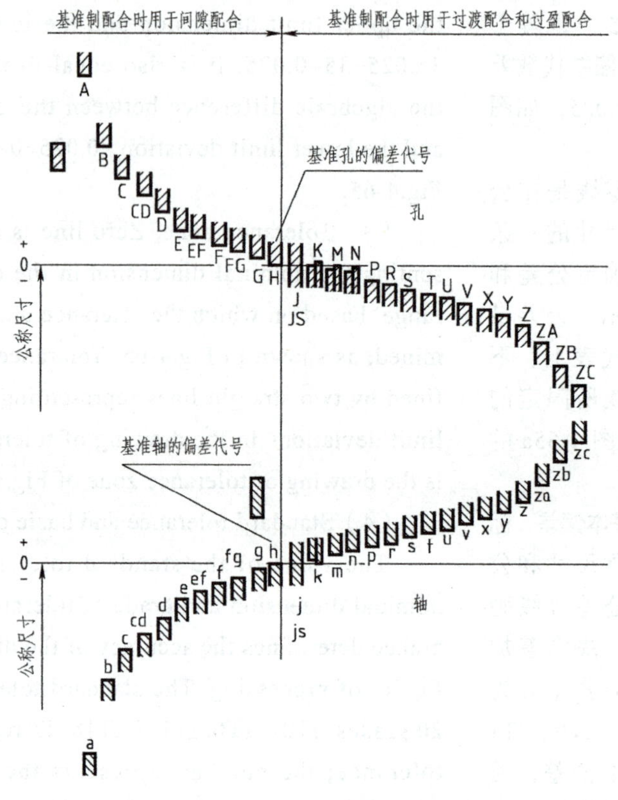

图 4-66 基本偏差系列

Fig.4-66 Basic deviation series

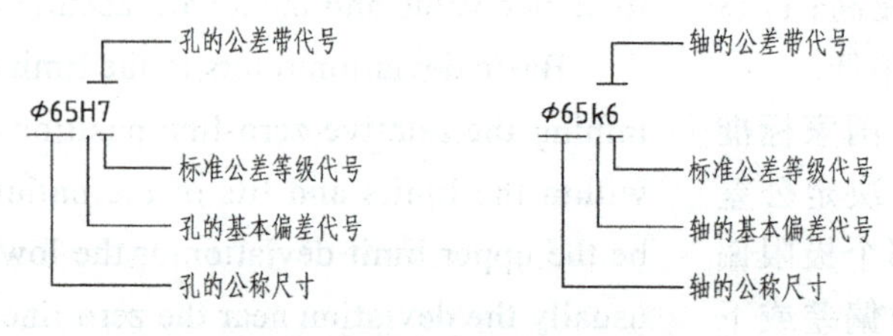

图 4-67 公差带代号标注含义

Fig.4-67 The meanings of tolerance zone codes

2. 配合

（1）配合种类 配合是指公称尺寸相同的相互结合的孔、轴公差带之间的关系。根据使用要求，孔与轴之间的配合有松有紧。国家标准规定配合分为三类：间隙配合、过盈配合和过渡配合。

2. Fits

（1）Types of fits. Fits refer to the relation between the tolerance zones of holes and shafts with the same nominal dimensions. According to different requirements in the application, the fits between holes and shafts could be loose or tight. Therefore, the national standard stipulates that fits have three types: clearance fit, interference fit and transition fit.

间隙配合指孔与轴装配时，有间隙（包括最小间隙等于零）的配合。如图 4-68a 所示，孔的公差带在轴的公差带之上。

Clearance fit: the fit has clearance (including zero clearance) in the assembling of holes and shafts. As shown in Fig.4-68a, the tolerance zone of the hole is above that of the shaft.

过渡配合指孔与轴装配时，可能有间隙或过盈的配合。如图 4-68b 所示，孔的公差带与轴的公差带互相交叠。

Transition fit: the fit might have clearance or interference in the assembling of holes and shafts. As shown in Fig.4-68b, the tolerance zone of the hole overlaps with that of the shaft.

过盈配合指孔与轴装配时，有过盈（包括最小过盈等于零）的配合。如图 4-68c 所示，孔的公差带在轴的公差带之下。

Interference fit: the fit has interference (including zero interference) in the assembling of holes and shafts. As shown in Fig.4-68c, the tolerance zone of the hole is under that of the shaft.

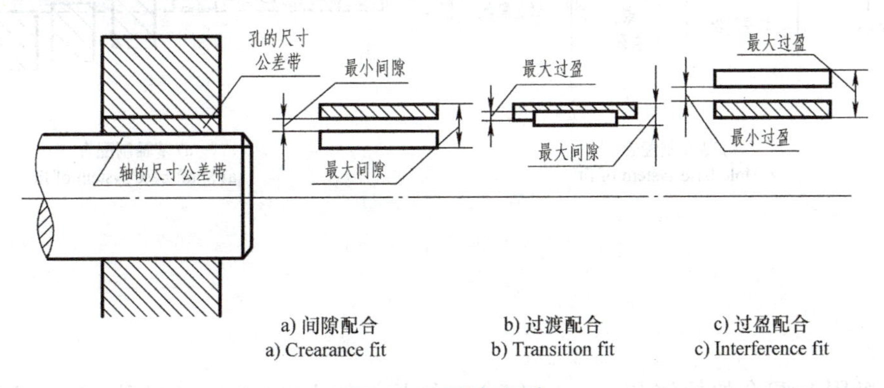

图 4-68 配合类别

Fig.4-68 Fit types

（2）配合制　在制造相互配合的零件时，使其中一种零件作为基准件，其基本偏差固定，通过改变另一零件的基本偏差来获得各种不同性质的配合制度称为配合制。根据生产实际需要，国家标准规定了两种配合制。

(2) Fit System. In the manufacture of matching parts, different properties could be realized by making one part as the benchmark part with fixed basic deviation and changing the basic deviation of the other part. This system is called the fit system. According to the actual needs in production, the national standard stipulates two kinds of fit systems.

基孔制配合指基本偏差为一定的孔的公差带，与不同基本偏差的轴的公差带形成各种配合的一种制度，如图 4-69a 所示。基准孔的下偏差为零，用代号 H 表示。

Hole-base system of fit: the system of various matches between the hole with the tolerance zone having a fixed basic deviation and the shaft with the tolerance zone having different basic deviations, as shown in Fig.4-69a. The lower deviation of the benchmark hole is zero and represented by the code H.

基轴制配合指基本偏差为一定的轴的公差带，与不同基本偏

Shaft-base system of fit: the system of various matches between the shaft with the tolerance zone having a fixed basic deviation and the hole with the tolerance zone having different basic deviations, as shown in Fig.4-69b. The upper

差的孔的公差带形成各种配合的一种制度，如图 4-69b 所示。基准轴的上偏差为零，用代号 h 表示。

deviation of the benchmark shaft is zero and represented by the code h.

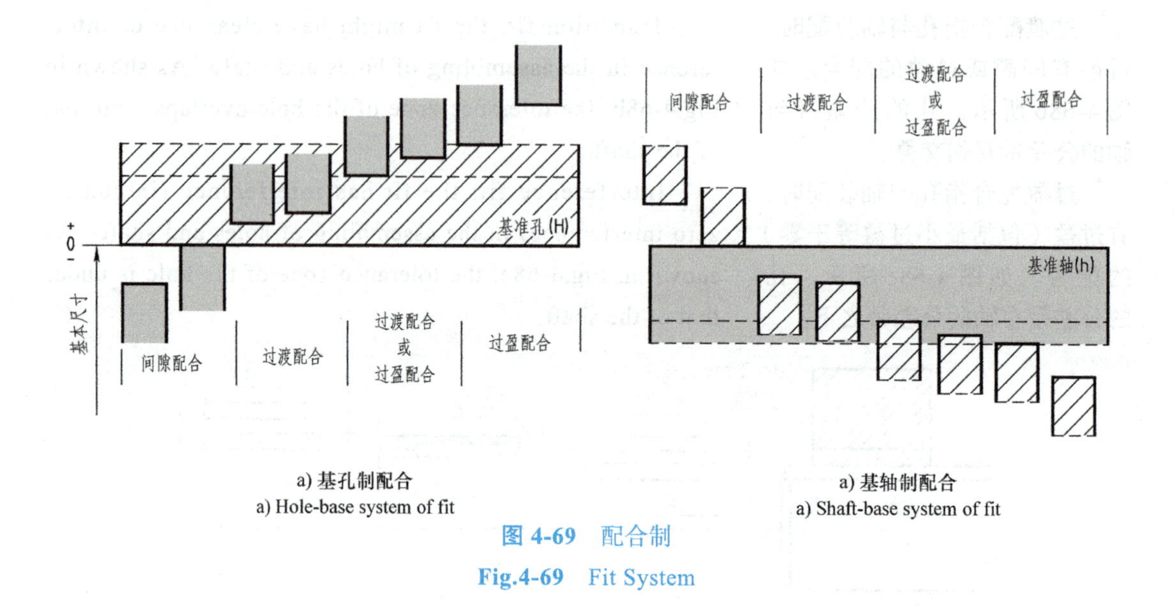

a) 基孔制配合
a) Hole-base system of fit

a) 基轴制配合
a) Shaft-base system of fit

图 4-69 配合制
Fig.4-69 Fit System

（3）极限与配合的标注与查阅 在装配图上标注极限与配合，采用组合式注法。它是在公称尺寸后面用一分数形式表示，分子为孔的公差带代号，分母为轴的公差带代号。通常分子中含 H 的为基孔制配合，分母中含 h 为基轴制配合，如图 4-70a 所示。

在零件图上标注公差的形式有三种：只注公差带代号，如图 4-70b 所示（用于成批生产的零件图上）；只注极限偏差数值，如图 4-70c 所示（用于单件或小批量生产的零件图上）；同时注公差带代号和极限偏差数值，如图 4-70d 所示（用于生产批量不定的零件图上）。

（3）Labeling and consulting of limits and fits. The limits and fits are labeled on the assembly drawing using the combined labeling method: a fraction after the nominal dimension; the numerator is the tolerance zone code of the hole; the denominator is the tolerance zone code of the shaft. Generally, if the numerator has H, it's the hole-base system of fit; if the denominator has h, it is the shaft-base system of fit, as shown in Fig.4-70a.

There are three types of tolerance labeling on the parts drawing: only label the tolerance zone code, as shown in Fig.4-70b (used for the parts drawings of batch production); only label the limit deviation value, as shown in Fig.4-70c (used for the drawings of a single part or small batch production);label the tolerance zone code and limit deviation value at the same time, as shown in Fig.4-70d (used for parts drawings of uncertain batch production).

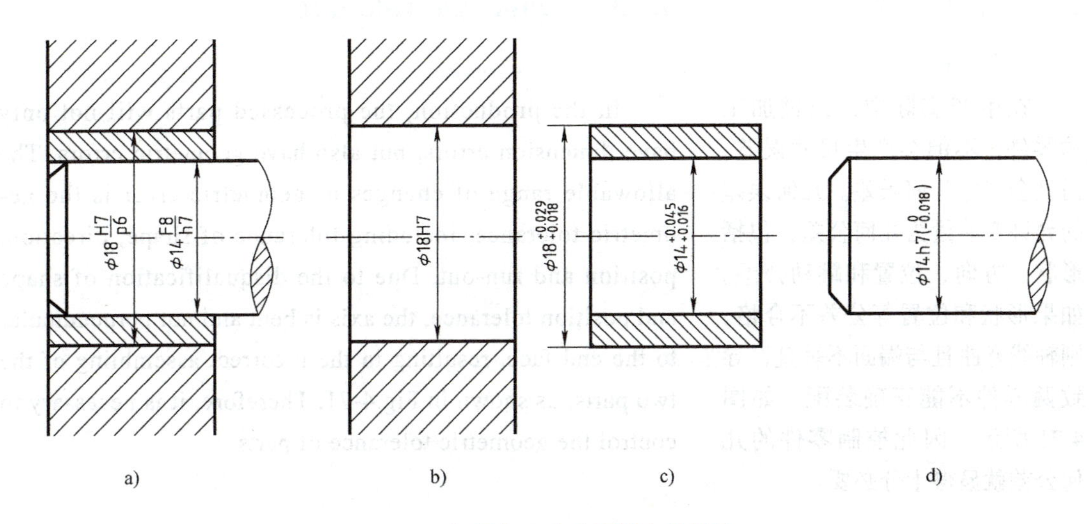

图 4-70 极限与配合在图样上的标注

Fig.4-70 Labeling of limits and fits on drawings

查表写出 ϕ18（H8/f7）的极限偏差数值。

解：对照基本偏差系列图 4-66 可知，H8/f7 是基孔制配合，其中 H8 是基准孔的公差带代号；f7 是配合轴的公差带代号。

1）ϕ18H8 基准孔的极限偏差，可由配套资源中的附表 E-5 中查得。在表中由公称尺寸从大于 14 至 18 的行和公差带 H8 的列相交处查得 $^{+27}_{0}$（即 +0.027 和 0mm）这就是基准孔的上、下极限偏差，所以，ϕ18H8 可写成 $\phi 18^{+0.027}_{0}$。

2）ϕ18f7 配合轴的极限偏差，可由配套资源中的附表 E-4 中查得。在表中由公称尺寸从大于 14 至 18 的行和公差带 f7 的列相交处查得 $^{-16}_{-34}$（−0.016，−0.034mm），它是配合轴的上、下极限偏差，所以 ϕ18f7 可写成 $\phi 18^{-0.016}_{-0.034}$。

Look up the table and write the limit deviation value.

According to the basic deviation series of Fig.4-66, H8/F7 is the hole-base system of fit, H8 is the tolerance zone code of the benchmark hole, f7 is the tolerance zone code of the shaft.

1）ϕ18H8 is the limit deviation of the benchmark hole, which can be found in the attached table E-5 of the supporting resources. In the table, (+0.027 and 0mm) can be found in the intersection of the nominal dimension row of 14-18 and tolerance zone column of H8. This is the upper and lower deviation of the benchmark hole. Therefore, ϕ18H8 can be written as $\phi 18^{+0.027}_{0}$.

2）ϕ18f7 is the limit deviation of the corresponding shaft, which can be found in the attached table E-4 of the supporting resources. In the table, $^{-16}_{-34}$ (−0.016, −0.034mm) can be found in the intersection of the nominal dimension row of 14-18 and tolerance zone column of f7. It is the upper and lower deviation of the corresponding shaft. Therefore, ϕ18f7 can be written as $\phi 18^{-0.016}_{-0.034}$.

4.3.3 几何公差

1. 基本概念

在生产实际中，经过加工的零件，不但会产生尺寸误差，而且会产生几何误差。几何误差的允许变动量为几何公差，包括形状、方向、位置和跳动公差。如果形状和位置等公差不合格，则轴线弯曲且与端面不垂直，导致两零件不能正确装配，如图4-71所示。因此控制零件的几何公差就显得十分必要。

4.3.3 Geometric Tolerance

1. Basic Concepts

In the production, the processed parts will not only have dimension errors, but also have geometric errors. The allowable range of changes of geometric error is the geometric tolerance, including tolerance of shape, direction, position and run-out. Due to the disqualification of shape and position tolerance, the axis is bent and not perpendicular to the end face, resulting in the incorrect assembling of the two parts, as shown in Fig.4-71. Therefore, it is necessary to control the geometric tolerance of parts.

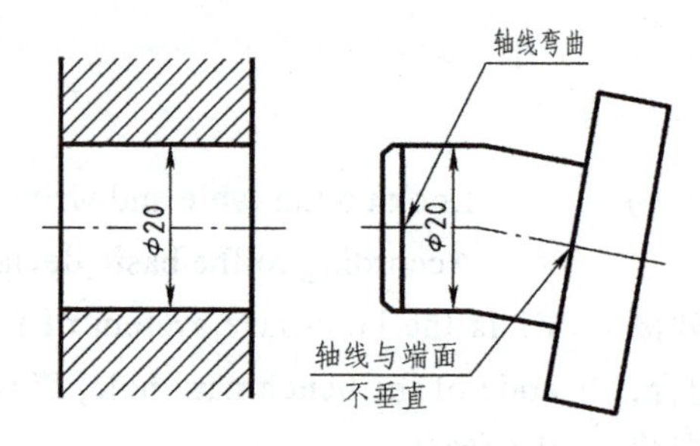

图 4-71 几何误差

Fig.4-71 Geometric error

2. 几何公差的几何特征和符号

表4-10为几何公差的类型、几何特征和符号（GB/T 1182—2018）

3. 几何公差的标注

几何公差采用公差框格、几何特征符号、公差值、基准、被测要素以及其他附加符号等标注，几何公差标注形式如图4-72所示。

第一格为几何公差项目的符号；第二格为几何公差数值和有关符号；第三格及以后各格为基准代号字母和有关符号。

2. Geometric features and symbols of geometric tolerances

Table 4-10 is the types, geometric features and symbols of geometric tolerances (GB/T 1182—2018).

3. Labeling of the geometric tolerance

The geometric tolerance is labeled with tolerance frames, geometric feature symbols, tolerance values, benchmark, measured elements and other additional symbols. The labeling patterns of geometric tolerance are shown in Fig.4-72.

The first frame: item symbol of geometric tolerance. The second frame: geometric tolerance value and related symbols. The third frame and subsequent frames: benchmark code letters and related symbols.

表 4-10　几何公差的类型、几何特征及符号

Table 4-10 Types, geometric features and symbols of geometric tolerances

公差类型 Specification	几何特征 Characteristics	符号 Symbol	有无基准 Datum needed
形状公差 Form tolerances	直线度 Straightness	—	无
	平面度 Flatness	⏥	无
	圆度 Roundness	○	无
	圆柱度 Cylindricity	⌭	无
	线轮廓度 Line profile	⌒	无
	面轮廓度 Surface profile	⌓	无
方向公差 Orientationtolerances	平行度 Parallelism	//	有
	垂直度 Perpendicularity	⊥	有
	倾斜度 Angularity	∠	有
	线轮廓度 Line profile	⌒	有
	面轮廓度 Surface profile	⌓	有

公差类型 Specification	几何特征 Characteristics	符号 Symbol	有无基准 Datum needed
位置公差 Location tolerance	位置度 Position	⌖	有或无
	同心度 （用于中心度） Concentricity	◎	有
	同轴度 （用于轴线） Coaxiality	◎	有
	对称度 Symmetry	⌯	有
	线轮廓度 Line profile	⌒	有
	面轮廓度 Surface profile	⌓	有
跳动公差 Run-out tolerance	圆跳动 Circle run-out	↗	有
	全跳动 Total run-out	⌰	有

指引线连接被测要素和公差框格，指引线的箭头指向被测要素的表面或其延长线，箭头方向一般为公差带的方向。

The leader line connects the measured element and tolerance frame. The arrowhead of the leader line points to the surface of the measured element and its extension line. The direction of the arrowhead is usually the direction of the tolerance zone.

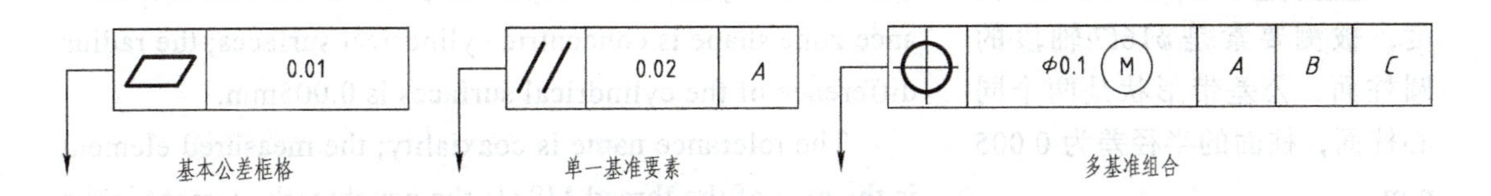

图 4-72　几何公差标注形式

Fig.4-72　The labeling patterns of geometric tolerance

GB/T 1182—2018 中基准符号的规定：由一个基准方框和一个涂黑或空白的基准三角形（等边），用细实线连接而构成。基准代号标注形式如图 4-73 所示。

The provision of benchmark symbols in GB/T 1182—2018: it is consisted of a benchmark frame and a black or blank benchmark triangle with the thin line connecting them. The labeling patterns of benchmark codes are shown in Fig.4-73.

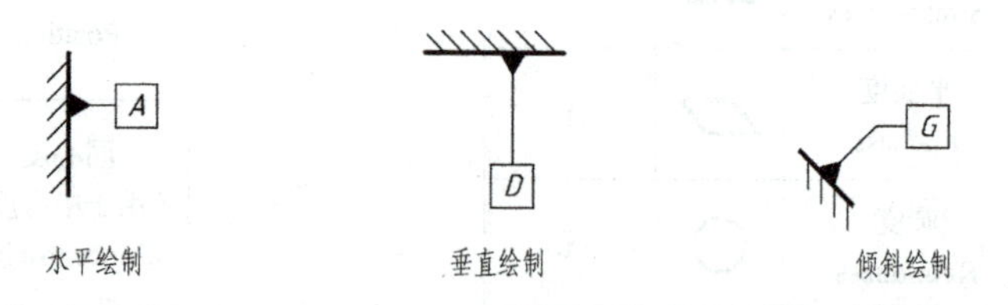

图 4-73 基准代号标注形式

Fig.4-73 The labeling patterns of benchmark codes

无论基准符号在图面上的方向如何，其方框中的字母都应水平书写。

The letter in the frame should be written horizontally regardless of the direction of the benchmark symbol in the drawing.

被测要素的基准在图样上用英文大写字母表示，为避免混淆，不得采用 *E*、*F*、*I*、*J*、*L*、*M*、*O*、*P*、*R* 九个字母。基准字母一般不许与图样中任何向视图的字母相同。

The benchmark of measured elements should be written in capital English letters in the drawing. To avoid confusion, nine letters *E*, *F*, *I*, *J*, *L*, *M*, *O*, *P* and *R* should not be used. The benchmark letters are generally not the same with the letters of any auxiliary views in the drawing.

4. 几何公差标注案例

4. The labeling examples of geometric tolerance

图 4-74 所示为螺纹量规的工作样图。

The working template of thread gauge is shown in Fig.4-74.

[↗ | 0.03 | A] 公差名称为圆跳动，被测要素是左球面，基准要素是 ϕ16f7 轴段轴线，公差带形状是以基准轴线为圆心的同心圆，同心圆的半径差为 0.03 mm。

The tolerance name is circle run-out; the measured element is the left sphere; the benchmark element is ϕ16f7 shaft end axis; the tolerance zone shape is the concentric circles with the benchmark axis as the center; the radius difference of the concentric circles is 0.03mm.

[⌭ | 0.005] 公差名称为圆柱度，被测要素是 ϕ16f7 轴段的圆柱面，公差带形状是两个同心柱面，柱面的半径差为 0.005 mm。

The tolerance name is cylindricity; the measured element is the cylindrical surface of ϕ16f7 shaft end; the tolerance zone shape is concentric cylindrical surfaces; the radius difference of the cylindrical surfaces is 0.005mm.

[◎ | ϕ0.1 | A] 公差名称为同轴度，被测要素是螺纹 M8x1 的

The tolerance name is coaxiality; the measured element is the axis of the thread M8x1; the benchmark element is the axis of the ϕ16f7 shaft end; the tolerance zone shape is the cylindrical surface with the benchmark axis as the axis; the

轴线，基准要素是 ϕ16f7 轴段的轴线，公差带形状是以基准轴线为轴线的圆柱面，圆柱面的直径差是 ϕ0.1mm。

radius difference of the cylindrical surface is ϕ0.1mm.

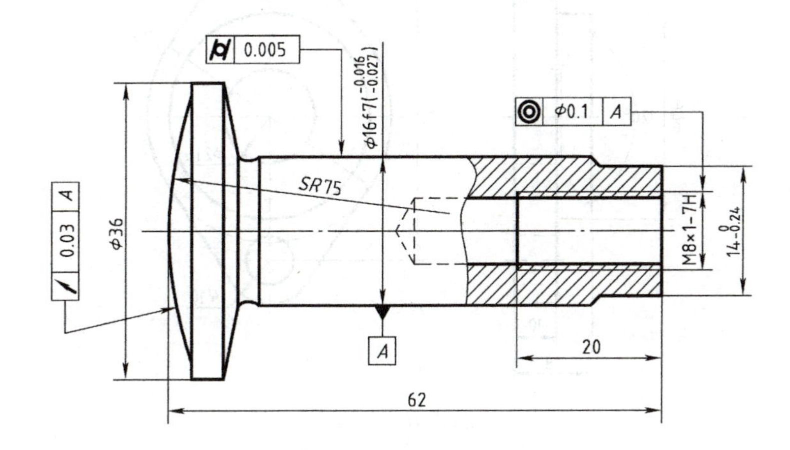

图 4-74 几何公差标注案例

Fig.4-74 The labeling examples of geometric tolerance

[本章习题]

[Chapter exercises]

1. 绘制如图 4-75 所示机件的全剖视图。

1. Draw the full view of the part as shown in Fig.4-75.

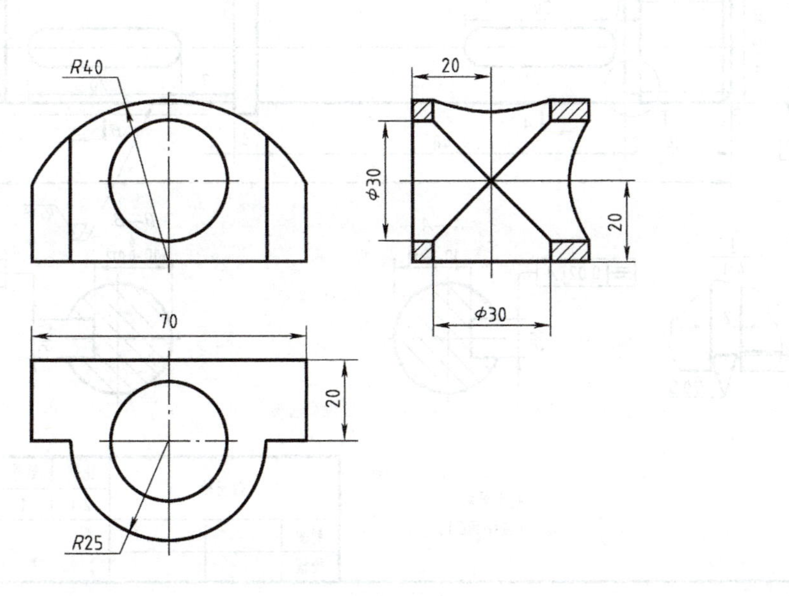

图 4-75 机件视图

Fig.4-75 view of machine parts

2. 请用合理地方法表达如图 4-76 所示压盖。

2. Express the grand as shown in Fig.4-76 reasonably.

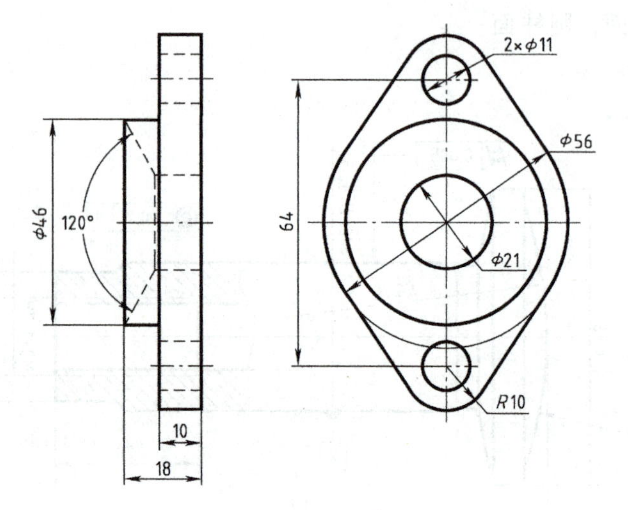

图 4-76 压盖

Fig.4-76 Gland

3. 绘制如图所示的输出轴

3. Draw the output shaft as shown in Fig.4-77.

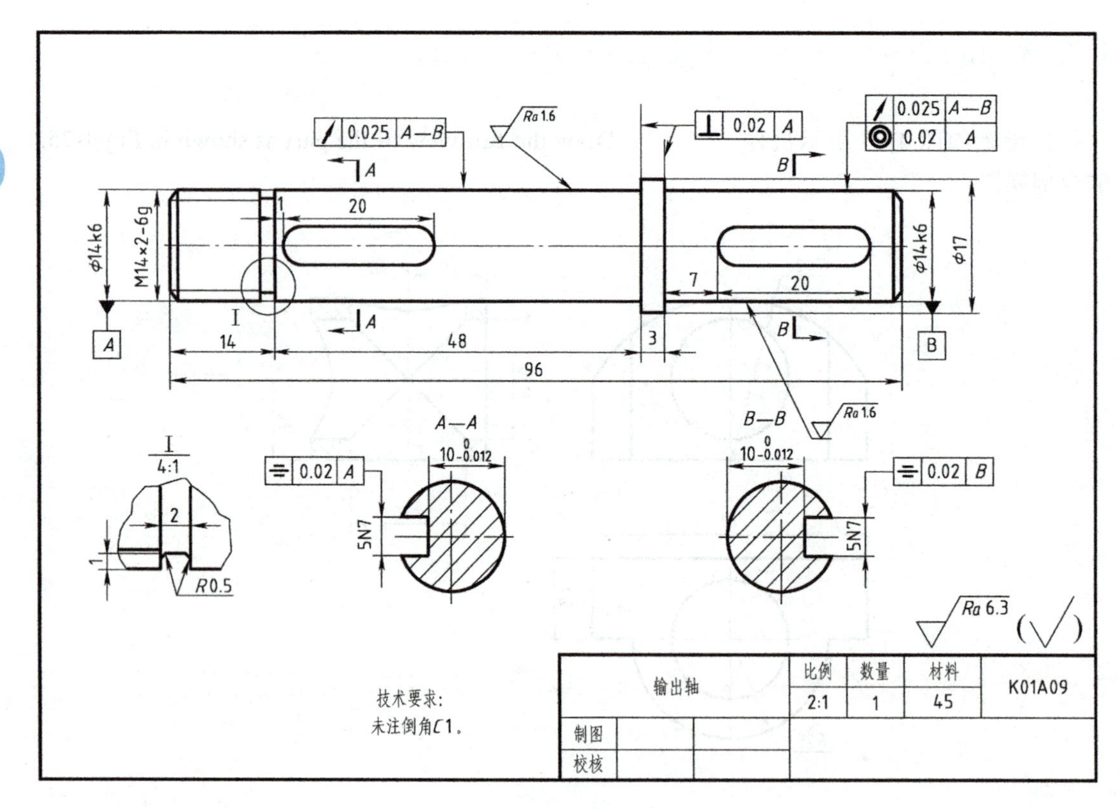

图 4-77 输出轴

Fig.4-77 Output shaft

第5章 常用机件及结构要素的表示

Chapter 5 The Representation of Commonly-used Machine Parts and Structural Elements

5.1 螺纹

5.1 Threads

5.1.1 螺纹概述

5.1.1 Threads Summary

5.1.1

1. 螺纹的概念

螺纹是指在圆柱或圆锥表面上，沿螺旋线所形成的，具有相同断面的连续凸起和沟槽。在圆柱外表面上形成的螺纹，称为外螺纹；在圆柱内表面上形成的螺纹，称为内螺纹。内外螺纹成对使用，可用于各种机械连接，传递运动和动力。

图 5-1 所示为在车床上车削加工外螺纹和内螺纹的情形。

1. The concept of the thread

A thread refers to the continuous convex and groove with the same fracture surface along the spiral line on the surface of a cylinder or circular cone. The thread on the external surface is called the external thread. The thread on the internal surface is called the internal thread. The internal and external threads are used in pairs and can be used for various mechanical connections to transfer motion and power.

Fig.5-1 is the graph of the processing of the external and internal threads on the lathe.

2. 螺纹的基本要素

螺纹的基本要素有牙型、直径、螺距、线数和旋向。

（1）牙型　在通过螺纹轴线的断面上，螺纹的轮廓形状称为牙型。常见的螺纹牙型有三角形、梯形、锯齿形等。

2. Basic elements of the thread

The basic elements of a thread include thread type, diameter, thread pitch, line number and screwing direction.

（1）The thread type is called the thread shape on the cross section of the thread axis, and the outline shape of the thread. The common thread types are triangular, trapezoidal, serrated, etc.

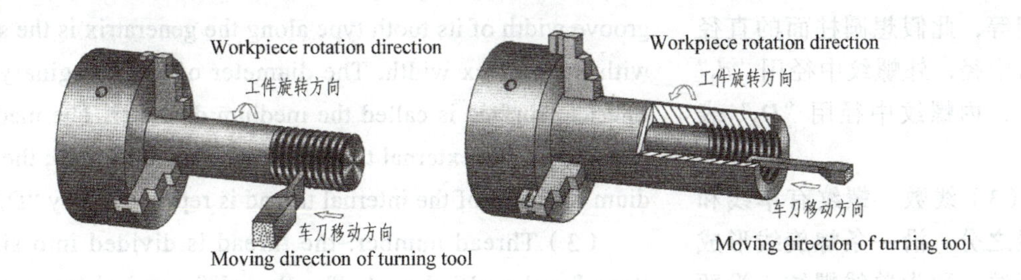

a) 车外螺纹
a) External thread lathing

b) 车内螺纹
b) Internal thread lathing

图 5-1　车削螺纹
Fig.5-1　Thread turning

（2）直径　根据螺纹的结构特点，螺纹的直径主要分为以下几种，如图 5-2 所示。

（2）According to the structural features of the thread, the diameter of the thread is mainly divided into the following types, as shown in Fig.5-2.

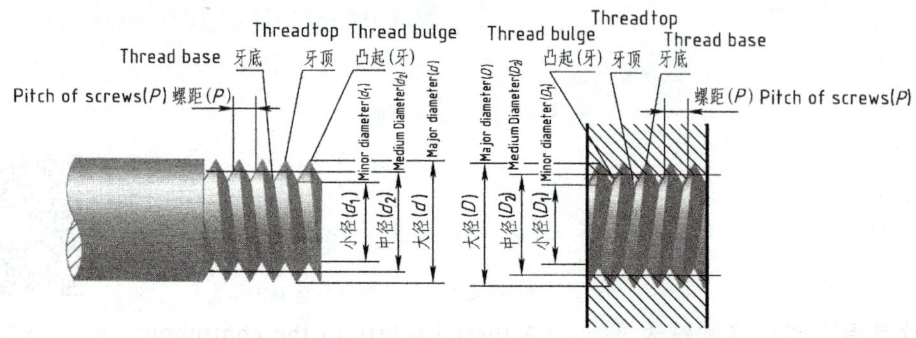

图 5-2　螺纹直径

Fig.5-2　Thread diameters

1）大径：螺纹的最大直径，又称公称直径，即与外螺纹的牙顶或内螺纹的牙底相重合的假想圆柱面的直径。外螺纹的大径用“d”表示，内螺纹的大径用“D”表示。

2）小径：螺纹的最小直径，即与外螺纹的牙底或内螺纹的牙顶相重合的假想圆柱面的直径。外螺纹的小径用“d_1”表示，内螺纹的小径用“D_1”表示。

3）中径：在大径和小径之间有一假想圆柱面，在其母线上牙型的沟槽宽度和凸起宽度相等，此假想圆柱面的直径称为中径，外螺纹中径用“d_2”表示，内螺纹中径用“D_2”表示。

（3）线数　螺纹有单线和多线之分。沿一条螺旋线形成的螺纹，称为单线螺纹；沿两条或两条以上，且在轴向等距离分布的螺旋线所形成的螺纹，称为多线螺纹，螺纹的线数用 n 来表示，如图 5-3 所示。

1）Major diameter: it is the maximum diameter of the thread, also called the nominal diameter. It is the diameter of the imaginary cylindrical surface where the thread top of the external thread and the thread base of the internal thread coincide with each other. The major diameter of the external thread is represented by “d”; the major diameter of the internal thread is represented by “D”.

2）Minor diameter: it is the minimum diameter of the thread. It is the diameter of the imaginary cylindrical surface where the thread top of the internal thread and the thread base of the external thread coincide with each other. The minor diameter of the external thread is represented by “d_1”, and the minor diameter of the internal thread is represented by “D_1”.

3）Medium diameter: there is an imaginary cylindrical surface between the major diameter and minor diameter, the groove width of its tooth type along the generatrix is the same with the convex width. The diameter of this imaginary cylindrical surface is called the medium diameter. The medium diameter of the external thread is represented by “d_2”; the medium diameter of the internal thread is represented by “D_2”.

（3）Thread number: the thread is divided into single thread and multi thread. The thread formed along a spiral line is called the single thread. The thread formed along two or more spiral lines which are located at the same distance along the axis is called the multi thread. The thread number is represented by n, as shown in Fig.5-3.

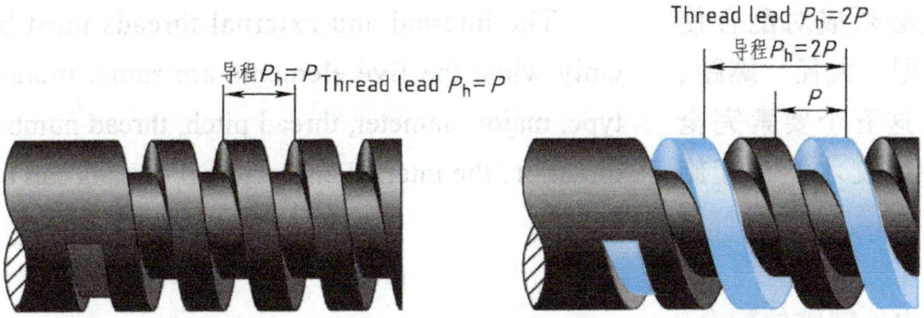

a) 单线螺纹
a) Single thread spiral

b) 双线螺纹
b) Double thread spiral

图 5-3　螺距与导程

Fig.5-3　Thread pitch and lead

（4）螺距和导程　相邻两牙在中径线上对应两点间的轴向距离，称为螺距，用“P”表示。在同一螺旋线上的相邻两牙在中径线上对应两点间的轴向距离，称为导程，用“P_h”表示。若螺旋线数为 n，则导程与螺距有如下关系：$P_h = nP$。

（5）旋向　螺纹分左旋和右旋两种，顺时针旋转时旋入的螺纹，称为右旋螺纹；逆时针旋转时旋入的螺纹，称为左旋螺纹，如图 5-4 所示。

旋向的判定：将外螺纹轴线垂直放置，螺纹的可见部分右高左低者为右旋螺纹，反之为左旋螺纹，如图 5-4 所示，工程上常用右旋螺纹。

（4）Screw pitch and lead: the axis distance between two corresponding points of two adjacent teeth along the pitch line is called the screw pitch, represented by “P”. The axis distance between two corresponding points of two adjacent teeth of the spiral line along the pitch line is called the lead, represented by “P_h”. If the thread number is n, the lead and the screw pitch have the following relation: $P_h = nP$.

（5）Screw direction: the thread is divided into left-hand thread and right-hand thread. The thread screwing clockwise is called right-hand thread. The thread screwing anticlockwise is called left-hand thread, as shown in Fig.5-4.

Determination of screwing direction: place the axis of the external thread vertically, if the visible part of the thread is higher in the right and lower in the left, the thread is right-hand thread, otherwise it is the left-hand thread, as shown in Fig.5-4. The right-hand thread is commonly-used in engineering.

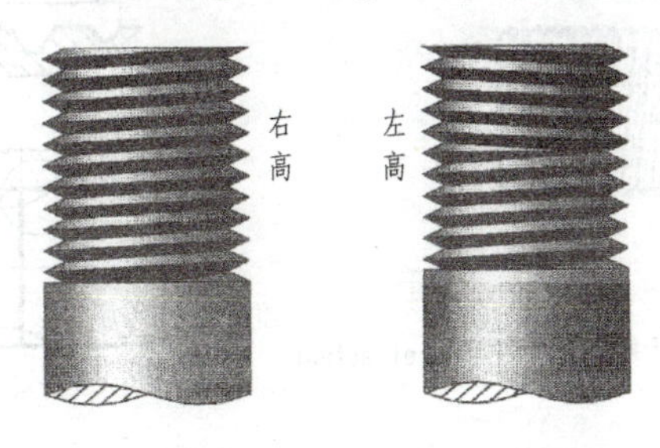

a) 右旋螺纹
a) Right-hand thread

b) 左旋螺纹
b) Left-hand thread

图 5-4　螺纹旋向

Fig.5-4

内外螺纹必须成对配合使用，螺纹的牙型、大径、螺距、线数和旋向，这五个要素完全相同时，内外螺纹才能相互旋合。

The internal and external threads must be used in pair. Only when the five elements are same, including the thread type, major diameter, thread pitch, thread number and screwing direction, the internal and external threads can be screwed.

3. 倒角、螺尾及退刀槽

3. Chamfering, washout thread and tool withdrawal groove

为了便于内外螺纹的旋合，在螺纹的端部制成 45° 的倒角，如图 5-5 所示。在制造螺纹时，由于退刀的缘故，螺纹的尾部会出现渐浅部分，这种不完整的牙型，称为螺尾，如图 5-6 所示。为了消除这种现象，应在螺纹终止处加工一个退刀槽，如图 5-7 所示。

In order to facilitate the screwing of the internal and external threads, the end of the thread is made into the chamfering of 45°, as shown in Fig.5-5. In the manufacturing of the thread, due to the tool withdrawal, the thread will be gradually shallower at the end. This incomplete thread type is called washout thread, as shown in Fig.5-6. To eliminate this phenomenon, a tool withdrawal groove shall be processed at the end of the thread, as shown in Fig.5-7.

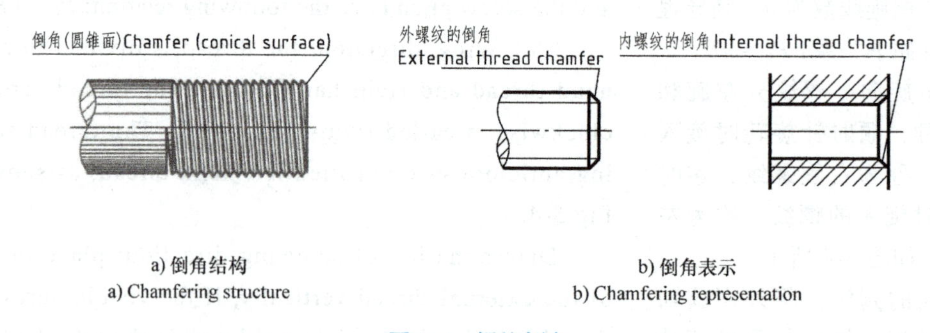

a) 倒角结构
a) Chamfering structure

b) 倒角表示
b) Chamfering representation

图 5-5 螺纹倒角

Fig.5-5 Thread chamfering

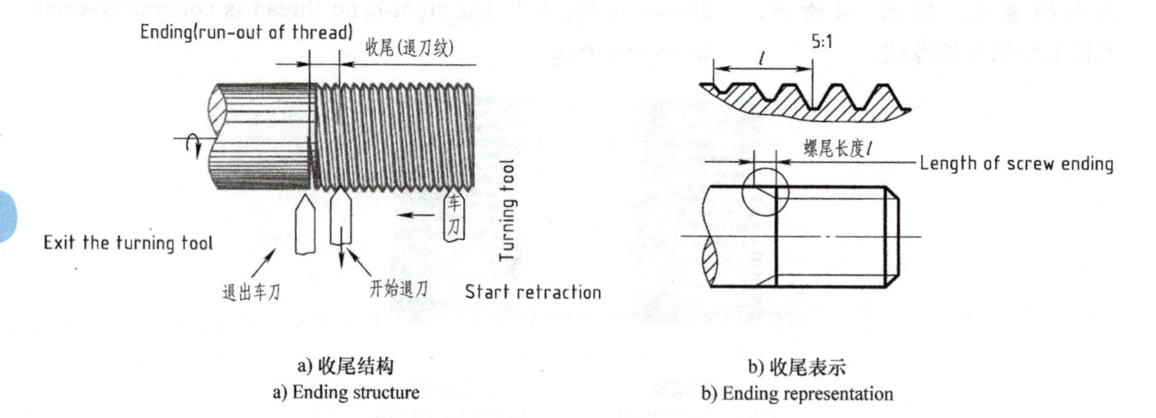

a) 收尾结构
a) Ending structure

b) 收尾表示
b) Ending representation

图 5-6 螺纹收尾

Fig.5-6 Thread ending

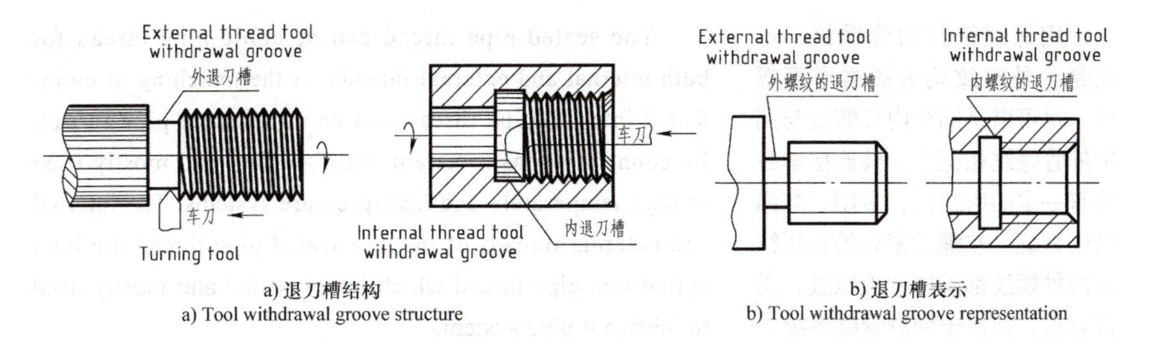

a) 退刀槽结构
a) Tool withdrawal groove structure

b) 退刀槽表示
b) Tool withdrawal groove representation

图 5-7　螺纹退刀槽

Fig.5-7　Spiral tool withdrawal groove

对于不穿通的螺纹孔，应将钻孔深度和螺纹孔深度分别画出，钻孔深度比螺纹孔深度深 0.5d。底部的锥顶角应画成 120°，但不必标注尺寸，如图 5-8 所示。

For non-penetrating screw holes, the drilling depth and the screw hole depth should be drawn respectively. The drilling depth is 0.5d deeper than that of the screw hole. The cone apex angle at the bottom shall be drawn as 120°, but needs not dimensioned, as shown in Fig.5-8.

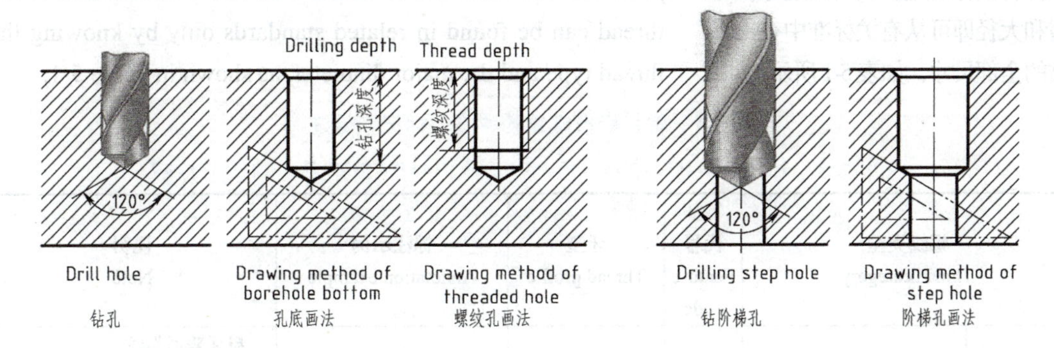

图 5-8　钻孔底部与阶梯孔的画法

Fig.5-8　The drawing methods of borehole bottom and stepped hole

5.1.2　螺纹的种类和标注

1. 螺纹的种类

（1）螺纹按用途可分为联接螺纹和传动螺纹两大类，如表 5-1 所示。

联接螺纹：起连接作用的螺纹。常用的联接螺纹有普通螺纹、管螺纹、锥管螺纹。其中普通螺纹又可分为粗牙普通螺纹和细牙普通螺纹两种，管螺纹又可分为用螺纹密封的管螺纹和非螺纹密封的管螺纹两种。

5.1.2　Types and Labeling of the Thread

1. Thread Types

（1）The thread can be divided into two categories: connecting thread and transmission thread, as shown in Table 5-1.

5.1.2-1

Connecting thread: thread having the connect function. Most commonly used connecting threads include common thread, pipe thread, taper pipe thread. The common thread can be divided into coarse thread and fine thread. The pipe thread can be divided into sealed pipe thread and non-sealed pipe thread.

用螺纹密封的管螺纹，可以是内外螺纹均为圆锥形管螺纹，也可以是圆柱内管螺纹与圆锥外管螺纹相配合。其连接本身具有一定的密封性，多用于高温高压系统。非螺纹密封的管螺纹其内外螺纹都是圆柱管螺纹，无密封性，常用于润滑管路系统。

The sealed pipe thread can be conic pipe thread for both internal and external threads, or the matching of cylindrical internal pipe thread and conic external pipe thread. Its connection has a certain sealing effect and mostly used in high-temperature and high-pressure systems. The internal and external threads of the non-sealed pipe thread are both cylindrical pipe thread which is not sealed and mostly used in lubricant pipe system.

传动螺纹：用于传递动力和运动的螺纹。最常用的有梯形和锯齿形螺纹，其中梯形螺纹应用最广。

Transmission thread: the thread used to transmit power and motion. The most commonly used transmission threads are trapezoid and zigzag threads of which the trapezoid thread is the most widely used.

（2）按符合国家标准的情况，螺纹可分为以下三类。

(2) According to the national standard, the thread can be divided into three types.

标准螺纹：牙型、大径、螺距符合国标规定，只需知道螺纹牙型和大径即可从有关标准中查出螺纹的全部尺寸，如表 5-1 所示。

Standard thread: thread type, major diameter and thread pitch meet the national standard. The entire dimensions of the thread can be found in related standards only by knowing the thread type and the major diameter, as shown in Table 5-1.

表 5-1　常用标准螺纹的种类、标记和标注

Table 5-1　Types, markers and labeling of the common standard thread

螺纹类型 Thread category		特征代号 Feature code	牙型 Thread profile	标注示例 Annotation example	说明 Note
联接和紧固用螺纹 Thread for connection and fastening	粗牙普通螺纹 Coarse pitch thread	M	60°	M16	粗牙普通螺纹 公称直径 16mm；中径公差带和顶径公差带均为 6g（省略不标）；中等旋合长度；右旋 Coarse pitch thread Nominal diameter: 16mm; medium diameter tolerance zone and top diameter tolerance zone: 6g (omitted); medium screwing length; right-handed
	细牙普通螺纹 Fine pitch thread			M16×1	细牙普通螺纹 公称直径 16mm，螺距 1mm；中径公差带和顶径公差带均为 6H（省略不标）；中等旋合长度；右旋 Fine pitch thread Nominal diameter is 16mm; pitch is 1mm; pitch tolerance zone and pitch tolerance zone are both 6h (omitted); medium screw length; right-handed

（续）

螺纹类型 Thread category			特征代号 Feature code	牙型 Thread profile	标注示例 Annotation example	说明 Note
55° 管螺纹 55°pipe thread	55° 非密封管螺纹 55°non sealing pipe thread		G		G1A G1	55° 非密封管螺纹 G——螺纹特征代号 1——尺寸代号 A——外螺纹公差等级代号 55°non sealing pipe thread G——thread feature code 1——dimension code A—— external thread tolerance grade code
	55° 密封管螺纹 55°sealing pipe thread	圆锥内螺纹 Conical internal thread	Rc	55°	$Rc1\frac{1}{2}$ $R_1 1\frac{1}{2}$	55° 密封管螺纹 Rc——圆锥内螺纹 Rp——圆柱内螺纹 R_1——与圆柱内螺纹相配合的圆锥外螺纹 R_2——与圆锥内螺纹相配合的圆锥外螺纹 $1\frac{1}{2}$——尺寸代号 55°sealing pipe thread Rc——taper internal thread Rp——cylindrical internal thread R_1——conical external thread matching with cylindrical internal thread R_2——Taper external thread matching with tapered internal thread
		圆柱内螺纹 Cylindrical internal thread	Rp			
		圆锥外螺纹	R_1 R_2			
传动螺纹 Drive thread	梯形螺纹 Trapezoidal thread		Tr	30°	Tr36×12(P6)−7H	梯形螺纹 公称直径 36mm，双线螺纹，导程 12mm，螺距 6mm；中径公差带为 7H；中等旋合长度；右旋 Trapezoidal thread Nominal diameter 36mm, double thread, lead 12mm, pitch 6mm; pitch tolerance zone 7H; medium screw length; right-handed

特殊螺纹：牙型符合国标规定，大径、螺距不符合国标规定。

非标准螺纹：牙型、大径、螺距均不符合国标规定。

Special thread: thread type meets the national standard; major diameter and thread pitch don't meet the national standard.

Non-standard thread: thread type, major diameter and thread pitch don't meet the national standard.

2. 螺纹的标注

由于螺纹采用规定画法，因此各种螺纹的画法都是相同的。不同螺纹的种类和要素只能通过标注来区分。螺纹标注的一般形式为：

2. Thread Labeling

As the drawing of the thread adopts the stipulated method, the drawing method of different threads is the same. Therefore, the types and elements of different threads can only be distinguished by labeling. The general form of thread labeling is:

5.1.2-2

螺纹特征代号公称直径 × 螺距(或Ph导程/P螺距) – 公差带代号 – 旋合长度代号 – 旋向

Thread feature code/ nominal diameter × thread pitch (or Ph lead/ P thread pitch) – Tolerance zone code – screwing length code – screwing direction

无论何种螺纹，左旋螺纹均应注写旋向代号LH，右旋螺纹不注旋向，单线螺纹只注螺距，多线螺纹则注Ph导程和P螺距。

公差带代号由中径公差带和顶径公差带两组组成，它们都是由表示公差等级的数字和表示公差带位置的字母组成。大写字母表示内螺纹，小写字母表示外螺纹。若两组公差带相同，则只标注一组。

旋合长度分为短（S）、中（N）、长（L）三种，中等旋合长度最为常用。当采用中等旋合长度时，不标注旋合长度代号。

No matter what kind of the thread is, the screwing direction code of the left-hand thread is LH; the screwing direction of the right-hand thread doesn't have to be labeled; the single thread shall be labeled only with the thread pitch; the multiple thread shall be labeled with Ph lead and P thread pitch.

The tolerance zone code is consisted of medium diameter tolerance zone and crest diameter tolerance zone which are composed of numbers indicating the tolerance grade and letters indicating the tolerance zone location. Capital letters represents the internal thread; the lowercase letters represent the external thread. If two tolerance zones are the same, only one of them needs to be labeled.

The screwing length is divided into three types: short (S), medium (N) and long (L). The medium screwing length is the most common one. When the medium screwing length is adopted, its code doesn't have to be labeled.

普通螺纹标记：

Common thread labels:

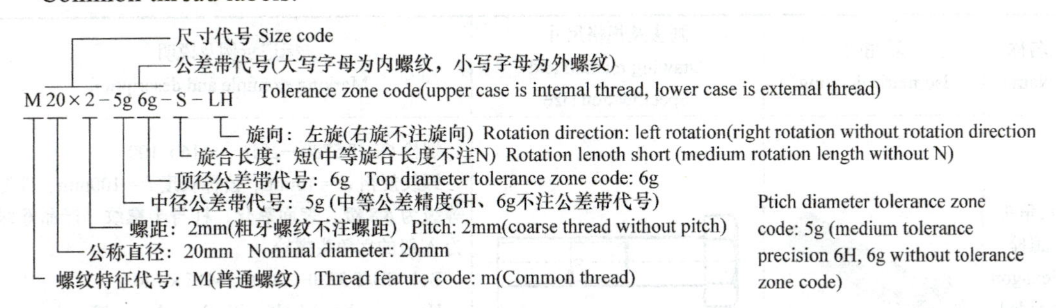

梯形螺纹标记：

Trapezoidal screw thread:

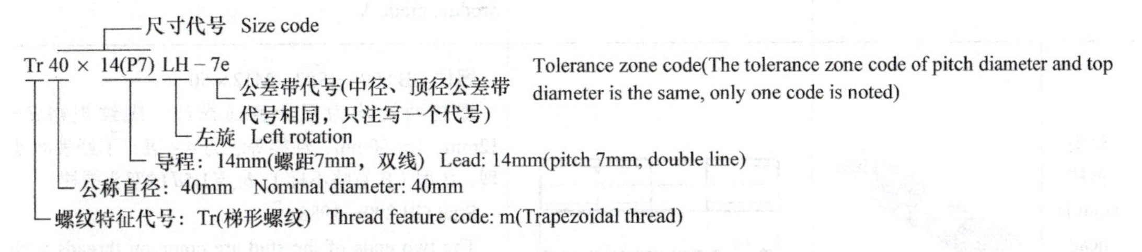

管螺纹标记：

Pipe screw thread:

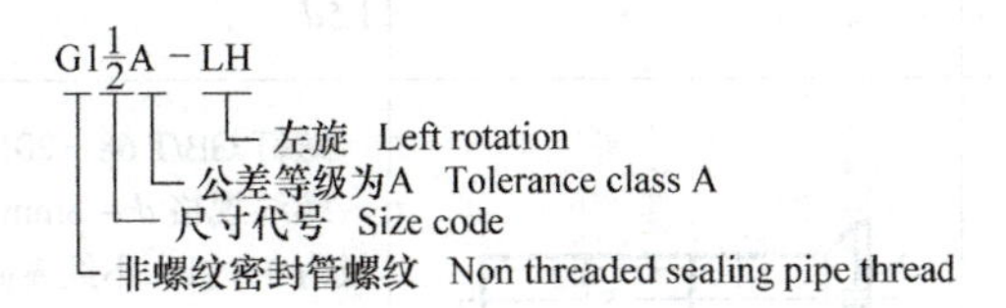

5.2 螺纹紧固件

5.2 Thread Fastener

螺纹紧固件包括螺栓、螺柱、螺钉、螺母和垫圈等。它们都是标准件，由专门的工厂生产，一般不画出它们的零件图，只要按规定进行标记，根据标记就可从国家标准中查到它们的结构形式和尺寸数据。

The thread fastener includes bolt, stud, screw, nut and washer. They are standard parts and produced in specialized factories, so they shall be labeled according to regulation instead of drawing the parts drawing. Their structural forms and dimension data could be found in the national standard based on the labels.

5.2.1 螺纹紧固件的标记

5.2.1 Labeling of Thread Fastener

螺纹紧固件的规定标记为：名称、标准、代号、型号、规格。表 5-2 列举出常用螺纹紧固件的标记和示例。

The stipulated label of the thread fastener is: name, standard, code, type and specification. The labels and examples of commonly-used thread fasteners are listed in Table 5-2.

表 5-2 常用螺纹紧固件的标记

Table 5-2 Labels of commonly-used thread fastener

名称 Name	轴测图 Isometric drawing	画法及规格尺寸 Drawing method and Specification size	标记示例及说明 Marking example and description
六角头螺栓 Hexagon headed bolt			螺栓 GB/T 5780—2016　M16 × 100 螺纹规格 d = 16mm、公称长度 l = 100mm、性能等级为 4.8 级、表面氧化、杆身半螺纹、产品等级为 A 级的六角头螺栓 Bolt GB/T 5780—2016　M16 × 100 Hexagon head bolt with thread specification d = 16mm, nominal length L = 100mm, performance grade 4.8, surface oxidation, half thread of rod body and product grade A
双头螺柱 Double head stud			螺柱 GB 899—1988　M12 × 50 螺柱两端均为粗牙普通螺纹、螺纹规格 d = 12mm、l = 50mm、性能等级为 4.8 级、不经表面处理、B 型（B 省略不标）、b_m = 1.5d 的双头螺柱 Stud GB 899—1988 The two ends of the stud are common threads with coarse teeth, d = 12mm, l = 50mm, performance grade 4.8, without surface treatment, type B (b omitted), b_m = 1.5d
螺钉 Screw			螺钉 GB/T 68—2016　M8 × 40 螺纹规格 d = 8mm、公称长度 l = 40mm、性能等级为 4.8 级、不经表面处理的开槽沉头螺钉 Screw GB/T 68—2016　M8 × 40 Slotted countersunk head screw with thread size d = 8mm, nominal length l = 40mm, performance grade 4.8 and without surface treatment
六角螺母 Hexagon nut			螺母 GB/T 41—2016　M16 螺纹规格 D = 16mm、性能等级为 5 级、不经表面处理、产品等级为 C 级的六角螺母 Nut GB/T 41—2016　M16 Hexagon nut with thread size D = 16mm, performance grade 5, no surface treatment and product grade C
垫圈 Washer			垫圈 GB/T 97.1—2002　16 标准系列、公称规格 d_1 = 16mm、硬度等级为 200HV 级、不经表面处理、产品等级为 A 级的平垫圈 Washer GB/T 97.1—2002　16 Standard series, nominal specification d_1 = 16mm, hardness grade 200HV, no surface treatment, product grade A flat washer

5.2.2 螺纹联接

常用螺纹紧固件的联接形式有：螺栓联接、双头螺柱联接、螺钉联接和螺钉紧定等。

1. 螺栓联接

螺栓联接适用于连接两个不太厚的零件。螺栓穿过两被联接件上的通孔，加上垫圈，拧紧螺母，就将两个零件联接在一起了，如图 5-9 所示。

5.2.2 Thread Connection

The connection forms of common thread fasteners include: bolt connection, stud connection, screw connection, screw set, etc.

1. Bolt Connection

The bolt connection is suitable for connecting two parts that are not too thick. Two parts are connected together by making the bolt penetrating the through hole of the two parts, adding the washer and tightening the nut, as shown in Fig.5-9.

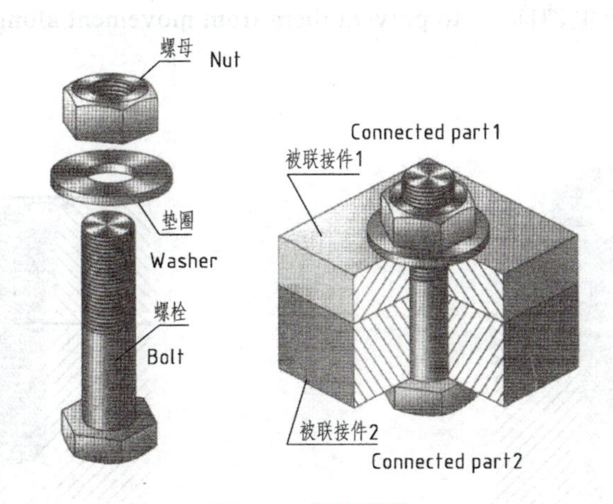

图 5-9 螺栓联接

Fig.5-9 Bolt connection

2. 双头螺柱联接

双头螺柱联接常用于被联接件之一太厚而不能加工成通孔的情况。双头螺柱两端都有螺纹，其中一端全部旋入被联接件的螺孔内，称为旋入端；另一端穿过另一被连接件的通孔，加上垫圈，旋紧螺母，如图 5-10 所示。此时采用的是弹簧垫圈，它依靠弹性增加摩擦力，防止螺母因受振动而松开。

3. 螺钉联接

螺钉联接一般用于受力不大而又不经常拆卸的地方。被

2. Stud Connection

The stud connection is commonly used when one of the parts is too thick to be processed with a through hole. The stud has thread in both ends. One end is entirely screwed into the hole of a part and called screwing end. The other end penetrates the through hole of the other part, add the washer, tighten the nut, as shown in Fig.5-10. The spring washer is used here which could increase the friction based on elasticity and prevent the nut from loosening due to vibration.

3. Screw Connection

The screw connection is generally used in places where the stress is small and it’s not frequently disassembled. One

连接的零件中一个为通孔，另一个为不通的螺纹孔。螺孔深度和旋入深度的确定与双头螺柱联接基本一致，螺钉头部的形式很多，应按规定画出，如图 5-11 所示。

of the connected parts has s through hole; the other one has a non-through screw hole. The screwing depth and screw hole depth are basically consistent with that of the stud connection. The head of the screw has a lot of forms and they should be drawn according to the regulation, as shown in Fig.5-11.

4. 螺钉紧定

紧定螺钉也是一种常用的螺钉联接，主要用于防止两个零件的相对运动。图 5-12 表示用锥端紧定螺钉限制轮和轴的相对运动，使其不能产生轴向移动。

4. Screw set

The screw set is a commonly-used screw connection form, mainly used to prevent the relative movement of two parts. Fig.5-12 shows that the relative movement of the gear and the shaft can be restrained by the cone-point screw set to prevent them from movement along the shaft direction.

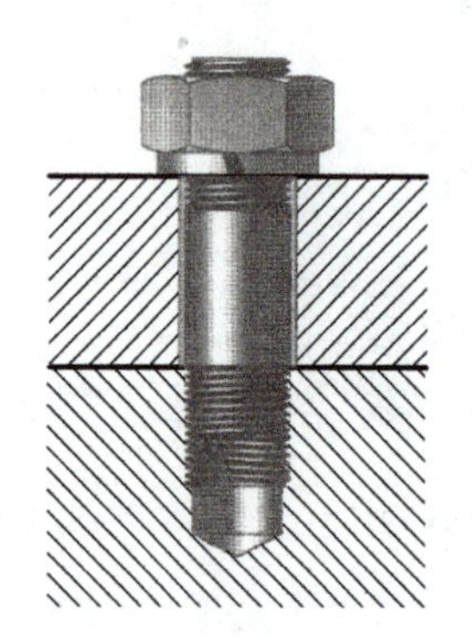

图 5-10　双头螺柱联接

Fig.5-10　Stud connection

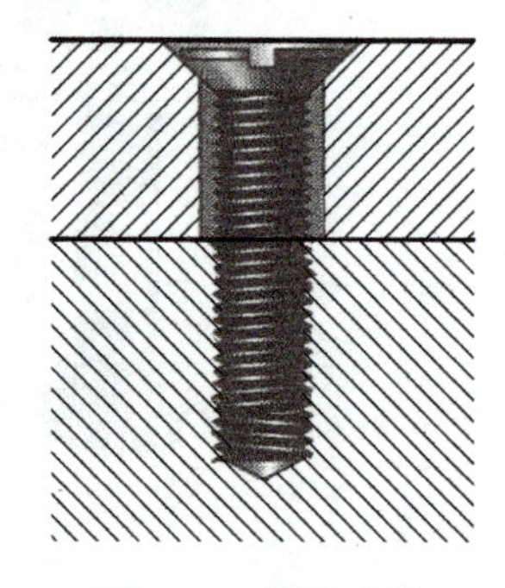

图 5-11　螺钉联接

Fig.5-11　Screw connection

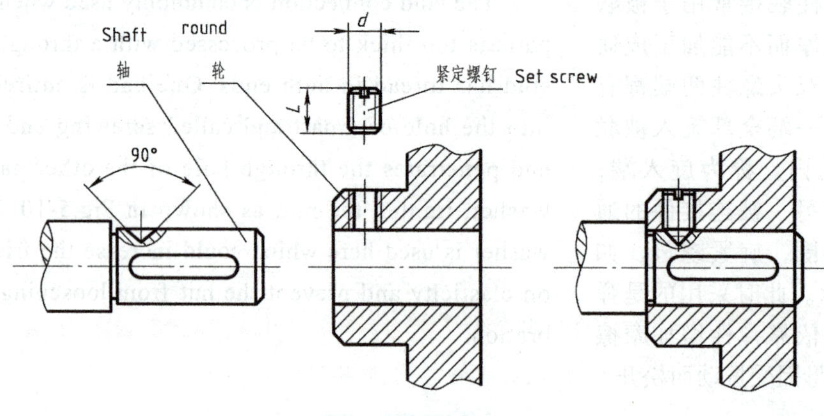

图 5-12　螺钉紧定

Fig.5-12　Screw set

5.3　键、销联接

5.3　Key and Pin Connection

5.3.1　键联接

键是标准件，它通常可以用来连接轴和轴上的传动零件，如齿轮、带轮等，起传递扭矩的作用。通常在（齿）轮和轴上分别加工出键槽，再将键装入键槽内，即可实现轮和轴的共同转动，如图 5-13 所示。

5.3.1　Key Connection

Key is a standard part, which can be used to connect the transmission parts on the shaft, such as gears and belt pulleys, to transfer the torque. Usually, the keyway is processed on the gear and shaft. Then the key is installed in the keyway to realize the common rotation of the gear and the shaft, as shown in Fig.5-13.

图 5-13　键联接

Fig.5-13　Key Connection

1. 常用键的标记

常用键有普通平键、半圆键和钩头楔键。其结构形式、规格尺寸及键槽尺寸等可从标准中查出。

普通平键应用最广，按轴槽结构可分为圆头普通平键（A 型）、方头普通平键（B 型）和单圆头普通平键（C 型）三种型式。键的选择一般根据轴的直径查表获得。

表 5-3 列出了几种常用键的标准代号和标记示例。

1. Labeling of common keys

Common keys include general flat key, woodruff key and gib head key. The structural form, specification, dimension and keyway dimension can be found in the standard.

The general flat key is the most widely used one. Based on the shaft groove structure, it can be divided into three types: round end general flat key (type A), square end general flat key (type B), single round end general flat key (type C). The selection of the key can be determined by searching the shaft diameter in the table.

The standard codes and labeling examples of several common keys are listed in Table 5-3.

2. 键联接的画法

在采用键联接时，轴和轮的零件图上都应画出键槽，并应标注出尺寸。图 5-14a 是轴上键槽的画法，图 5-14b 是轮上键槽的画法。

2. Drawing method of key connection

When using the key connection, the parts drawings of the shaft and gear shall be drawn with the keyway and labeled with dimensions. Fig.5-14a is the drawing method of shaft keyway; Fig.5-14b is the drawing method of gear keyway.

表 5-3　键及其标记示例

Table 5-3　Keys and Labeling Examples

名称（标准号） Name (standard number)	图例 Legend	标记示例 Tag example
普通平键 GB/T 1096—2003 General flat key	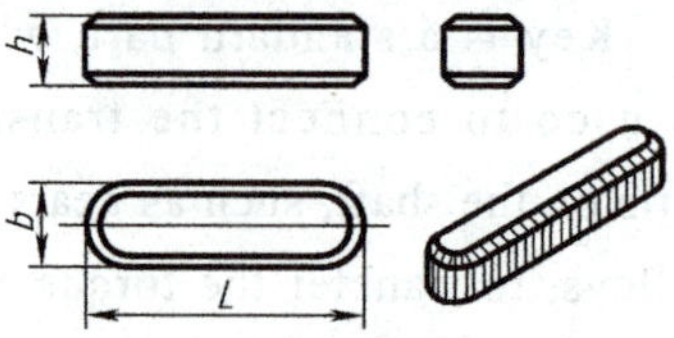	$b=8$、$h=7$、$L=25$ 的普通平键（A 型）标记为： GB/T 1096 键 8×7×25 The general flat key (type A) with $b=8$, $h=7$ and $L=25$ is marked as: GB/T 1096 key 8×7×25
半圆键 GB/T 1099.1—2003 Woodruff key	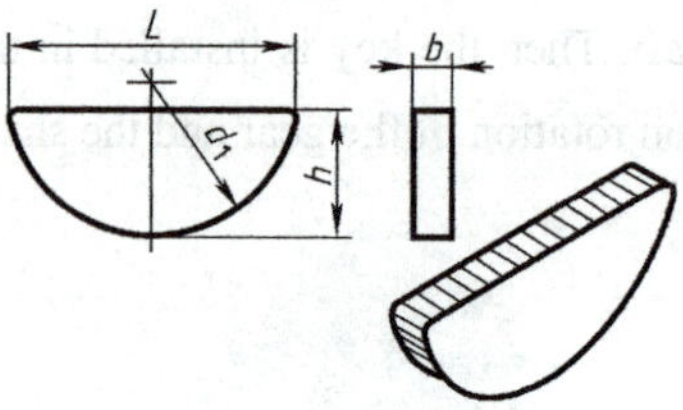	$b=6$、$h=10$、$D=25$ 的半圆键标记为： GB/T 1099.1 键 6×10×25 The woodruff key of $b=6$, $h=10$, $D=25$ is marked as: GB/T 1099.1 key 6×10×25
钩头楔键 GB/T 1565—2003 Gib head key	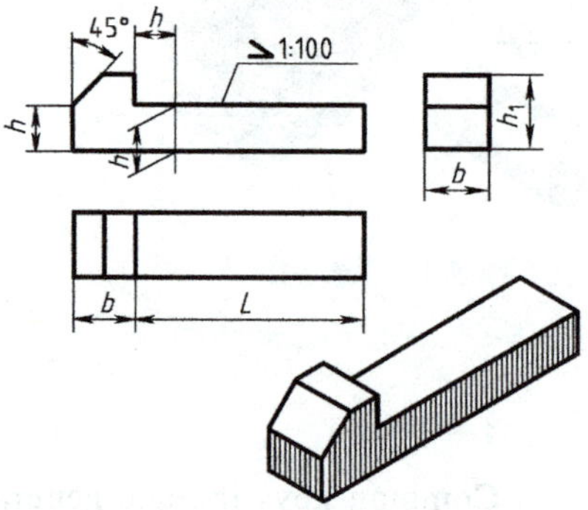	$b=6$、$L=25$ 的钩头楔键标记为： GB/T 1565 键 6×25 The gib head key with $b=6$, $L=25$ is marked as: GB/T 1565 key 6×25

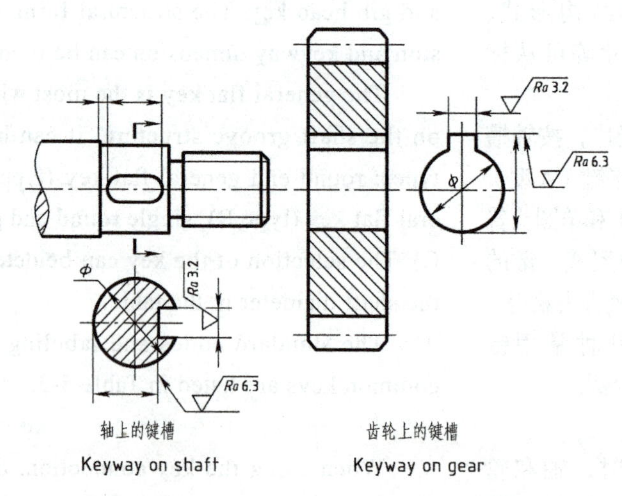

a) 轴上键槽画法　　b) 轮上键槽画法

a) Drawing method of shaft keyway　b) Drawing method of gear keyway

图 5-14　键槽的表达方法和尺寸标注

Fig.5-14　Representation and dimensioning of keyway

画普通平键联接图时，一般采用一个主视图和一个左视图来表达它们的装配关系，如图 5-15 所示。在主视图中，键和轴均按不剖来绘制。为了表达键在轴上的装配情况，主视图又采用了局部剖。在左视图上，由于键的两个侧面是工作面，只画一条线。键的顶面与键槽顶面不接触，应画两条线。

In the drawing of the general flat key connection, the assembly relation shall be represented by a front view and a left view, as shown in Fig.5-15. In the front view, the key and shaft shall be drawn as non-section view. In order to describe the assembly situation of the key on the shaft, the front view shall adopt the local section view. In the left view, as both sides of the key are working surfaces, only one line needs to be drawn. The top surface of the key is not in contact with the top surface of the keyway, two lines shall be drawn.

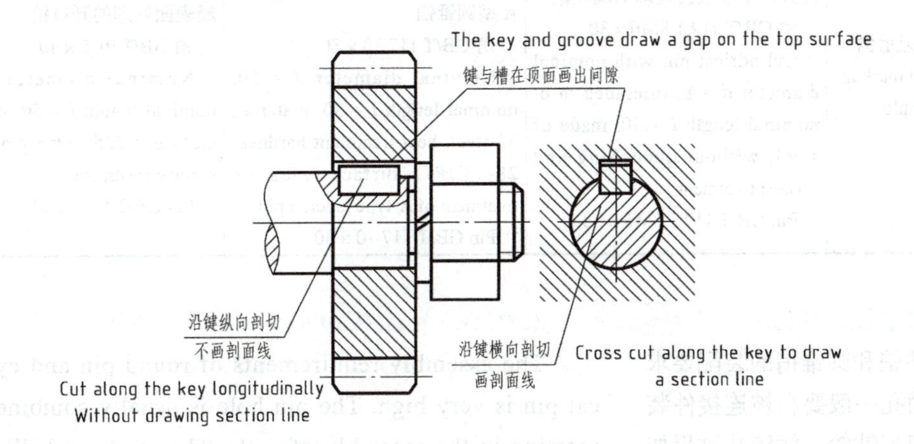

图 5-15　键联接的画法

Fig.5-15　Drawing method of the keyway

5.3.2　销联接

销是标准件，常用的销有圆柱销、圆锥销、开口销等。圆柱销和圆锥销主要用于零件间的连接或定位；开口销用来防止螺母松动或固定其他零件。

5.3.2　Pin Connection

The pin is the standard part. The commonly used pins include round pin, taper pin, split pin, etc. The round pin and taper pin are mostly used for the connection and positioning of parts. The split pin is used to prevent the nut from loosening and tighten other parts.

1. 销的标记

各种销的尺寸可以根据连接零件的大小以及受力情况由查表获得。表 5-4 列出了三种销的简图和标记。

1. Labeling of the pin

The dimensions of different pins could be found in the table based on the size of the part and stress situation. The diagrams and labeling of three pins are listed in Table 5-4.

表 5-4 销的简图和标记

Table 5-4 The diagrams and labeling of three pins

种类 Type	圆柱销 Cylindrical pin	圆锥销 Taper pin	开口销 Cotter pin
结构和规格尺寸 Structure and Specification size			
简化标记示例 Simplified markup example	公称直径 $d=8$，公差 m6，公称长度 $l=30$，材料为钢，不经淬火，不经表面处理的圆柱销 销 GB/T 119.1 8M6 × 30 Cylindrical pin with nominal diameter d = 8, tolerance m 6, nominal length l = 30, made of steel, without quenching and surface treatment Pin GB/T 119.1 8M6 × 30	公称直径 $d=10$，公称长度 $l=30$，材料为 35 钢，热处理硬度 28~38HRC，表面氧化处理的 A 型圆锥销 销 GB/T 117 10 × 30 Nominal diameter d = 10, nominal length l = 30, material: 35 steel, heat treatment hardness 28-38HRC, surface oxidation treatment of A-type tapered pin Pin GB/T 117 10 × 30	公称直径 $d=5$，公称长度 $l=50$，材料为 Q215 或 Q235，不经表面处理的开口销 销 GB/T 91.5 × 40 Nominal diameter d = 5, nominal length l = 50, material: Q215 or Q235, cotter pin without surface treatment Pin GB/T 91 5 × 40

2. 销联接的画法

圆柱销和圆锥销的装配要求较高，销孔一般要在被连接件装配时一起配钻铰，钻铰孔过程如图 5-16 所示。因此，在零件图上标注销孔尺寸时，应注明“配作”字样，如图 5-17a 所示。销联接的画法如图 5-17b 所示。锥销孔的公称直径指小端直径，如图 5-17b 的 $\phi6$。开口销联接的画法如图 5-17c 所示。

2. Drawing method of pin connection

The assembly requirements of round pin and cylindrical pin is very high. The pin hole is usually combined with reaming in the assembly of parts. The reaming drilling process is shown in Fig.5-16. Therefore, when the pin hole is dimensioned on the parts drawing, it shall be labeled with the word of “matching”, as shown in Fig.5-17a. The drawing method of the pin connection is shown in Fig.5-17b. The nominal diameter of the taper pin hole refers to the small-end diameter, such as $\phi6$ in Fig.5-17b. The drawing method of the cotter pin connection is shown in Fig.5-17c.

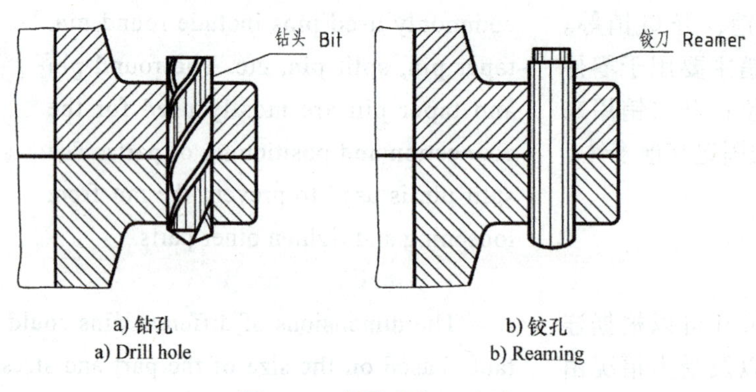

a) 钻孔
a) Drill hole

b) 铰孔
b) Reaming

图 5-16 销孔的加工方法

Fig.5-16 Processing method of the pin hole

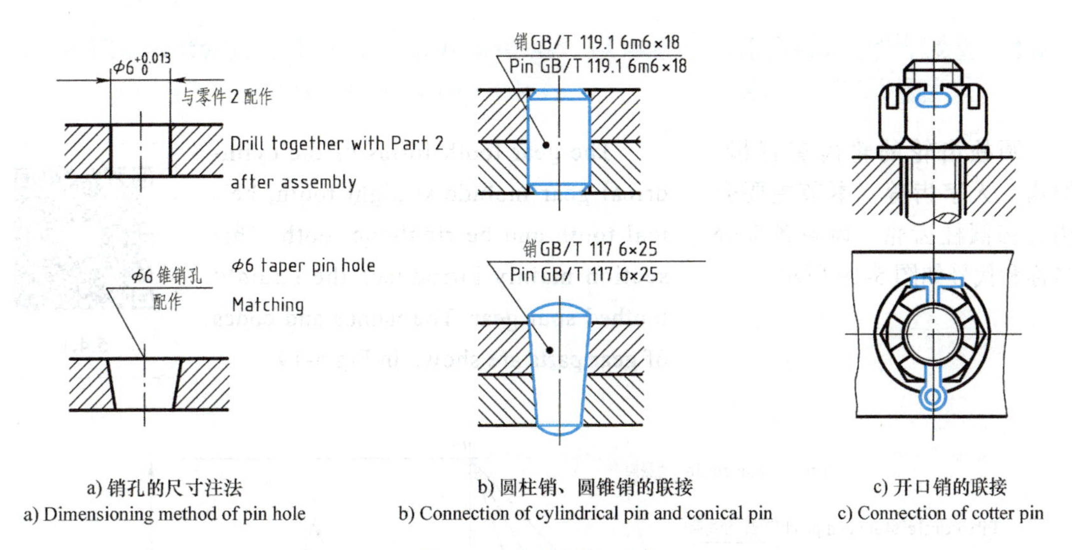

a) 销孔的尺寸注法
a) Dimensioning method of pin hole

b) 圆柱销、圆锥销的联接
b) Connection of cylindrical pin and conical pin

c) 开口销的联接
c) Connection of cotter pin

图 5-17　销联接的画法

Fig.5-17　Drawing method of the pin connection

5.4　齿轮

齿轮是机械传动中广泛应用的传动零件，它可以用来传递动力、改变转速和旋转方向。其常用的传动形式有：圆柱齿轮传动，如图 5-18a 所示；圆锥齿轮传动，如图 5-18b 所示；蜗杆、蜗轮传动，如图 5-18c 所示。

5.4　Gear

Gear is widely used in mechanical transmission. It can be used to transfer power, change speed and screwing direction. The commonly-used transmission forms include: cylindrical gear transmission, as shown in Fig.5-18a; cone gear transmission, as shown in Fig.5-18c. worm and gear transmission, as shown in Fig.5-18c.

a) 圆柱齿轮
a) Cylindrical gear

b) 圆锥齿轮
b) Cone gear

c) 蜗轮蜗杆
c) Worm and gear

图 5-18　常见的齿轮传动

Fig.5-18　Common gear transmission

5.4.1 直齿圆柱齿轮各部分名称和主要参数

圆柱齿轮的轮齿有直齿、斜齿、人字齿等，本节主要介绍直齿圆柱齿轮，齿轮各部分名称和代号如图 5-19 所示。

5.4.1 Names and Major Parameters of Parts of Straight Toothed Spur Gear

The gear tooth forms of the cylindrical gear include straight tooth, helical tooth and herringbone tooth. This section mainly introduces the straight toothed spur gear. The names and codes of gear parts are shown in Fig.5-19.

5.4.1

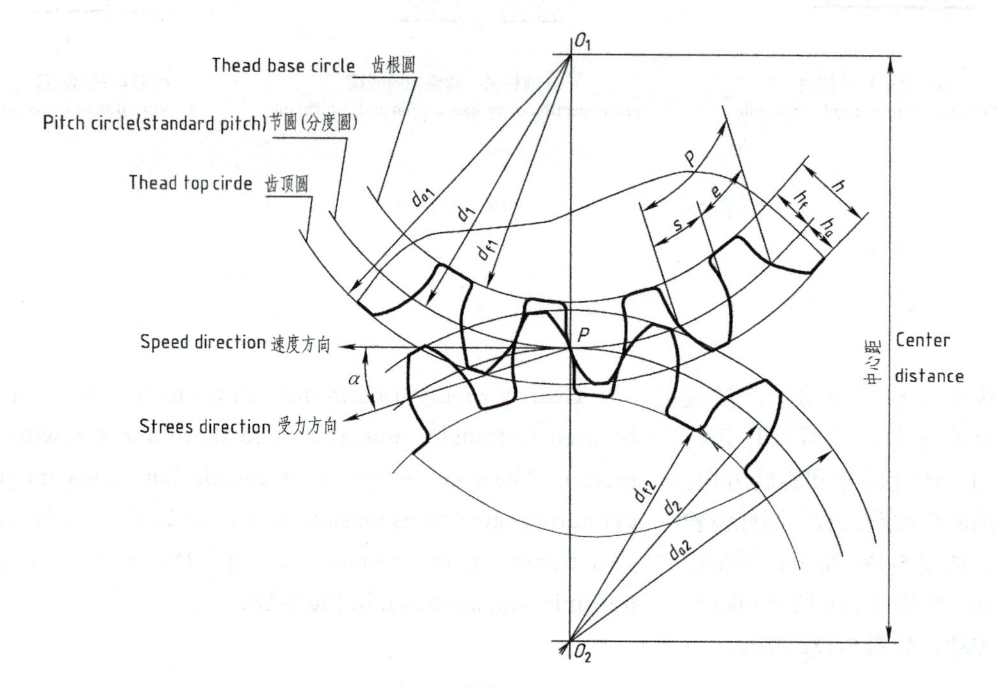

图 5-19 齿轮各部分名称和代号

Fig.5-19 The names and codes of gear parts

（1）齿数 z 轮齿的数量。

（2）齿顶圆 d_a 通过齿轮轮齿顶端的圆称为齿顶圆。

（3）齿根圆 d_f 通过齿轮轮齿根部的圆称为齿根圆。

（4）分度圆 d 在齿轮上有一个设计和加工时计算尺寸的基准圆，它是一个假想圆，在该圆上，齿厚 s 与齿槽宽 e 相等。

（5）齿顶高 h_a 分度圆到齿顶圆之间的径向距离，称为

Teeth number z: number of the gear teeth.

Addendum circle d_a: the circle at the top of the gear is called the addendum circle.

Dedendum circle d_f: the circle at the root of the gear is called the dedendum circle.

Reference circle d: there is an imaginary circle on the gear which is the basic circle for dimension calculation in the design and processing. On this circle, the tooth thickness s is equal to the space width e.

Tooth addendum h_a: the radial distance between the reference circle and the addendum circle is called the tooth

齿顶高。

（6）齿根高 h_f　分度圆到齿根圆之间的径向距离，称为齿根高。

（7）齿高 h　齿顶圆到齿根圆之间的径向距离，称为齿高，$h = h_a + h_f$。

（8）齿厚 s　在分度圆上，同一齿两侧齿廓之间的弧长，称为齿厚。

（9）齿间 e　在分度圆上，齿槽宽度的一段弧长，称为齿间，也称为齿槽宽。

（10）齿距 p　在分度圆上，相邻两齿同侧齿廓之间的弧长，称为齿距。

（11）压力角 α　一对啮合的齿轮分度圆上啮合点 P 的受力方向与该点瞬时运动方向的夹角，称为压力角，如图 5-19 所示。压力角已经标准化，国家标准规定为 20° 或 15°，一对相互啮合的齿轮，其压力角应该相等。

（12）中心距 a　两齿轮回转中心的连线称为中心距。

（13）模数 m　模数是设计、制造齿轮的重要参数。

如图 5-19 所示，分度圆大小与齿距和齿数有关，即：

$\pi d = zp$ 或 $d = zp/\pi$

令 $m = p/\pi$，则 $d = mz$

m 称为模数，单位为毫米，模数的大小直接反映出轮齿的大小。一对相互啮合的齿轮，其模数必须相等。为了便于设计和制造齿轮，减少齿轮加工的刀具，模数已标准化，其系列值如表 5-5 所示。

addendum.

Tooth dedendum h_f: the radial distance between the reference circle and the dedendum circle is called the tooth dedendum.

Tooth depth h: The radial distance between the addendum circle and the dedendum circle is called the tooth depth, $h = h_a + h_f$.

Tooth thickness s: on the reference circle, the arc length between the two sides of the tooth is called the tooth thickness.

Intertooth space e: on the reference circle, the arc length of the space width is called the intertooth space, also known as the space width.

Tooth pitch p: on the reference circle, the arc length between the tooth forms of the same side of two adjacent teeth is called the tooth pitch.

Pressure angle α: the angle between the force direction of the meshing point P of two meshing gear reference circles and the instantaneous motion direction of the same point is called the pressure angle, as shown in Fig.5-19. The pressure angle has been standardized. The national standard stipulates that the angle shall be 20° or 15°. A pair of meshing gears should have equal pressure angles.

Center distance a: the connection between the centers of rotation of two gears is called the center distance.

Modulus m: the modulus is an important parameter in the design and manufacturing of gears.

As shown in Fig.5-19, the size of the reference circle is related to the tooth pitch and number, that is:

$\pi d = zp$ or $d = zp/\pi$

Let: $m = p/\pi$, then: $d = mz$

m represents the modulus. Its unit is millimeter. The size of the modulus directly reflects the size of the gear tooth. A pair of meshing gears must have the same modulus. In order to facilitate the design and manufacturing of gears and reduce the number of gear cutting tools, the modulus has been standardized, and the series values are shown in Table 5-5.

5.4.2 各部分尺寸计算公式

直齿圆柱齿轮各部分尺寸计算公式及计算举例见表 5-5。

5.4.2 Computational Formula of Part Dimensions

The computational formulas and examples of part dimensions of the straight toothed spur gear are shown in Table 5-5.

表 5-5 直齿圆柱齿轮的尺寸公式及计算举例 （单位：mm）

Table 5-5 The computational formulas and examples of part dimensions of the straight toothed spur gear （Unit：mm）

基本参数：模数 m，齿数 z Basic parameters: modulus m, tooth number z			已知：$m=3$，$z_1=21$，$z_2=42$ Known: $m=3, z_1=21, z_2=42$	
名称 Name	代号 Code	尺寸公式 Dimensional formula	计算举例 Computational example	
分度圆 Reference circle	d	$d=mz$	$d_1=63$	$d_2=126$
齿顶高 Tooth addendum	h_a	$h_a=m$	$h_a=3$	
齿根高 Tooth dedendum	h_f	$h_f=1.25m$	$h_{f=}3.75$	
齿高 Tooth depth	h	$h=h_a+h_f=2.25m$	$h=6.75$	
齿顶圆直径 Tip diameter	d_a	$d_a=d+2h_a=m(z+2.5)$	$d_{a1}=70.5$	$d_{a2}=132$
齿根圆直径 Root diameter	d_f	$d_f=d-2h_f=m(z-2.5)$	$d_{f1}=55.5$	$d_{f2}=118.5$
齿距 Tooth pitch	p	$p=\pi m$	$p=9.42$	
齿厚 Tooth thickness	s	$s=p/2$	$s=4.71$	
中心距 Center distance	a	$a=(d_1+d_2)/2=m(z_1+z_2)/2$	$a=94.5$	

5.4.3 齿轮规定画法

单个齿轮的表达一般只采用两个视图，主视图画成剖视图，可采用半剖。投影为圆的视图应将键槽的位置和形状表达出来，如图 5-20 所示。

5.4.3 Stipulated Drawing Method of the Gear

The description of a single gear usually uses two views, the front view and the section view (or half section view). The view whose projection is round shall draw the position and shape of the keyway, as shown in Fig.5-20.

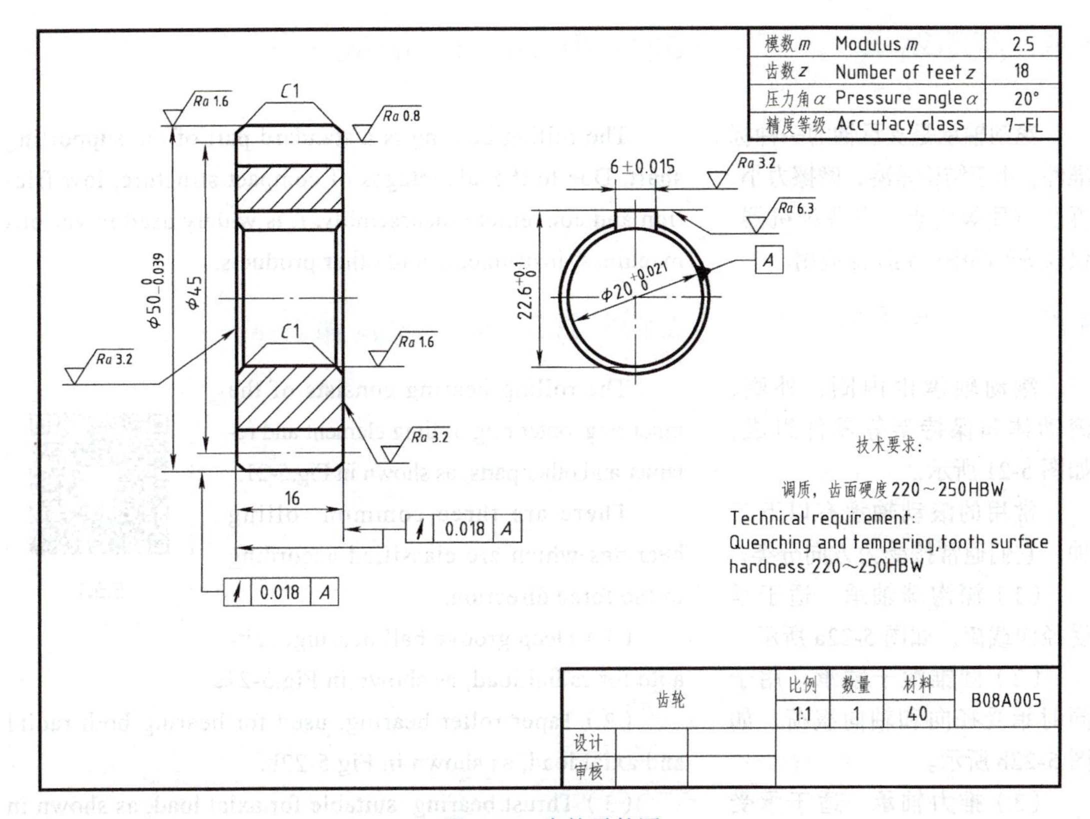

图 5-20 齿轮零件图

Fig.5-20 Gear part diagram

单个齿轮的表达也可采用一个视图和一个局部视图。当需要表示斜齿轮和人字齿轮的齿线方向时，可用三条与齿线方向一致的细实线表示。

The description of a single gear could also use one view and one local view. When the tooth direction of the helical gear and the herringbone gear need to be described, it can be described with three fine lines in the same direction with the tooth.

齿顶线和齿顶圆用粗实线绘制，分度线和分度圆用细点画线绘制，在剖视图中，当剖切平面通过齿轮轴线时，齿根线用粗实线绘制，轮齿按不剖处理，即轮齿部分不画剖面线。

The addendum line and circle are drawn with heavy lines; the reference line and circle are drawn with thin dash-dotted line. In the section view, when the section plane passes the gear axis, tooth dedendum line is drawn with the heavy line, the gear tooth is processed as non-section view. That is, the section line shall not be drawn for the gear teeth.

齿轮的零件图应按零件图的全部内容绘制和标注完整，并且在其零件图的右上角画出有关齿轮的啮合参数和检验精度的表格并注明有关参数，如图 5-20 所示。

The parts drawing of gear teeth shall be drawn completely with all labels. The table of meshing parameters and examination precision of gear teeth shall be drawn on the top right corner of the parts drawing and labeled with related parameters, as shown in Fig.5-20.

5.5 滚动轴承

滚动轴承是支承轴的一种标准件。由于结构紧凑，摩擦力小，拆装方便等优点，在各种机器、仪表等产品中得到广泛应用。

5.5.1 滚动轴承的种类

滚动轴承由内圈、外圈、滚动体和保持架等零件组成，如图 5-21 所示。

常用的滚动轴承有以下三种，它们通常按受力方向分类。

（1）深沟球轴承　适于承受径向载荷，如图 5-22a 所示。

（2）圆锥滚子轴承　用于同时承受径向和轴向载荷，如图 5-22b 所示。

（3）推力轴承　适于承受轴向载荷，如图 5-22c 所示。

5.5 Rolling Bearing

The rolling bearing is a standard part of the supporting shaft. Due to the advantages of compact structure, low friction and convenient disassembly, it is widely used in various machines, instruments and other products.

5.5.1 Types of Rolling Bearing

The rolling bearing consists of the inner ring, outer ring, rolling element and retainer and other parts, as shown in Fig.5-21.

There are three common rolling bearings which are classified according to the force direction.

（1）Deep groove ball bearing: suitable for radial load, as shown in Fig.5-22a.

（2）Taper roller bearing: used for bearing both radial and axial load, as shown in Fig.5-22b.

（3）Thrust bearing: suitable for axial load, as shown in Fig.5-22c.

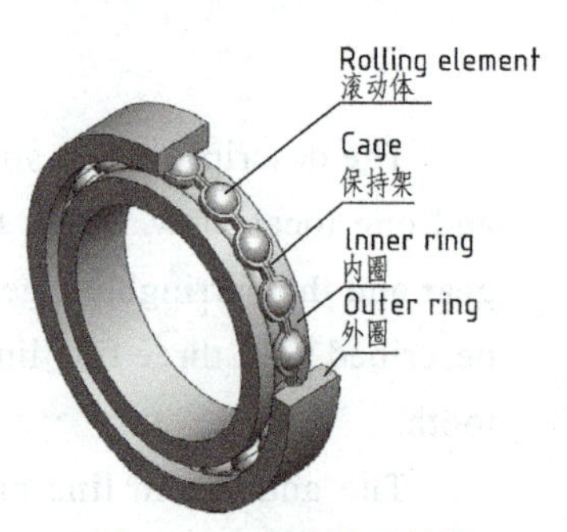

图 5-21　滚动轴承结构

a) 深沟球轴承
a) Deep groove ball bearing

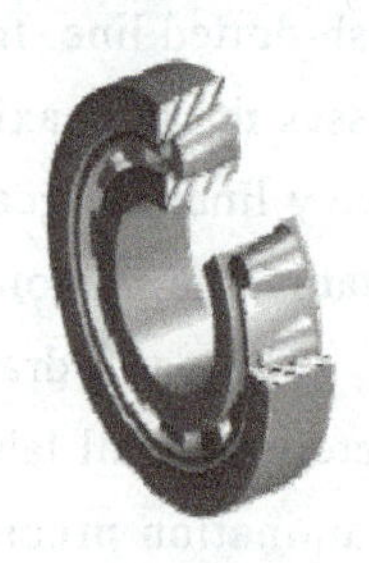

b) 圆锥滚子轴承
b) Taper roller bearing

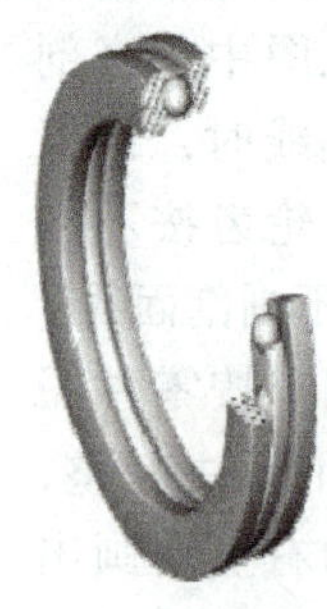

c) 推力轴承
c) Thrust bearing

图 5-22　滚动轴承类型

5.5.2 滚动轴承的代号（GB/T 272—2017）

5.5.2 Codes of Rolling Bearing (GB/T 272—2017)

滚动轴承的代号由前置代号、基本代号和后置代号三部分组成，各部分的排列如下。

The code of the rolling bearing is composed of three parts: pre-positional code, basic code and post-positional code. The arrangement of these three codes are as follows.

前置代号 基本代号 后置代号

Pre-positional code Basic code Post-positional code

1. 基本代号

滚动轴承的基本代号表示轴承的基本类型、结构和尺寸，是滚动轴承代号的基础。滚动轴承基本代号由轴承类型代号、尺寸系列代号、内径代号构成。

（1）类型代号由数字或字母表示，其含义如表 5-6 所示。

1. Basic code

The basic code represents the basic type, structure and size of the rolling bearing which is the basis of the entire code. The basic code of the rolling bearing consists of bearing type code, size series code and inner diameter code.

（1）Type code: consists of numbers or letters. Their meanings are shown in Table 5-6.

表 5-6 滚动轴承类型代号

Table 5-6 Codes of Rolling Bearing Types

代号 Code	轴承类型 Bearing type	代号 Code	轴承类型 Bearing type
0	双列角接触球轴承 Double row angular contact ball bearing	7	角接触球轴承 Angular contact ball bearing
1	调心球轴承 Self aligning ball bearing	8	推力圆柱滚子轴承 Thrust cylindrical roller bearing
2	调心滚子轴承和推力调心滚子轴承 Self aligning roller bearings and thrust self-aligning roller bearings	N	圆柱滚子轴承 Cylindrical roller bearing
3	圆锥滚子轴承 Tapered roller bearing	NN	双列或多列圆柱滚子轴承 Double row or multi row cylindrical roller bearing
4	双列深沟球轴承 Double row deep groove ball bearing	U	外球面球轴承 Outer spherical ball bearing
5	推力球轴承 Thrust ball bearing	QJ	四点接触球轴承 Four point contact ball bearing
6	深沟球轴承 Deep groove ball bearing		

（2）尺寸系列代号由轴承宽（高）度系列代号和直径系列代号组合而成，用两位数字表示，其中左边一位数字为宽（高）度系列代号，右边一位数字为直径系列代号，如表 5-7 所示。

（3）内径代号表示轴承的公称内径，如表 5-8 所示。

（2）Dimension series code: consists of width series code and diameter serial code of the bearing which are represented by a double-digit number. The left number is the width series code; the right number is the diameter series code, as shown in Table 5-7.

（3）Inner diameter code: represents the nominal diameter of the bearing, as shown in Table 5-8.

表 5-7　向心轴承、推力轴承尺寸系列代号

Table 5-7　The dimension series codes of radial bearing and thrust bearing

直径系列代号 Diameter series code	向心轴承 Radial bearing								推力轴承 Thrust bearing			
	宽度系列代号 Width series code								高度系列代号 Height series code			
	8	0	1	2	3	4	5	6	7	9	1	2
	尺寸系列代号 Size series code											
7	—	—	17	—	37	—	—	—	—	—	—	—
8	—	08	18	28	38	48	58	68	—	—	—	—
9	—	09	19	29	39	49	59	69	—	—	—	—
0	—	00	10	20	30	40	50	60	70	90	10	—
1	—	01	11	21	31	41	51	61	71	91	11	—
2	82	02	12	22	32	41	52	62	72	92	12	22
3	83	03	13	23	33	—	—	—	73	93	13	23
4	—	04	—	24	—	—	—	—	74	94	14	24
5	—	—	—	—	—	—	—	—	—	95	—	—

表 5-8　滚动轴承内径代号

Table 5-8　Inner diameter code of the rolling bearing

轴承公称内径 /mm Nominal inner diameter of bearing/mm	内径代号 Inner diameter code	示例 Examples
0.6~10（非整数） 0.6-10 (Non integer)	用公称内径毫米数直接表示，内径与尺寸系列代号之间用"/"分开 The nominal inner diameter is directly expressed in mm, and the inner diameter and the size series code are separated by "/"	深沟球轴承　618/2.5　$d = 2.5$mm Deep groove ball bearing 618/2.5
1~9（整数） 1-9(Integer)	用公称内径毫米数直接表示，对深沟及角接触球轴承 7、8、9 直径系列，内径与尺寸系列代号之间用"/"分开 The nominal inner diameter is directly expressed in mm. For 7, 8 and 9 diameter series of deep groove and angular contact ball bearings, the inner diameter and dimension series codes are separated by "/"	深沟球轴承　625　$d = 5$mm Deep groove ball bearing 625 深沟球轴承　618/5　$d = 5$mm Deep groove ball bearing 618/5

（续）

轴承公称内径 /mm Nominal inner diameter of bearing/mm		内径代号 Inner diameter code	示例 Examples
10-17	10	00	深沟球轴承　6200　$d=10$mm Deep groove ball bearing 6200 深沟球轴承　6201　$d=12$mm Deep groove ball bearing 6201 深沟球轴承　6202　$d=15$mm Deep groove ball bearing 6202 深沟球轴承　6203　$d=17$mm Deep groove ball bearing 6203
	12	01	
	15	02	
	17	03	
20~480（22、28、32 除外） 20-480(Except 22,22,32)		公称内径除以 5 的商数，商数为个位数，需在商数左边加“0”，如 08 The quotient of the nominal inner diameter divided by 5, which is a single digit, needs to add “0” to the left of the quotient, such as 08	圆锥滚子轴承　30308　$d=40$mm Tapered roller bearing 30308 深沟球轴承　6215　$d=75$mm Deep groove ball bearing 6215
≥ 500 以及 22、28、32 ≥ 500 and 22, 28, 32		用公称内径毫米数直接表示，但在与尺寸系列之间用“/”分开 It is directly expressed in mm of nominal inner diameter, but separated from dimension series by “/”	调心滚子轴承　230 × 500　$d=500$mm Self aligning roller bearing 230/500 深沟球轴承　62/22　$d=22$mm Deep groove ball bearing 62/22

前置代号和后置代号是轴承在结构形式、尺寸、公差和技术要求等有改变时，在其基本代号前后添加的补充代号。

The pre-positional code and post-positional code are the supplementary codes added before and after the basic code when the structural form, dimension, tolerance and technical requirement are changed.

滚动轴承的代号举例：

Code examples of the rolling bearing:

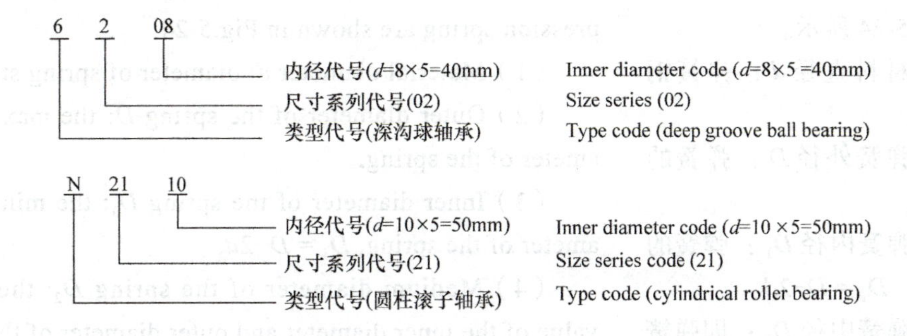

5.6 弹簧

5.6 Spring

弹簧是一种常用件，它通常用来减振、夹紧、测力和储存能量。弹簧的种类多，常见的有螺旋弹簧和涡卷弹簧等。根据受力情况不同，螺旋弹簧又可分为压缩弹簧、拉伸弹簧和扭转弹簧等，常用的各种弹簧如图 5-23 所示。弹簧的用途很广，本节只介绍圆柱螺旋压缩弹簧。

The spring is a commonly-used part for absorbing shock, clamping, measuring the force and storing power. There are many kinds of springs and the most common ones are the spiral spring and scroll spring. The spiral spring is divided into compression spring, extension spring and torsion spring according to different force conditions. Commonly-used springs are shown in Fig.5-23. The spring is widely applied and this section only introduces the cylindrical helical compression spring.

a) 压缩弹簧 b) 拉伸弹簧 c) 扭转弹簧 d) 涡卷弹簧 e) 板弹簧

a) Compression spring b) Extension spring c) Torsion spring d) Scroll spring e) Blade spring

图 5-23 常用弹簧

Fig.5-23 Common springs

5.6.1 圆柱螺旋压缩弹簧各部分名称及尺寸计算

5.6.1 Names and Dimension Calculation of Different Parts of Cylindrical Helical Compression Spring

圆柱压缩弹簧相关名称与代号如图 5-24 所示。

（1）材料直径 d：弹簧钢丝的直径。

（2）弹簧外径 D：弹簧的最大直径。

（3）弹簧内径 D_1：弹簧的最小直径，$D_1 = D-2d$。

（4）弹簧中径 D_2：即弹簧内径和外径的平均值。

Related names and codes of cylindrical helical compression spring are shown in Fig.5-24.

（1）Material diameter d: diameter of spring steel wire.

（2）Outer diameter of the spring D: the maximum diameter of the spring.

（3）Inner diameter of the spring D_1: the minimum diameter of the spring, $D_1 = D-2d$.

（4）Medium diameter of the spring D_2: the average value of the inner diameter and outer diameter of the spring.

$$D_2 = (D+D_1)/2 = D_1+d = D-d。$$

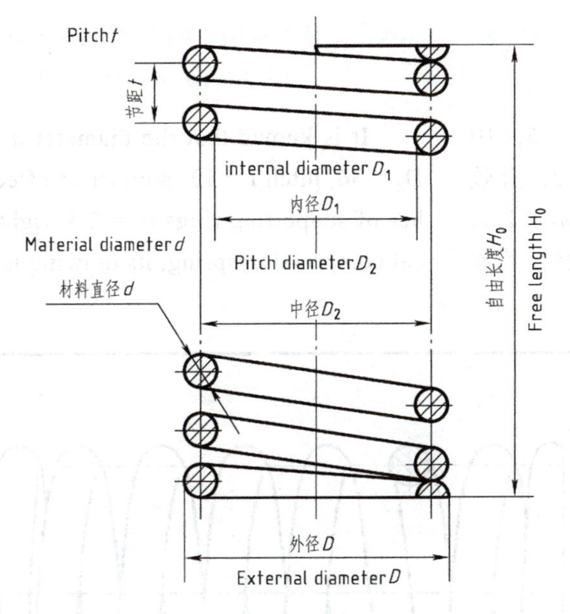

图 5-24　圆柱压缩弹簧相关名称与代号

Fig.5-24　Related names and codes of cylindrical helical compression spring

（5）节距 t：除支承圈外，相邻两圈沿轴向的距离。

（6）有效圈数 n：弹簧能保持相同节距的圈数。

（7）支承圈数 n_0：为了使压缩弹簧工作时受力均匀，保证轴线垂直于支承面，通常将弹簧的两端并紧磨平。这部分圈数只起支承作用，叫支承圈数，常见的有 1.5 圈、2 圈、2.5 圈三种，其中 2.5 圈用得最多。

（8）总圈数 n_1：有效圈数与支承圈数之和，称为总圈数，$n_1 = n+n_0$

（9）自由高度 H_0：弹簧没有负荷时的高度。

（10）展开长度 L：弹簧丝展开后的长度。

$$L = n_1\sqrt{(\pi d)^2 + t^2}$$

（11）旋向：分为左旋和右旋两种。

（5）Pitch t: the distance between two adjacent rings along the axial direction except the supporting ring.

（6）Number of effective rings n: the number of rings the spring could maintain the same pitch.

（7）Number of support rings n_0: in order to make the compression spring work with uniform force and ensure that the axis is perpendicular to the support surface, both ends of the spring usually are tightened together and ground flat. This part of rings only plays a supporting role and called the supporting rings. There are three common types: 1.5 rings, 2 rings and 2.5 rings, among which the 2.5 rings type is mostly used.

（8）Number of total rings n_1: the sum of the number of effective rings and supporting rings; $n_1 = n+n_0$.

（9）Free height H_0: the height of the spring without load.

$$H_0 = nt+(n_0-0.5)d.$$

（10）Developed length L: the length of the spring after expansion.

（11）Screwing direction: divided into left-hand screwing and right-hand screwing.

5.6.2 圆柱螺旋压缩弹簧的绘制

已知材料直径 $d = 6$，中径 $D_2 = 36$，节距 $t = 12$，有效圈数 $n = 6$，支承圈数 $n_0 = 2.5$，右旋圆柱螺旋压缩弹簧，其零件图如图 5-25 所示。

5.6.2 Drawing of Cylindrical Helical Compression Spring

It is known that the diameter $d = 6$, medium diameter $D_2 = 36$, pitch $t = 12$, number of effective rings $n = 6$, number of supporting rings $n_0 = 2.5$, right-hand cylindrical helical compression spring, its drawing is shown in Fig.5-25.

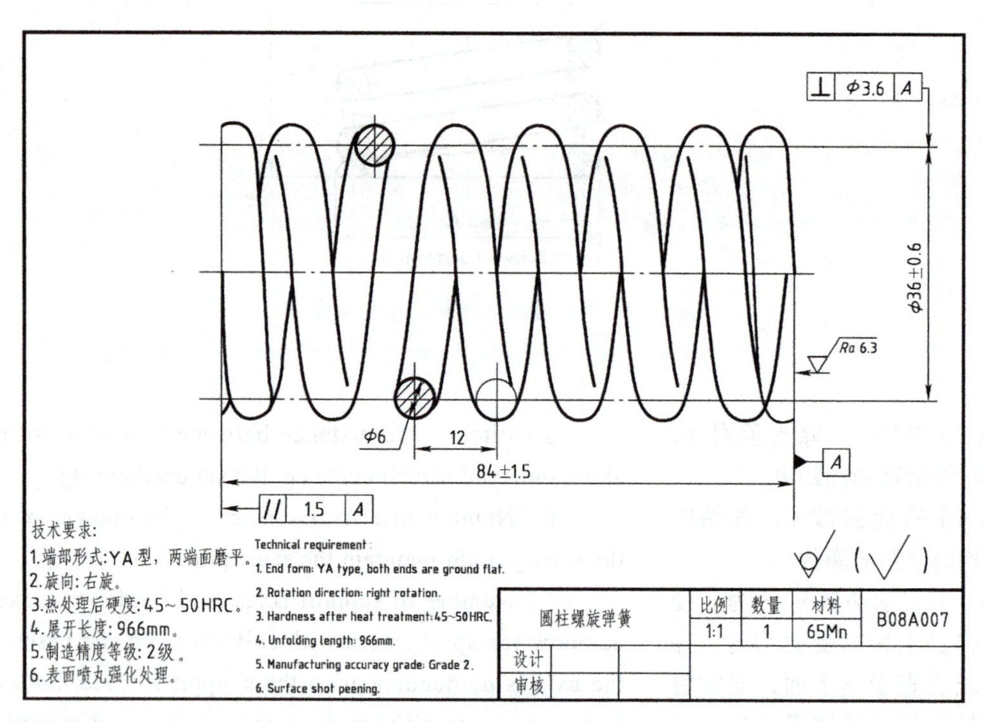

图 5-25 圆柱压缩弹簧零件图

Fig.5-25 Parts drawing of cylindrical helical compression spring

[本章习题]

1. 已知一对啮合的直齿圆柱齿轮，模数 $m = 2$，小齿轮齿数 = 20，大齿轮齿数 = 55，齿宽 = 20mm，压力角 = 20°，请用 Inventor 软件中的设计加速器完成其建模。

2. 抄画图 5-25 圆柱压缩弹簧零件图（提示：先建模再生成零件图）。

[Chapter exercises]

1. Given a pair of meshing spur gears with module $m = 2$, pinion teeth number = 20, large teeth number = 55, tooth width = 20 mm, and pressure Angle = 20 degrees, please use the design accelerator in Inventor software to complete the modeling.

2. Copy the part drawing of cylindrical compression spring as shown in Fig.5-25. (Hint: First modeling regenerate into part drawing)

第 6 章　零件图的识读与绘制

Chapter 6　The Reading and Drawing of the Detail Drawings

零件图是加工、制造和检验零件的依据，也是设计和生产过程中的重要技术资料。

本章重点讲述零件图绘制和读图的基本方法。先用 Inventor 软件进行造型设计，然后把零件或部件模型转换成二维的工程图。Inventor 软件可实现快速生成视图，通过比较，选择最优方案。

Detail drawings are the basis for machining, manufacturing and inspection, as well as important technical information in the process of design and production.

This chapter focuses on the fundamental methods of drawing and reading detail drawings. Inventor will be used to design model and then transform the part or the subassembly model into a two-dimensional (2D) engineering drawing. Inventor can generate views quickly. The optimal solution can be selected by comparison.

6.1　零件图的作用和内容

6.1　Function and Contents of Detail Drawings

6.1.1　零件图的作用

6.1.1　Function of Detail Drawings

一台机器是由若干个零件按一定的装配关系和技术要求装配而成的。构成机器的最小单元称为零件。表达零件的结构、形状、大小和技术要求的图样称为零件图，如图 6-1 所示。零件图是用于指导加工、检验和生产零件的依据，是设计和生产部门的重要技术文件。

A machine is assembled from a number of parts according to the assembly relationships and technical requirements, the smallest unit that makes up the machine is called a part. Drawings that represent the structure, shape, size, and technical requirements of the part are called part drawings, as shown in Fig.6-1. Detail drawings are the basis for guiding the processing, inspection and production of parts, and are important technical documents in the design and production departments.

6.1

6.1.2　零件图的内容

6.1.2　Contents of Detail Drawings

一张完整的零件图，应包括下列内容。

A full part drawing should include the following.

1. 一组视图

用恰当的视图、剖视图、断面图及其他表达方法，正确、完整、清晰地表达零件内外结构。

1. A group of views

Use the appropriate views, section views, cut views and other representations to describe the inner and outer structure of the part correctly, completely and clearly.

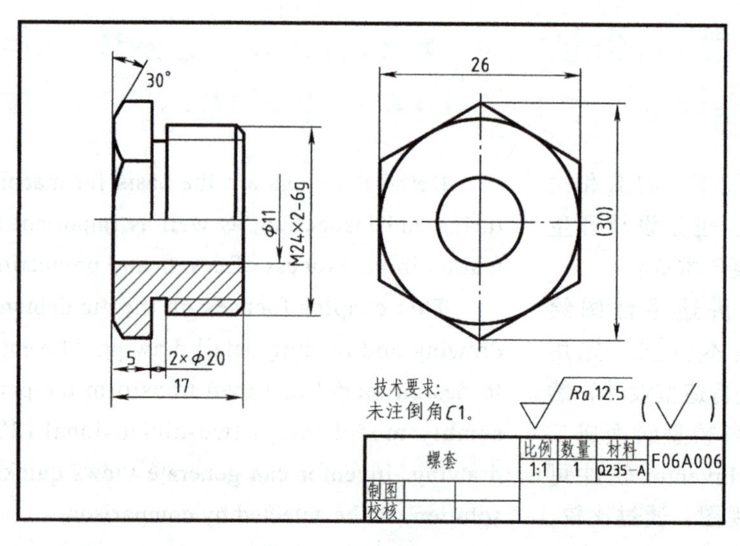

图 6-1 螺套零件图

Fig.6-1 Screw sleeve drawing

2. 完整尺寸

完整尺寸即合理、完整、清晰地标注出制造零件所需的全部尺寸。

2. Dimensioning

Dimensioning part reasonably, completely and clearly manufacture the part as required.

3. 技术要求

技术要求是用规定的代号、符号和文字注释，简明准确地给出零件在制造、检验时应达到的技术要求。如表面粗糙度、尺寸公差、几何公差、表面处理及其他特殊要求等。

3. Technical requirements

Technical requirements illustrate the technical requirements for manufacturing and inspection by using prescribed code, symbol and text comment, which should be met. Technical requirements include surface roughness, dimensional tolerances, geometric tolerances, surface treatment and other special requirements.

4. 标题栏

图纸右下角的标题栏中填写零件的名称、材料、数量、比例、图号以及设计人员的签名等。

4. Title block

The name, material, quantity, scale, drawing number of the part and the signature of the designer should be filled in the title block in the lower right corner of drawing.

6.2 零件图的视图选择

6.2 Choosing Views of Detail Drawings

零件图的视图选择应分析零件结构形状特点，了解零件的用途及主要加工方法，以最少数量的图形表达出零件各部分的结构形状。

If the view of detail drawing should be selected, the structural shape characteristic of parts should be analyzed, and understand the usage of parts and the main processing methods, and express the structure shape of parts with a minimum number of graphs.

6.2.1　主视图的选择

主视图是一组图形的核心，能较清楚和较详细地反映该零件的结构形状，在画图和看图中起主导作用。选择主视图时主要考虑以下几点。

1. 形状特征

主视图的投影方向，应最能反映零件结构形状特征和相对位置，如图6-2和图6-3所示。

6.2.1　Choosing the Front View

The front view is the core in a group of graphs, which can reflect the structural shape of the part clearly. It plays a leading role in drawing and reading. The following should be considered when selecting the front view.

1. Form features

The projection direction of the front view should reflect the structural shape characteristics and relative position of the part, as shown in Fig.6-2 and Fig.6-3.

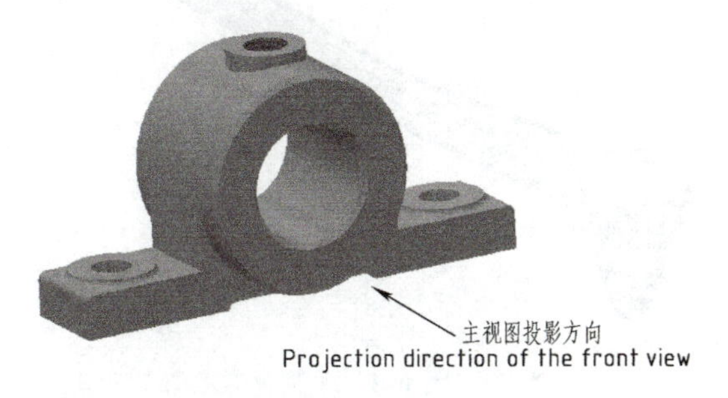

图 6-2　选择主视图

Fig.6-2　Choosing the front view

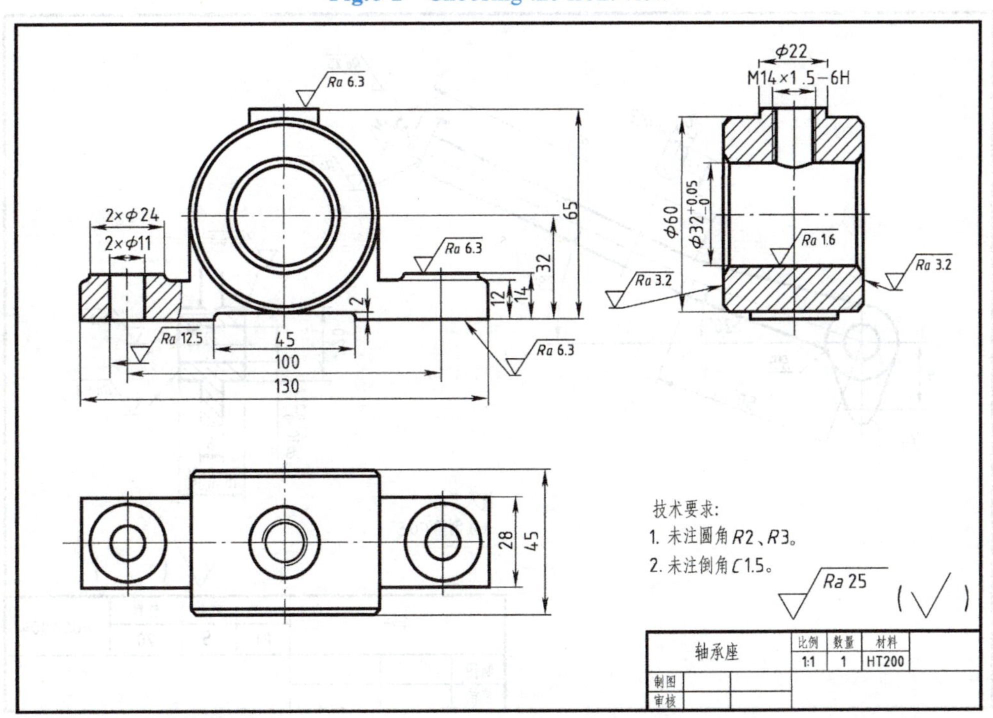

图 6-3　轴承座零件图

Fig.6-3　Drawing a bearing housing

2. 工作位置

如图 6-4 所示手柄，主视图的表达应尽量与零件在机器或部件中的工作位置一致，还应尽可能地和装配图中的位置保持一致，这样读图比较直观，也便于安装零件，如图 6-5 所示。

2. Working position

As shown in Fig.6-4, the selection of the front view representation should be consistent with the working position of part in the machine or subassembly. It could help read intuitionally and is convenient to install parts, as shown in Fig.6-5.

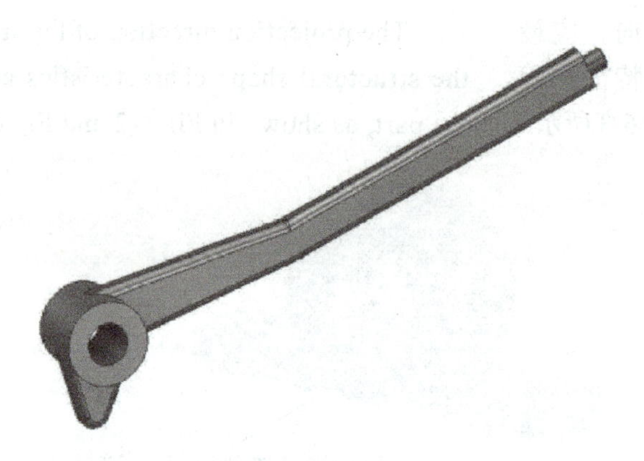

图 6-4 手柄

Fig.6-4 Handle

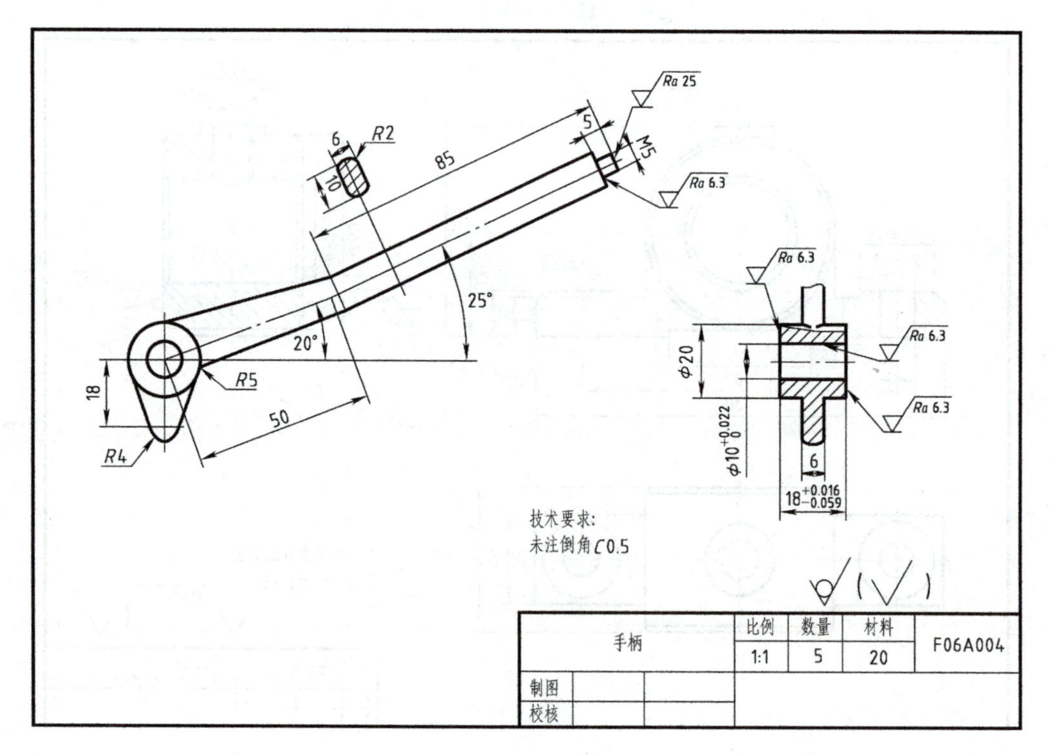

图 6-5 手柄零件图

Fig.6-5 Handle drawing

3. 加工位置

如图 6-6 所示阀杆，加工位置是指零件加工时在机床上的装夹位置，为便于工人生产，主视图应尽量和零件的主要加工位置保持一致。对轴套、轮盘类等回转体零件，选择主视图时，一般应遵循这一原则，如图 6-7 所示。

3. Machining position

As the valve stem shown in Fig.6-6, the machining position refers to the clamping position on the machine tool when the part is machined. For the convenience of workers in production, the front view should be consistent with the main machining position of the part. When selecting the front view for the rotate parts (sleeves and roulettes), the principle should be followed, as shown in Fig.6-7.

图 6-6　阀杆

Fig.6-6　Valve stem

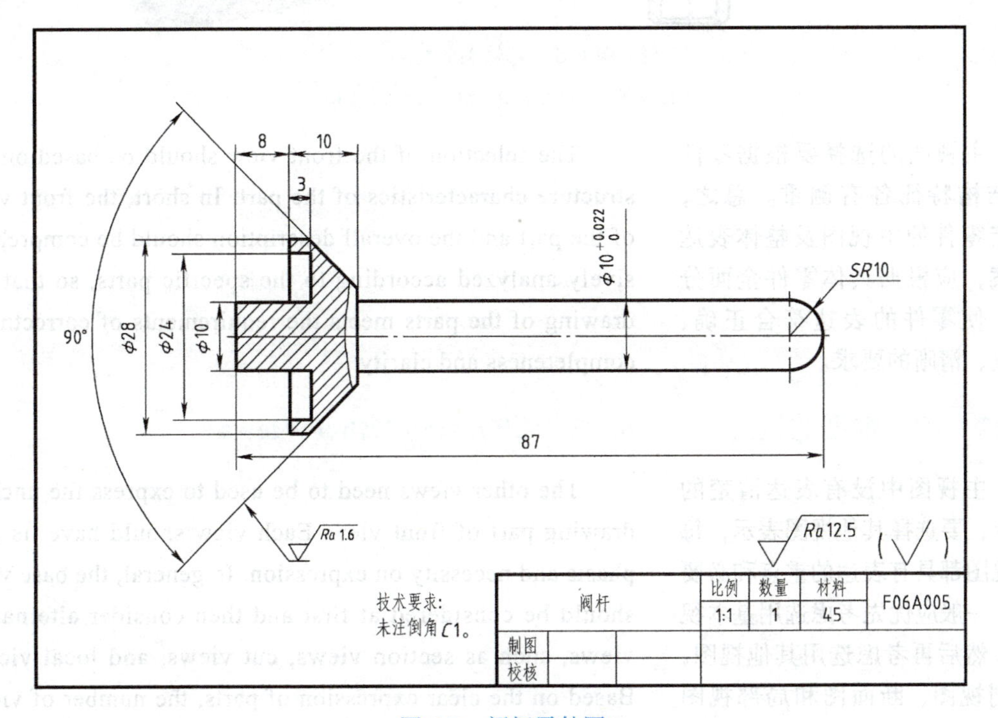

图 6-7　阀杆零件图

Fig.6-7　Valve stem drawing

4. 自然安放位置

如果零件的工作位置是斜的，或加工位置各有不同，又不便按加工位置放置，可将零件的主要部分放正，按自然安放位置放置。此外，还应兼顾其他视图的选择，考虑视图的合理布局，充分利用图幅。如图 6-8 所示。

4. Natural installation position

If the work position of the part is inclined or oblique, or the processing position is different so that it is inconvenient to place it according to the processing position, the main component of the part can be placed in a positive position according to the natural placement position. In addition, consider the rational layout of the view and make full use of the working sheet to select. As shown in Fig.6-8.

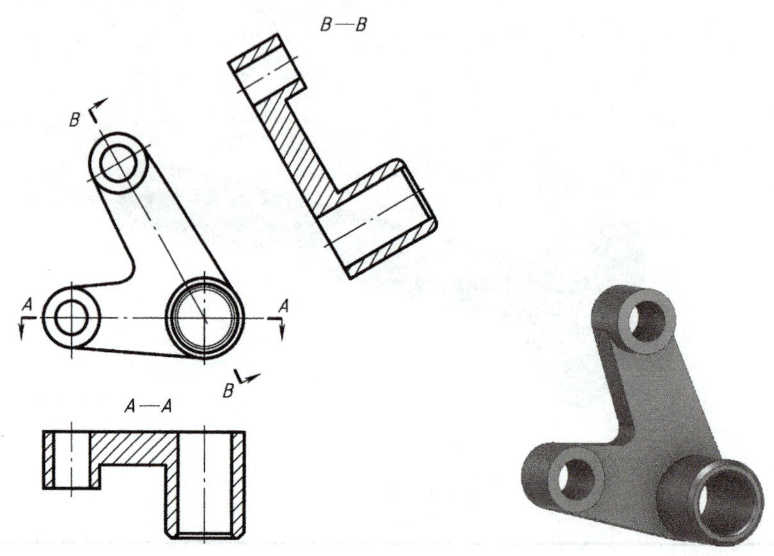

图 6-8 自然安放位置

Fig.6-8 Natural installation position

主视图的选择要根据零件的结构特征各有侧重。总之，确定零件的主视图及整体表达方案，应根据具体零件全面分析，使零件的表达符合正确、完整、清晰的要求。

The selection of the front view should be based on the structure characteristics of the part. In short, the front view of the part and the overall description should be comprehensively analyzed according to the specific parts, so that the drawing of the parts meets the requirements of correctness, completeness and clarity.

6.2.2 其他视图的选择

主视图中没有表达清楚的部分，要选择其他视图表示，每个视图都具有表达的重点和必要性。一般应优先考虑选用基本视图，然后再考虑选用其他视图，如剖视图、断面图和局部视图等。在明确表达零件的前提下，视图的数量要尽量少。

6.2.2 Choosing Other Views

The other views need to be used to express the unclear drawing part of front view. Each view should have its emphasis and necessity on expression. In general, the base view should be considered at first and then consider alternative views, such as section views, cut views, and local views. Based on the clear expression of parts, the number of views should be minimized.

如图 6-5 的手柄零件图中，主视图表达手柄外形，用移出断面图表达手柄杆部截面形状；左视图用局部视图表达手柄安装孔的形状。

Fig.6-5 shows the handle part drawing, the front view expresses the shape of the handle, and the removed cut view of the handle is used to express the cross-sectional shape of the handle. The left view uses a local view to express the shape of the mounting hole of the handle.

6.2.3　典型零件分析

6.2.3　Analysis of Typical Parts

由于零件的用途不同，其结构形状也是多种多样的，根据其结构形状，大致可分为轴套类零件、轮盘类零件、支架类零件和箱体类零件。它们在视图表达方面虽然有共同原则，但各有不同特点。

6.2.3

Due to the different usages of the parts, the structure and shape are also various. According to their structural shapes, the type of parts can be roughly divided into shafts and sleeves, wheel and plate, rod and bracket, box and casing. Although they share same principles in expression of views, they have different characteristics.

1. 轴套类零件

（1）结构与用途分析　轴套类零件包括各种转轴、丝杆、套筒等，主要用来支承传动零件和传递动力。轴套类零件的基本形状是回转体，沿轴线方向通常有轴肩、倒角、退刀槽、键槽、砂轮越程槽等结构要素。如图 6-9 所示。

（2）视图选择分析　轴套类零件一般是在车床或磨床上加工，按加工位置和反映轴向特征原则，将其轴线水平放置。视图的选择是画一个主视图，再根据各部分结构特点，选用断面图或局部放大图，如图 6-10 所示。

（3）尺寸标注分析　在加工和测量径向尺寸时，均以轴线作为基准，沿轴线方向分别注出各段轴的直径尺寸。轴长度方向的尺寸一般都以重要的定位面作为主要的尺寸基准，再按加工、测量要求选取辅助面为辅助基准。

1. Shafts and sleeves

6.2.3-1

(1) Structure and application analysis: these parts include various shafts, screws, sleeves, etc., which are mainly used to support transmission parts and deliver power. The basic shape of these parts is a rotated body. There are general structural elements along the axial direction include shoulder, chamfer, escape, keyway, and grinding, which are shown in Fig.6-9.

(2) Views analysis: shafts and sleeves are generally machined on a lathe or grinder, and their axes are placed horizontally according to the machining position and the principle of reflecting the axial characteristics. The choice of view is to draw a front view, and then select the cut view or partial enlarged view according to the structural characteristics of each part, as shown in Fig.6-10.

(3) Dimension analysis: when machining and measuring the radial dimension, take the axis as the reference, then mark the diameter of each segment along the axis. The dimension in the axis direction generally uses the important positioning surface as the main dimension reference, and then selects the auxiliary surface as the secondary reference according to the machining and measurement requirements.

图 6-9　输出轴

Fig.6-9　Output shaft

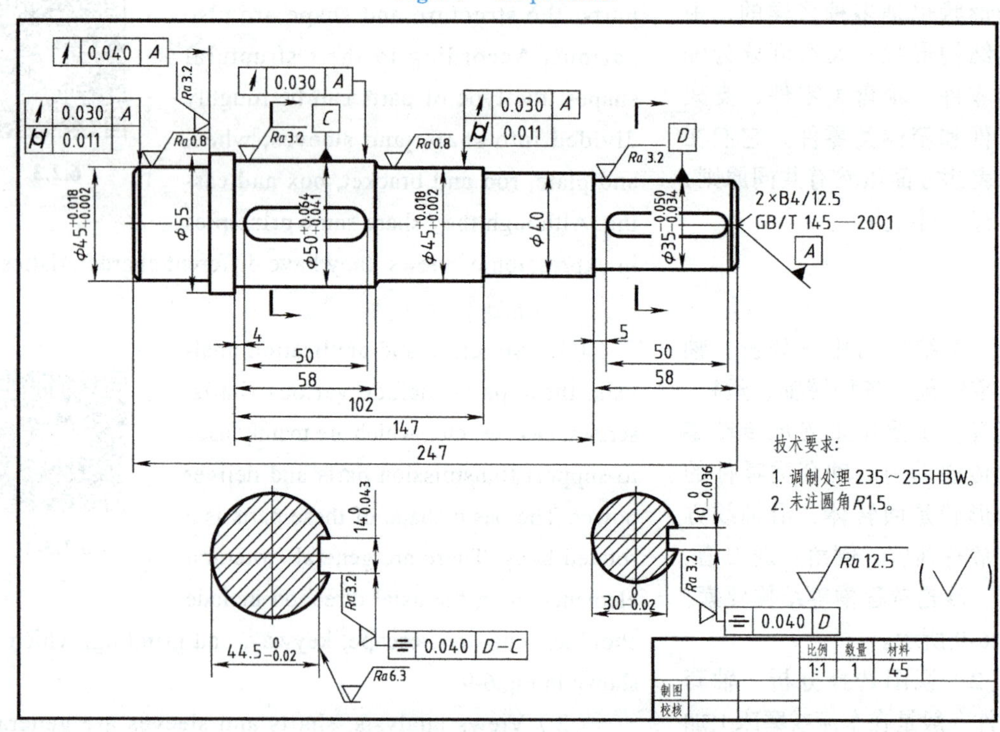

图 6-10　输出轴零件图

Fig.6-10　Output shaft drawing

2. 轮盘类零件

（1）结构与用途分析　轮盘类零件包括端盖、法兰盘、箱盖和各种轮子等，多用于传递扭矩和起支承、轴向定位及密封等作用。轮盘类零件的基本形状特点是扁平的盘状，上面通常有均匀分布的凸缘、孔和筋板等结构，如图 6-11 所示。

2. Wheels and plates

(1) Structure and application analysis: this category of parts includes end caps, flanges, case lids and a variety of wheels. Wheels play an important part on torque transmission and support, axial positioning and sealing, etc. The basic shape of these parts is characterized by a flat disc shape, which usually has a uniform distribution of flanges, holes and ribs, as shown in Fig.6-11.

6.2.3-2

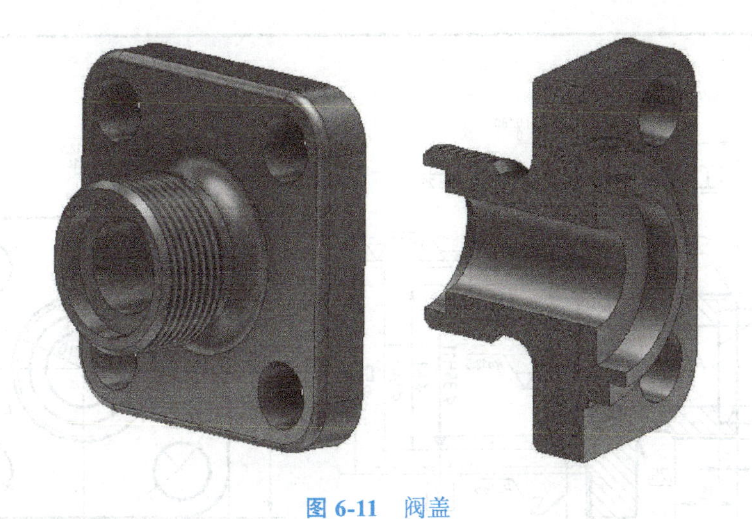

图 6-11　阀盖

Fig.6-11　Valve cover

（2）视图选择分析　轮盘类零件主视图的选择可以是外形视图，但选用剖视图更好，层次分明，可表达各台阶与内孔的形状及相对位置，也符合主要加工位置。一般轮盘类零件多数带有各种形状的凸缘、孔槽和筋等结构，用一个主视图不能完整表达零件，需要其他基本识图。如图 6-12 阀盖零件图所示，主要加工表面以车削为主，主视图将轴线水平放置。主视图选用全剖视图，表达内孔的形状；左视图采用基本视图，可表达四个分布的通孔。

（3）尺寸标注分析　轮盘类零件的宽度和高度方向的基准都是回转轴线，长度方向的主要基准是经过加工的较大端面。圆周上均匀分布的小孔的定位圆直径是这类零件的定位尺寸，对结合面（工作面）的有关尺寸精度、表面粗糙度和几何公差有比较严格的要求。

（2）Views selection Analysis: the front view of these parts can be a shape view, but the section view is better with distinct layers. In this way, it’s easy to show the shape and relative position of the step and the inner hole. It also fits the main position for machining. Generally, most of the plate parts have various shapes of flanges, holes and ribs, etc. The part cannot be expressed fully by front view so that the other view should be drawn to express. As shown the part drawing (bonnet) in Fig.6-12, the main machined surface is turned, and the axis should be placed horizontally in the front view. The section view should be used for the front view to express the shape of inner hole. The basic view is used for left view to show the four-distributed through hole.

（3）Dimension analysis: the reference for the width and height direction of these parts is the axis of rotation, and the main reference of length direction is the larger end face that has been machined. The positioning circle diameter of the small holes uniformly distributed on the circumference is the positioning size of such parts. There are strict requirements on the dimensional accuracy, surface roughness and geometric tolerance of the joint surface (working surface).

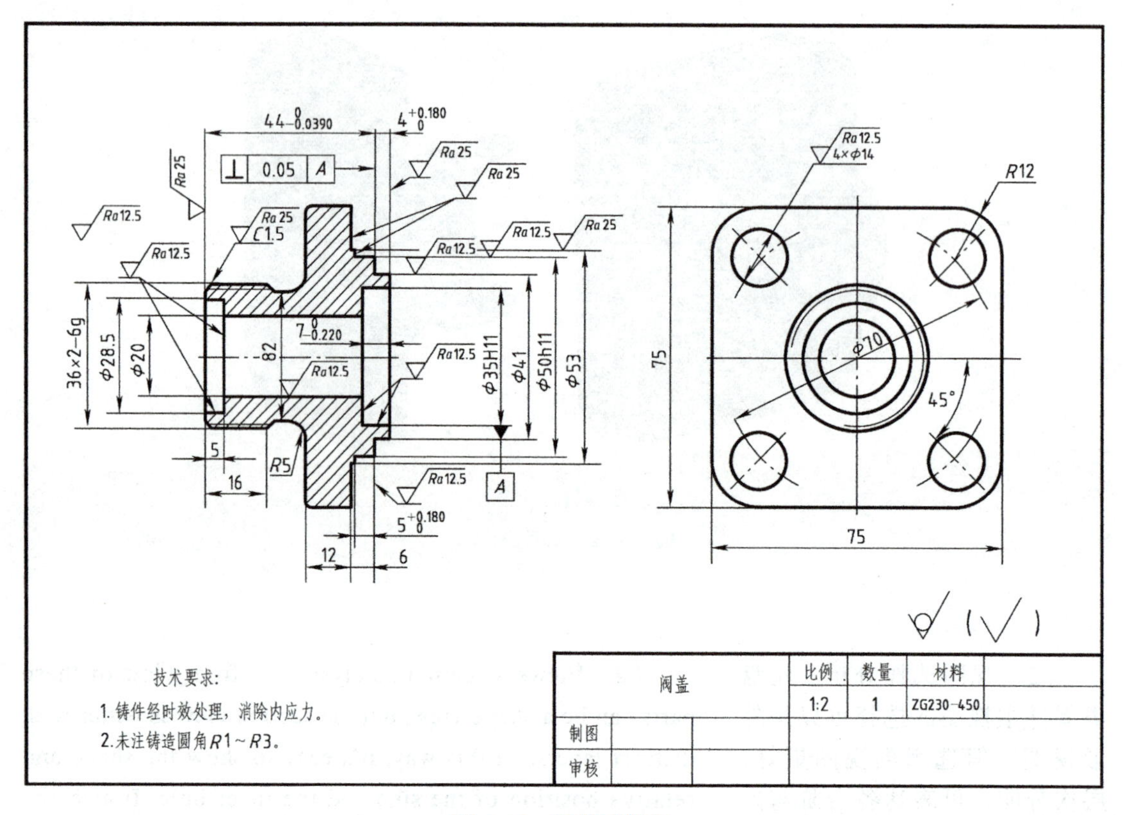

图 6-12　阀盖零件图

Fig.6-12　Valve cover drawing

3. 叉架类零件

（1）结构与用途分析　叉架类零件主要包括拨叉、连杆、支架、支座等。叉架类零件在机器或部件中主要是起操纵、连接、传动或支承作用。叉架类零件形状不规则，外形比较复杂，常有弯曲或倾斜结构，并带有底板、筋板、轴孔、螺孔等结构，加工位置较多。根据零件结构形状和作用不同，一般叉架类零件的结构可看成是由支承部分、工作部分和连接部分组成，如图 6-13 所示。

3. Brackets

（1）Structure and application analysis: these parts mainly include fork, connected rod, bracket, support and bearing, etc. These parts play the important role on manipulation, connection, transmission or support in a machine or subassembly. These parts are irregular in appearance, and the shape is complicated with curved or inclined structures and have structures such as a bottom plate, a rib plate, a shaft hole, and a screw hole,etc. There are also lots of machining positions. Based on the structural shape and function of the part, the structure of these type parts consists of the part of support,working and connection, as shown in Fig.6-13.

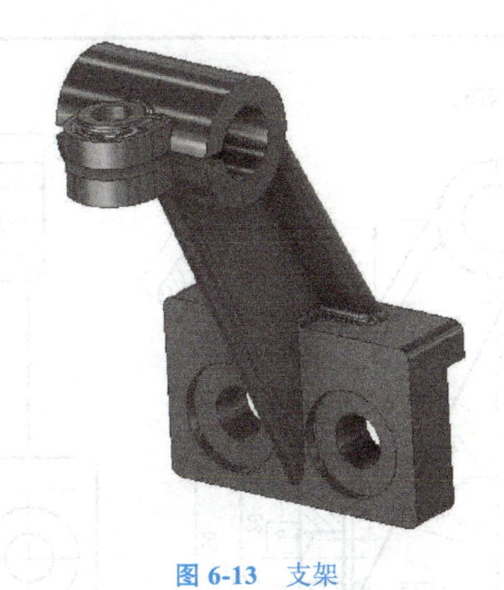

图 6-13　支架

Fig.6-13　Support

（2）视图选择分析　因叉架类零件一般都是锻件或铸件，往往要在多种机床上加工，各工序的加工位置不尽相同。所以在选择主视图时，主要按形状特征和工作位置确定。这类零件的结构形状较为复杂且不太规则，一般都需要两个以上视图。某些不平行于投影面的结构形状，常采用斜视图、斜剖视图和断面图表达；对一些内部结构形状可采用局部剖视图；也可采用局部放大图表达其较小结构。如图 6-14 所示。

（2）Views selection Analysis: because these parts are generally forged or casted so that they may often machined on a variety of machine tools and each working step of machining positions is different. Therefore, when selecting the front view, it is mainly determined by the structure feature and the working position. The structural shape of these parts is complex and irregular so it is required more than two views to be used to express them. Some structural shapes that are not parallel to the projection surface are often represented by oblique views, oblique section views, and section views. Local section views could be used to express the internal structural shape or partial enlarged views could be used to describe the smaller structures, which is shown in Fig.6-14.

（3）尺寸标注分析　叉架类零件在长、宽、高三个方向的主要基准一般为孔的中心线（或轴线）、对称平面和较大的加工面。定位尺寸较多，孔的中心线（或轴线）之间、孔的中心线（或轴线）到平面或平面到平面间的距离一般都要注出。

（3）Dimension analysis: the main reference of these parts on the three directions (length, width and height)is generally the center line (or axis) of the hole, the symmetry plane and the larger machined surface. There are a lot of positioning dimensions, so the distance between the center line (or axis) of the hole, the center line (or axis) of the hole to the plane or the plane to the plane need to be marked.

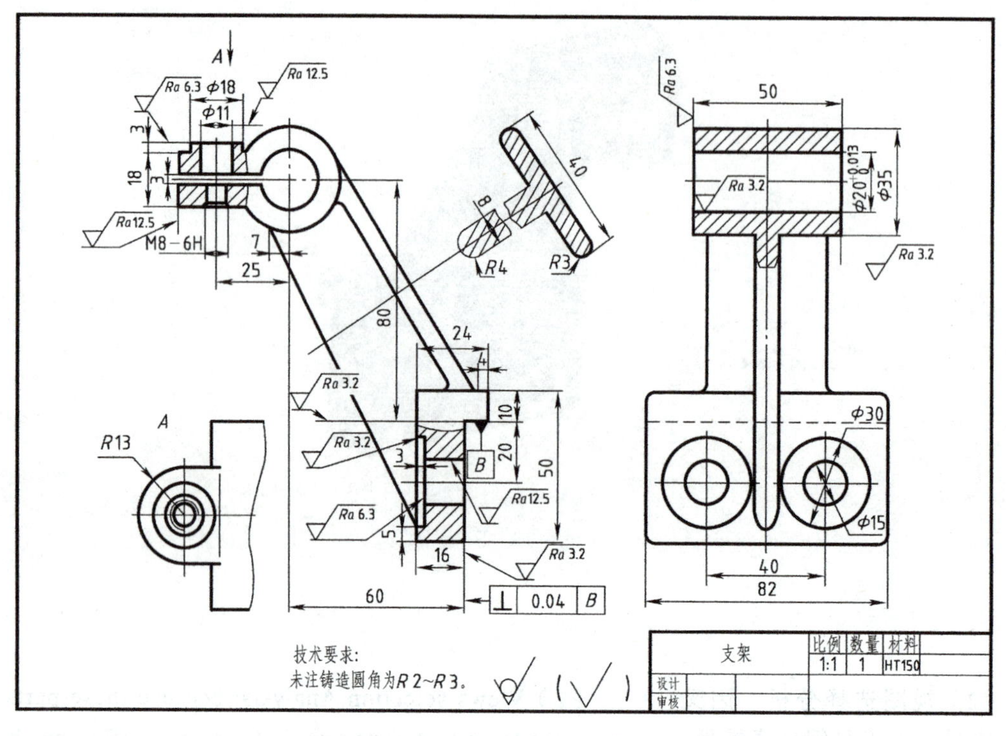

图 6-14 支架零件图

Fig.6-14 The part drawing of support

4. 箱体类零件

（1）结构与用途分析 箱体类零件一般是机器或部件的主体部分，常见的箱体类零件有箱体、泵体、阀体、机座等。箱体零件起着支承、包容其他零件的作用，又是保护机器中其他零件的外壳，其周围一般分布有连接螺孔等，结构形状复杂，一般多为铸件，也有焊接件。如图 6-15 所示。

（2）视图选择分析 箱体类零件结构比较复杂，加工工序较多，装夹位置又不固定，因此一般均按工作位置和形状特征原则选择主视图，其他视图至少在两个或两个以上。如果外部结构形状简单，内部形状复杂，且具有对称平面，可

4. Boxes

（1）Structure and application analysis: these parts are the main part of the machine or subassembly. There are some common box parts, which include box body, pump body, valve body, base part, etc. Box parts play the role of supporting and containing other parts and also protect the outer shells of other parts in the machine. Their surroundings are generally distributed with the connected screw holes. The structure is complex and most of them are casted and welded, as shown in Fig.6-15.

6.2.3-4

（2）Views Analysis: the structure of these parts is complicated, involving many machining steps and the clamping position is usually unfixed. Therefore, the front view is generally selected according to the working position and structure characteristic. In addition, at least two or more other views need to be used. If the structure of the external structure is simple and the internal shape is complex with

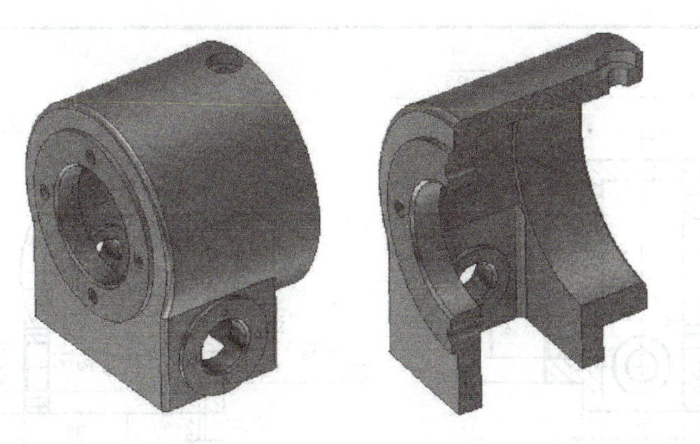

图 6-15　箱体

Fig.6-15　Box

采用全剖；如果外部形状复杂，内部形状简单，且具有对称平面，可采用局部剖视或用虚线表示；如果内外部结构形状都较复杂，投影不重叠，可采用局部剖视图；重叠时，内、外部结构形状应分别表达；对局部内、外部结构形状可采用局部视图、局部剖视和剖面来表达。箱体零件上常常会出现一些截交线和相贯线；由于该类零件多为铸件，所以经常会出现过渡线，应认真分析，如图 6-16 所示。

（3）尺寸标注分析　箱体类零件的长、宽、高三个方向的主要基准采用中心线、轴线、对称平面和较大的加工平面。因结构形状复杂，定位尺寸多，各中心线（或轴线）间的距离一定要直接标注出来，如重要轴孔对基准的定位尺寸，与其他零件有装配关系的尺寸。箱体类零件的技术要求，主要是支承传动轴的轴孔部分，其轴孔的尺寸精度、表面粗糙度和几何公差，都将直接影响装配质量和使用性能，因此尺寸精度必须严格要求。

symmetrical plane, the full section view could be used. If the external shape is complex and the internal shape is simple with symmetrical plane, the local section view or dash line could be used. If the internal and external structure are both complicated, when the projections do not overlap, the local section view can be used, when the projections overlap, the inner and outer structural shapes should be expressed separately. The local views, local section views and the section can be used to express the partial inner and outer structural shapes. Some lines of section and intersection often appear on the box parts. Since the parts are mostly casted, transition lines often appear so that it should be analyzed carefully, as shown in Fig.6-16.

（3）Dimensional analysis: the main reference of these parts on the three directions (length, width and height) is the centerline, the axis, the symmetry plane and the larger machining plane. Due to the complicated structure and many positioning dimensions, the distance among the center lines (or axes) of each hole must be marked directly, such as the positioning dimension of the important shaft hole to the reference, and the dimension of other parts with assembly relationship. Because these parts are mainly to support the shaft hole of the drive shaft in the aspect of technical requirements, the dimension accuracy, surface roughness and geometric tolerance of the shaft hole will directly affect the assembly quality and performance, so the dimension accuracy must be strictly.

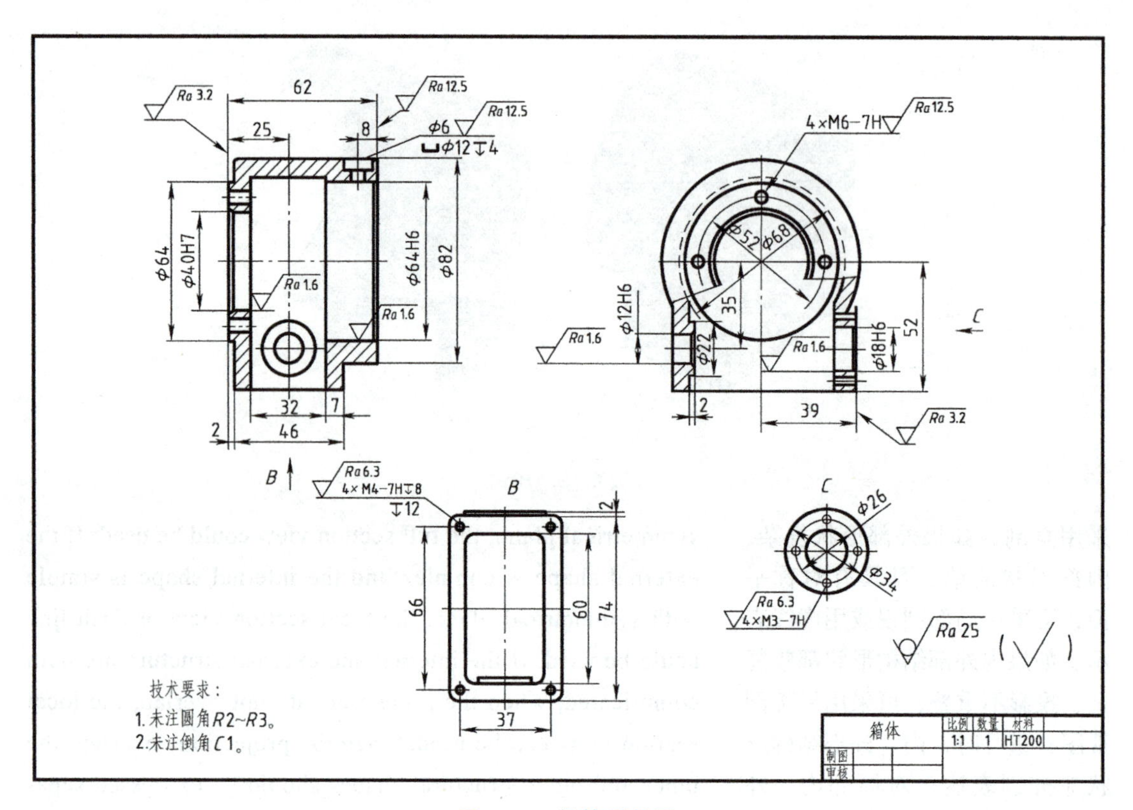

图 6-16　箱体零件图

Fig.6-16　Box drawing

除了上述类型零件外，还有一些其他类型的零件，例如冲压件、注塑件和镶嵌件等，它们的表达方法与上述类型零件的表达方法类似。

In addition to the above types of parts, there are some other types of parts, such as stampings, injection molded parts and inserts, which are expressed in a similar way to the above types of parts.

6.3　零件图的尺寸标注

6.3　Dimensioning of Detail Drawings

零件图的尺寸标注，除了要求正确、完整、清晰，还应考虑合理性。既要满足设计要求，又要符合生产实际，便于加工测量，并有利于装配。

The consideration of the dimensioning of detail drawings requires not only correctness, completion and clarity, but also rationality. In addition, it also needs not only meet the design requirements, but also facilitate machining, measurement and inspection, and facilitate assembly.

6.3.1　尺寸基准的选择

零件的尺寸基准是标注尺寸的起点，选择尺寸基准时，必须考虑零件在机器或部件中的位置，作用、零件之间的装配关系及零件在加工过程中的定位和测量要求。根据基准的用途不同，可分为设计基准和工艺基准。

设计基准是根据机器的结构和设计要求用以确定零件在机器中的一些点、线、面位置的基准。如图 6-17 所示，轴承座通常来支承轴，故轴承孔的轴线一般应与底面平行，轴承孔高度是影响轴承座工作性能的功能尺寸。尺寸 40 ± 0.02 以底面为基准，以保证轴承孔到底面的高度。其他高度方向的尺寸，如 10、12、58 均以底面为基准。

在标注底板上两孔的定位尺寸时，长度方向应以底板的对称面为基准，以保证底板上两孔的对称关系，如俯视图中尺寸 65。其他长度方向的尺寸，如主视图中 ϕ10、45、35，俯视图中 90、8 均以对称面为基准。

左视图中尺寸 30、15、5 均以后端面为基准。零件的长、宽、高三个方向上都各有一个主要基准，还可有辅助基准，如图 6-17 所示，注油孔 M8 的深度尺寸，以 ϕ10 的上端面作为辅助基准。主要基准和辅助基准之间必须有尺寸联系。

6.3.1　Selection of Dimensioning Datum

As mentioned before, a datum provides important reference for dimensioning. When selecting a datum, one must be consider is the part position, function, assembling relationship among parts and positioning and measurement requirements for parts during processing. According to the different applications for datum, it can be divided into design datum and process datum.

A design datum is a reference to the location of points, lines, and planes that determine the position of the part in the machine according to the structure and design requirements of the machine. As shown in Fig.6-17, the bearing seat is usually to support the shaft, so the axis of the bearing hole should generally be parallel to the bottom surface. The height of bearing hole is the functional size that affects the performance of the bearing housing. The dimension 40±0.02 is based on the base surface to ensure the height of the bearing bole. Other height-oriented dimensions (10, 12, and 58) are based on the base surface.

When labeling the location dimensions of the two holes, the symmetry plane of the chassis should be used as the length direction datum to ensure the symmetrical relationship of the two holes on the chassis. The dimensions of other length orientations are based on symmetrical plane, such as hole ϕ10, 45, 35 in the front view, and length 90, 8 in the top view.

The dimension 30, 15, 5 in the left view is the base of the end face. The part has each main datum in three directions (length, width and height) and also can have auxiliary datum. The Fig.6-17 shows the depth dimension of oil injection hole (M8) is based on the upper end surface of ϕ10 as auxiliary datum. Dimensions must be linked between main and auxiliary datum.

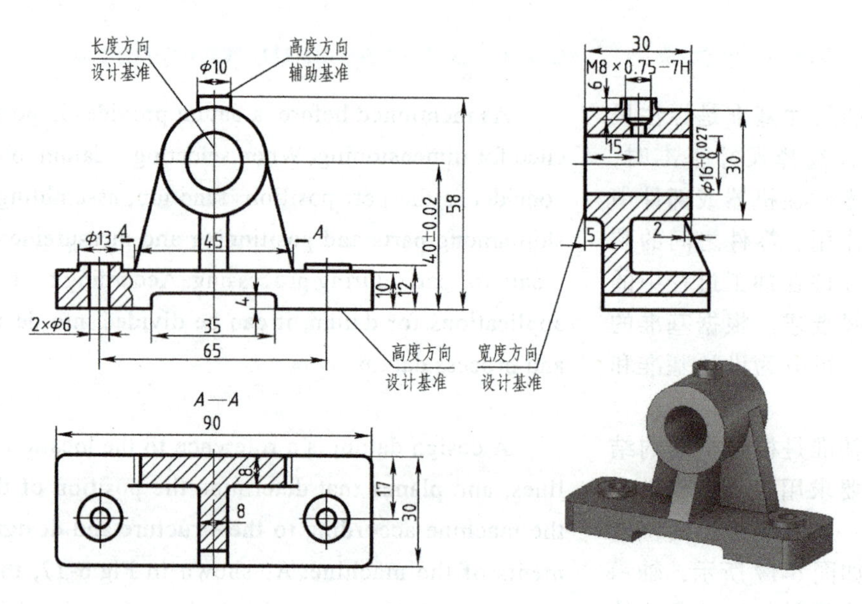

图 6-17 轴承座的尺寸基准

Fig.6-17 Dimension datum of a bearing seat

工艺基准是根据零件加工制造、测量和检验等工艺要求所选定的一些点、线、面位置的基准，如图 6-18a 所示阶梯轴。图 6-18b 是其加工顺序图，第一步，先加工 ϕ17 和总长 64；第二步，以右端面为基准加工 ϕ13 和 47；第三步，加工 ϕ9 和 25；第四步，从右端面钻孔 ϕ4 和 14；第五步，加工左端面及倒角并切断。从图中可看出阶梯轴以右端面为基准标注尺寸，符合加工顺序。在标注尺寸时，设计基准与工艺基准应尽量统一，以减少加工误差，提高加工质量。

Process datum includes datum (point, line and plane) used to process requirement (machining, measurement and examination, etc.) parts. The fig.6-18a shows the step shaft and Fig.6-18b reports machining sequence drawing. Firstly, ϕ17 and length 64 should be machined and then ϕ13 and length 47 should be machined by the datum of right end face. The following step is to process ϕ9 and 25. The step 4 is to drill holes ϕ4 and 14 from the right end face. The final step 5 is to process left end face and chamfer and cut off. It can be seen from the figure that it takes the right end face as the standard dimension and coincide with the processing sequence. In dimensioning, the design datum and process datum should be consistent with each other to reduce machining tolerance and improve the machining quality.

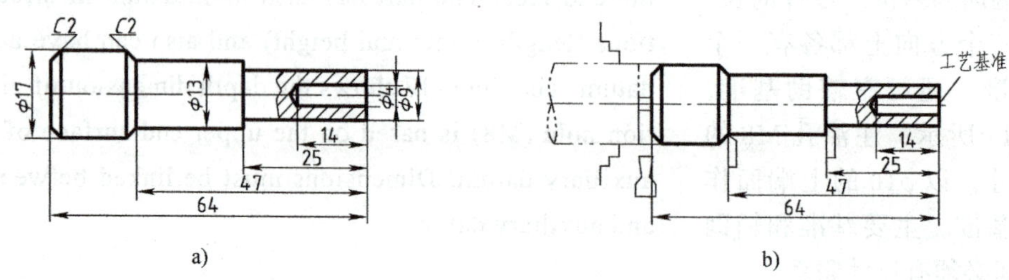

图 6-18 阶梯轴的尺寸

Fig.6-18 Dimensions of a step shaft

6.3.2　尺寸标注方法

6.3.2　Dimensioning Method

1. 重要尺寸要直接注出

重要尺寸指与其他零件相配合的尺寸、重要的相对位置尺寸、影响零件使用性能的其他尺寸，这些尺寸应在零件图上直接注出。如图 6-19a 所示，尺寸 *b* 和 *c* 的累积误差，会使孔中心高度不能满足设计要求。如图 6-19b 所示，孔轴线的高度尺寸 *a* 应直接注出。

1. Direct specifications of important dimensions

Important dimensions refer to dimensions that match other parts, important relative position dimensions, and other dimensions that affect the performance of the part, which should be specified directly during drawing parts. As shown in Fig.6-19a, the accumulation tolerance of dimension between *b* and *c* could make it hard to meet the design requirement. The height of the holes axis should be specified directly, As shown in Fig.6-19b.

2. 避免出现封闭尺寸链

在机器装配或零件加工过程中，由互相联系的尺寸按一定顺序首尾相接排列形成的封闭尺寸组，称为尺寸链。组成尺寸链的各个尺寸称为尺寸链的环。封闭尺寸链是指尺寸同一方向串联并头尾相接组成封闭的图形，如图 6-20a 所示轴的尺寸标注。这种标注会给加工带来困难，所以应采用图 6-20b 所示标注方法。

2. Avoiding close dimension chains

In the process of machine assembly or parts processing, the close dimension group is formed by interconnecting dimensions in a certain order, which called the dimension chain. Each dimension that makes up a dimension chain is called a ring of dimension chain. Dimensions are in series in the same direction and are composed of a closed graph, which called the closed dimension chain. Fig.6-20a shows the dimensions of the Axis. This labeling is difficult to process, so the labeling method shown in Fig.6-20b should be used.

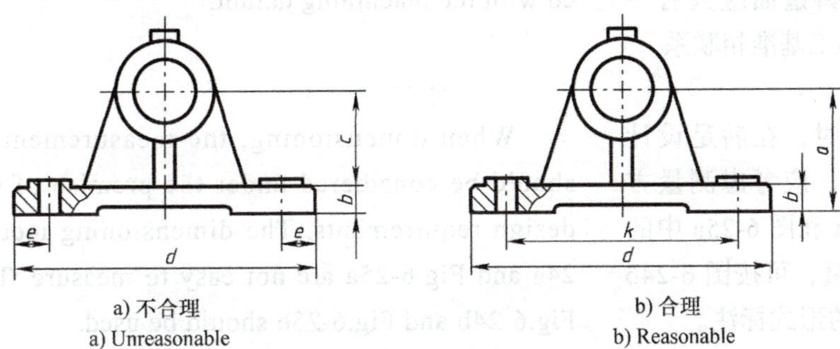

a) 不合理
a) Unreasonable

b) 合理
b) Reasonable

图 6-19　重要尺寸的标注

Fig.6-19　Dimensioning of important dimensions

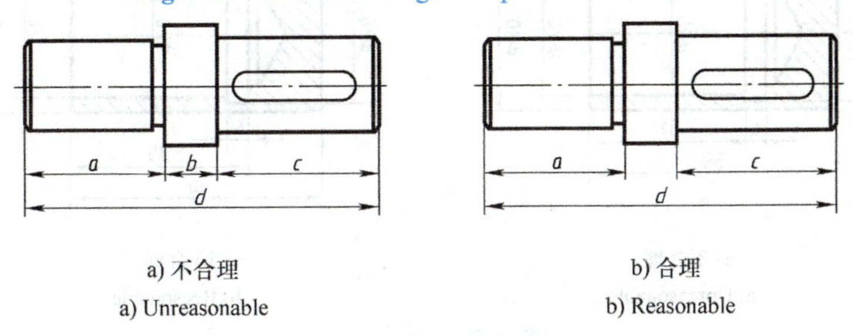

a) 不合理
a) Unreasonable

b) 合理
b) Reasonable

图 6-20　尺寸链

Fig.6-20　Dimension chains

3. 考虑加工方法，符合加工顺序

零件在加工时，都有一定的顺序。标注的尺寸应尽量与加工顺序一致，便于加工时看图、测量，且易于保证加工顺序。以轴套的尺寸注法为例，图 6-21a 中内孔的轴向尺寸按加工工序标注，所以是合理的；图 6-21b 中尺寸 2 和 31 的注法不符合加工工序，尺寸 31 也不易直接测量，所以是不合理的。

在加工过程中，会有不同的阶段，如铸、锻后再经机械加工，因此对不同工种所需的尺寸应分别标注，以利于看图，如图 6-22、图 6-23 所示。图 6-22a 标注不合理，当加工右端面时，由于铸造误差，要同时满足 20、88 和 108 是困难的。图 6-22b 标注合理，其中 88、8 为铸造尺寸，铸造面应只有一个尺寸 20 与加工基准相联系。

4. 便于测量

标注尺寸时，在满足设计要求的前提下，应考虑测量方便。如图 6-24a 和图 6-25a 中的注法不便于测量，可按图 6-24b 和图 6-25b 中的形式标注。

3. The processing method must conform to the processing order

Parts have the certain order during process. Dimensions should be consistent with the order of processing, so it is easy to see the diagram, measure and keep the processing sequence. Taking the size injection method of shaft sleeve as an example, the axial size of the inner hole in Fig.6-21a is dimensioned according to the processing procedure, which is reasonable. The injection method of sizes 2 and 31 in Fig.6-21b does not conform to the processing procedure, and size 31 is not easy to be directly measured, so it is unreasonable.

In the machining process, there are several different stages, such as casting, forging and then machining, so the size of different types of work should be dimensioned accordingly and separately, which is easy to read the drawing, as shown in Fig.6-22 and 6-23. The dimensions in Fig.6-22a are labeled incorrectly. When the right end surface is machined, it is difficult to meet the dimension 20, 88, and 108 at the same time due to casting tolerance. Fig.6-22b is labeled correctly, of which 88 and 8 are casting dimensions, and the casting surface should only be one size 20 associated with the machining datum.

4. Easy to measure

When dimensioning, the measurement convenience should be considered under the premise of satisfying the design requirements. The dimensioning method of Fig.6-24a and Fig.6-25a are not easy to measure. The method in Fig.6-24b and Fig.6-25b should be used.

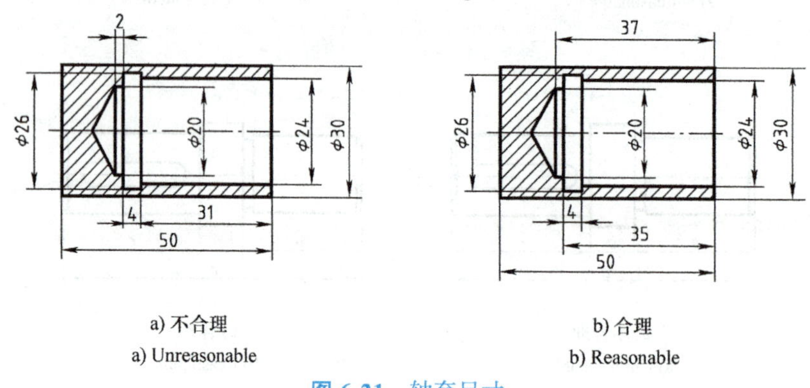

a) 不合理
a) Unreasonable

b) 合理
b) Reasonable

图 6-21 轴套尺寸
Fig.6-21 Sleeve size

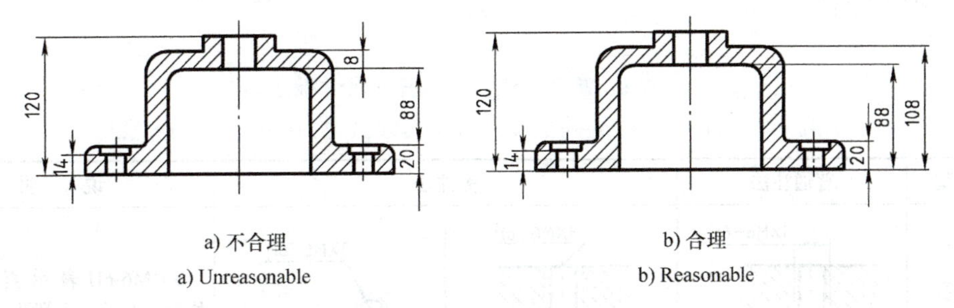

a) 不合理
a) Unreasonable

b) 合理
b) Reasonable

图 6-22 毛坯面的尺寸标注

Fig.6-22 The rough surface dimensioning

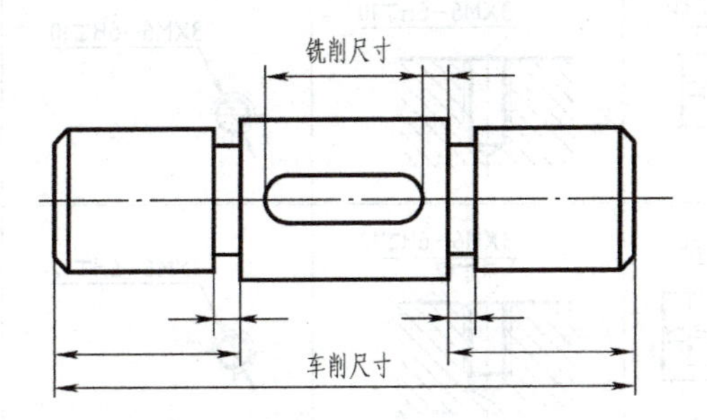

图 6-23 不同工种的尺寸分开标注

Fig.6-23 Separate dimensioning of different types of work

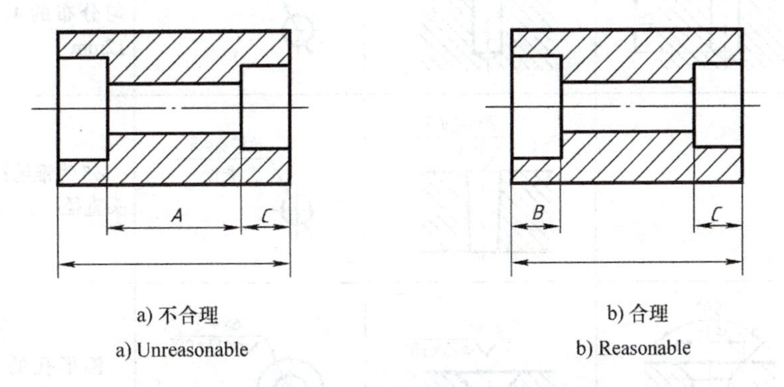

a) 不合理
a) Unreasonable

b) 合理
b) Reasonable

图 6-24 套筒的轴向尺寸

Fig.6-24 Axial dimensions of the sleeve

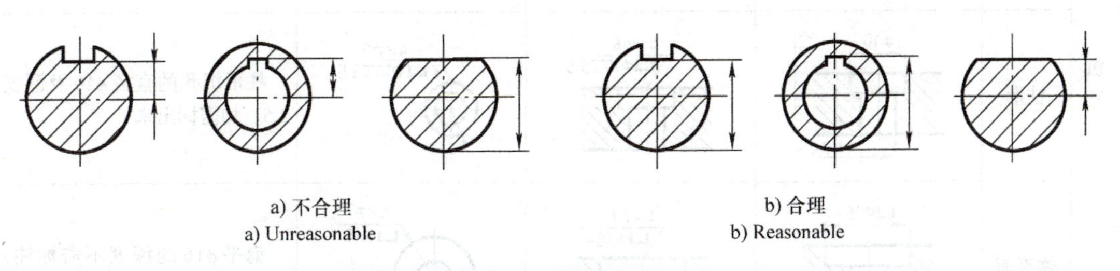

a) 不合理
a) Unreasonable

b) 合理
b) Reasonable

图 6-25 测量方便

Fig.6-25 Easy to measure

6.3.3 零件上常见结构的尺寸标注

6.3.3 Dimensioning of Common Structures on Parts

表 6-1 螺孔、光孔和沉孔的尺寸标注

Table 6-1 Dimensioning of screw hole, simple hole and sinking hole

结构类型		普通注法	旁注法		说 明
螺孔	通孔	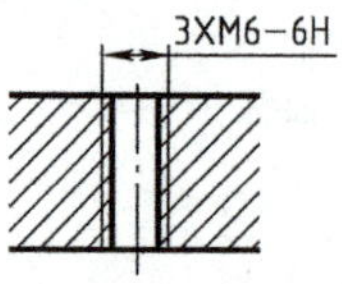 3XM6−6H	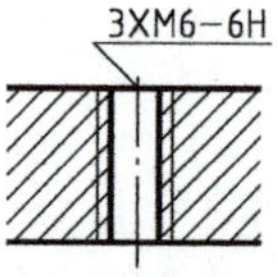 3XM6−6H	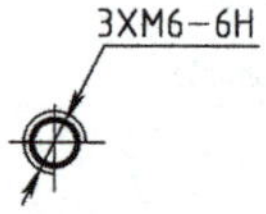 3XM6−6H	3×M6-6H 表示直径为 6，均匀分布的 3 个螺孔
	不通孔	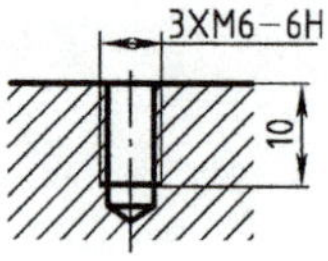 3XM6−6H 10	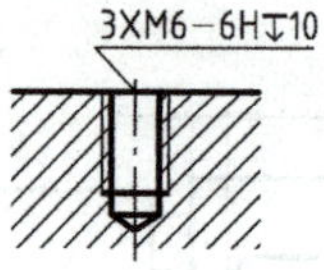 3XM6−6H↧10	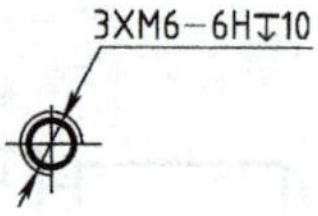 3XM6−6H↧10	螺孔的深度可以与螺孔的直径连注，也可以分开标注
		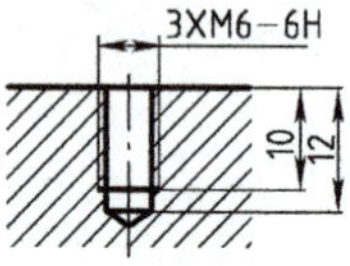 3XM6−6H 10 12	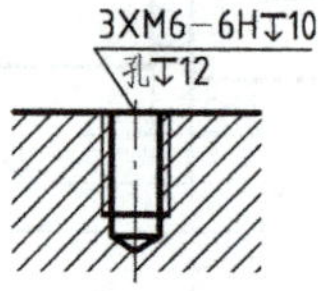 3XM6−6H↧10 孔↧12	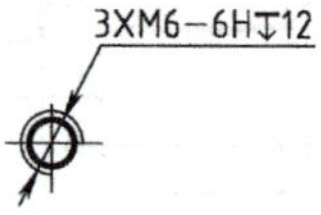 3XM6−6H↧12	3 个 M6-6H 螺纹不通孔，螺纹部分深 10mm，加工螺纹前钻孔深 12mm
光孔	一般孔	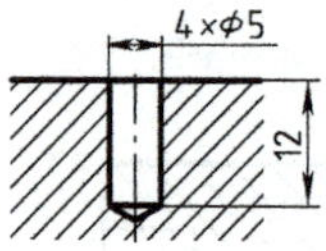 4×ϕ5 12	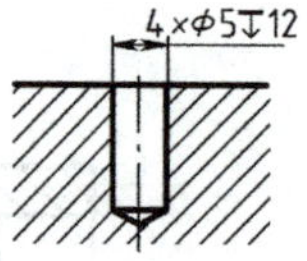 4×ϕ5↧12	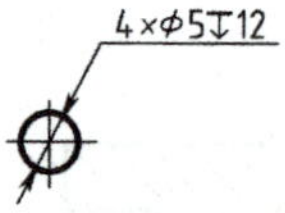 4×ϕ5↧12	4×ϕ5 表示直径为 5、均匀分布的 4 个光孔，孔深为 12mm
	锥销孔		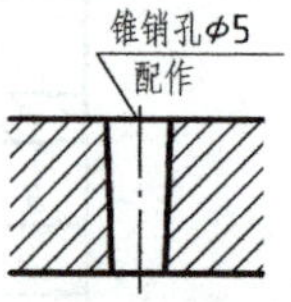 锥销孔ϕ5 配作	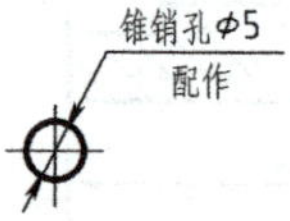 锥销孔ϕ5 配作	ϕ5 为锥销孔相配的圆锥销小头直径
沉孔	锥形	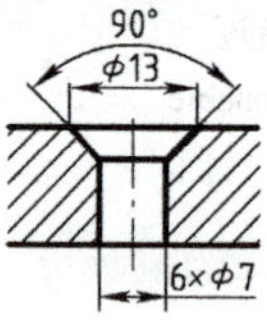 90° ϕ13 6×ϕ7	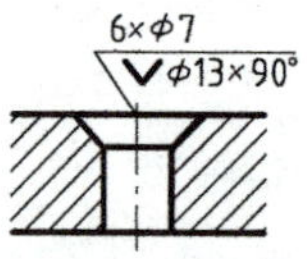 6×ϕ7 ⌵ϕ13×90°	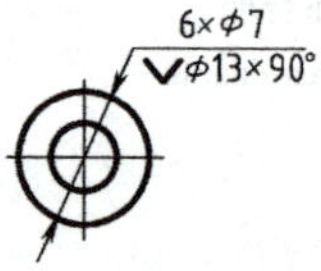 6×ϕ7 ⌵ϕ13×90°	锥形孔的直径 ϕ13 及锥角 90°，均需标出
	柱形	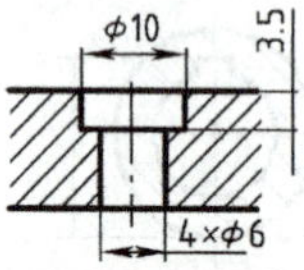 ϕ10 3.5 4×ϕ6	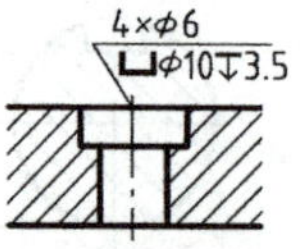 4×ϕ6 ⌴ϕ10↧3.5	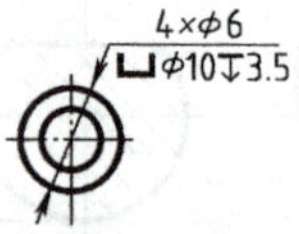 4×ϕ6 ⌴ϕ10↧3.5	柱形沉孔的直径 ϕ10 及深度 3.5，均需标出来
	锪平面	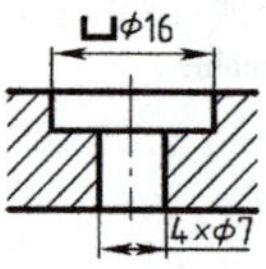 ⌴ϕ16 4×ϕ7	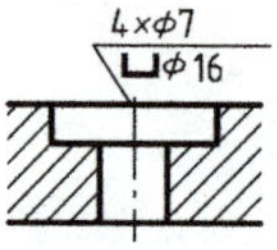 4×ϕ7 ⌴ϕ16	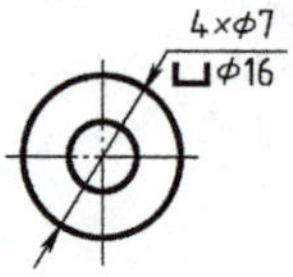 4×ϕ7 ⌴ϕ16	锪平 ϕ16 的深度不需标注，一般锪至不出现毛面为止

6.4 零件上常见的工艺结构

零件的结构形状，不仅要满足零件在机器中使用的要求，还要考虑零件在加工、测量、装配过程的一系列工艺要求，使零件具有合理的工艺结构。下面介绍一些常见的工艺结构。

6.4 Common Process Structure on Parts

The structure and shape of parts are not only meet the requirements of parts used in machines, but also consider a series of process requirements of parts in processing, measuring and assembling, so as to make the parts have reasonable technological structure. Processing structure is introduced below as a reference.

6.4.1 铸造结构

铸造时，先制造模样，将模样放置于型砂中，当型砂压紧后，取出模样，再在型腔内浇入铁水或钢水，待冷却后取出铸件毛坯。对零件上有配合关系的接触表面，还应切削加工，才能使零件达到最后的技术要求。

1. 起模斜度和铸造圆角

铸件造型时，为便于取出模样，沿脱模方向设计出起模斜度（1∶20），如图 6-26 所示。起模斜度较小时，在图上可以不标注；必要时，可以在技术要求中用文字说明。

铸件各表面相交处应做成圆角，便于铸件造型时起模，同时防止冷却时产生缩孔和裂纹，这种圆角称为铸造圆角。同一铸件的圆角应尽量相同，并可以在技术要求中统一说明。

无论塑料件、压铸件、五金件，考虑到生产问题，都要拔模，下面举例说明 Inventor 拔模斜度工具的使用。如图 6-27 所示，对实体进行拔模操作。首先，找到“修改”栏的“拔模”工具。

6.4.1 Casting Structure

When casting, the mould should be made at first and then placed in moulding sand. After pressing the moulding sand, remove the mould, and then pour hot liquid iron or steel into the cavity. When it come cool , casting blank could be taken out. The contact surface with a fitting relationship on the part should also be machined in order to make the parts meet the technical requirements.

1. Draft and fillets in castings

The draft (1∶20) along the drawing direction of the pattern facilitates to take out the pattern while molding, which is shown in Fig.6-26. If the draft is very small, it can be omitted in the drawing. If necessary, it can be marked in technical requirements.

The various surfaces of the castings should be made with fillet. In order to avoid sand drops while removing the mould, and avoid shrinkage cavity and the fissures while cooling. This fillet is called fillets in casting. The same casting radius round should be the same as far as possible, and can be unified description in the technical requirements.

Considering the production, whatever plastic parts, press castings and hardware parts all need draft process. The following example illustrates the use of Inventor draft slope tool. As shown in Fig.6-27, a draft operation is performed on the entity. Firstly, find the “draft” command in the “modify” bar.

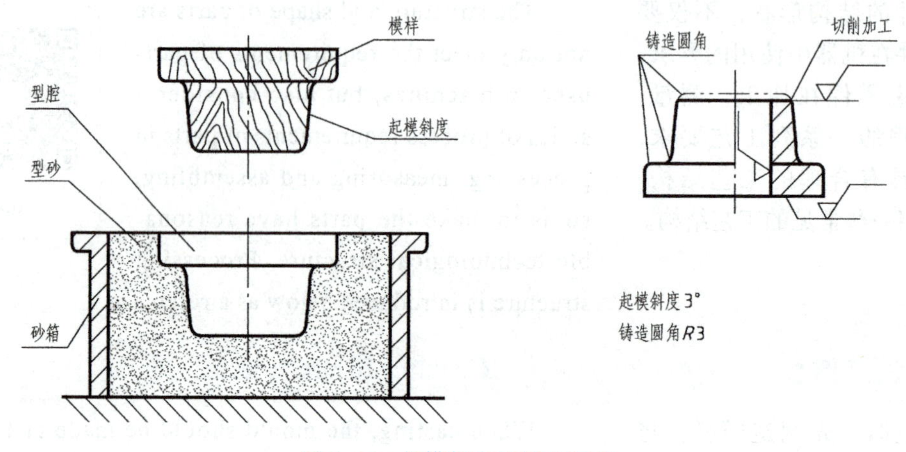

图 6-26　起模斜度和铸造圆角

Fig.6-26　Draft slope and fillets in castings

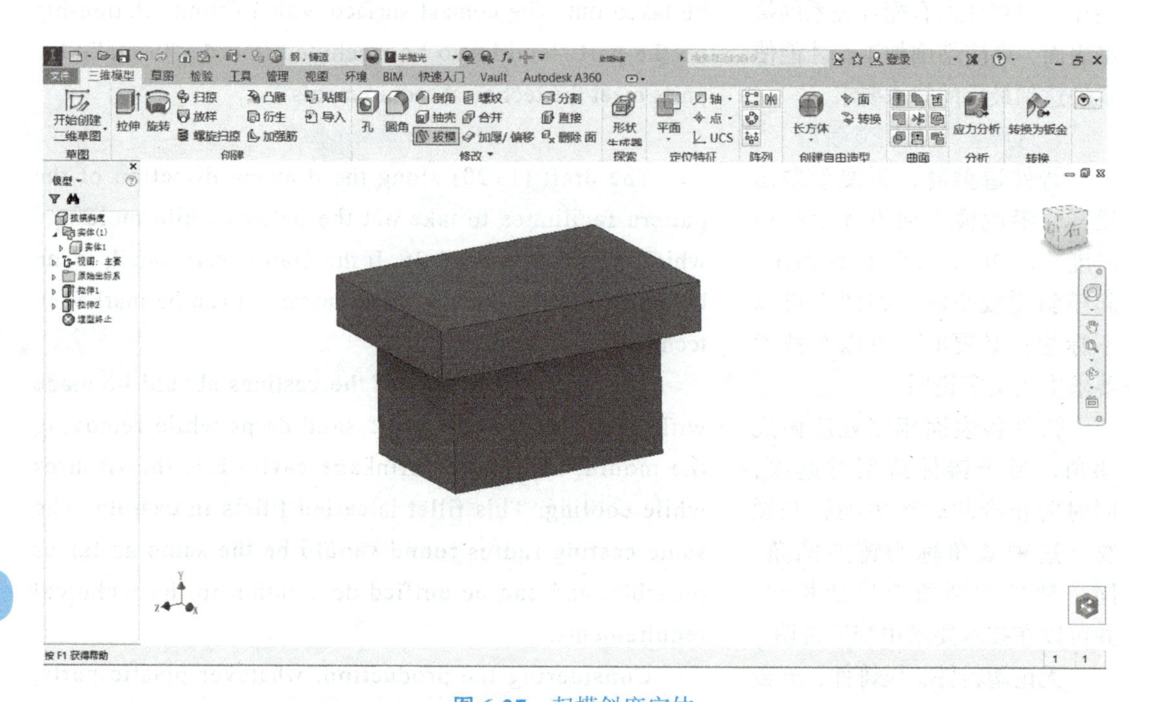

图 6-27　起模斜度实体

Fig.6-27　Draft slope entity

选择方式为“固定平面”，选择上面为固定平面，这个平面是在拔模过程中保持不变的固定的平面，如图 6-28 所示。

选择要拔模的面。输入需要拔模的斜度，调整拔模方向，如图 6-29 所示。

“fixed plane” command should be used, select upper plane as fixed plane, which remains unchanged in the process of draft, as shown in Fig.6-28.

Select planes which need the draft operation. Input the slope of draft to adjust direction of draft, as shown in Fig.6-29.

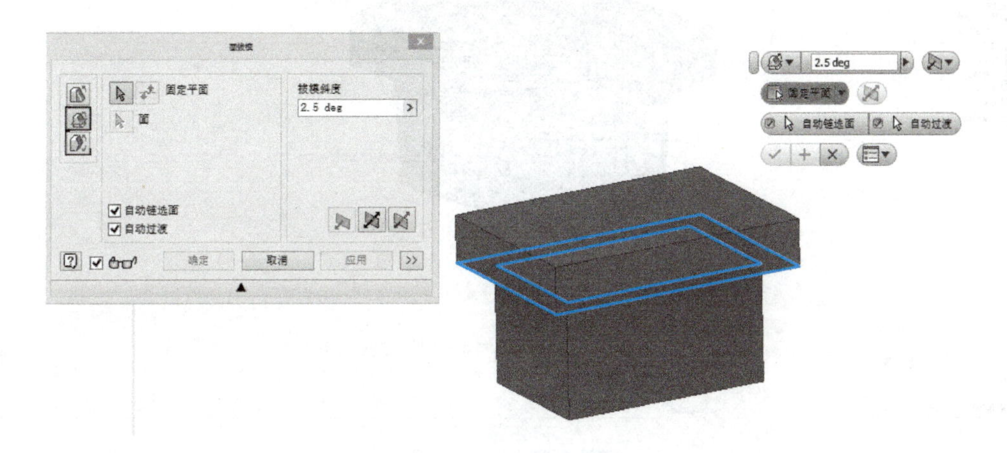

图 6-28 选择固定平面

Fig.6-28 Select fixed plane

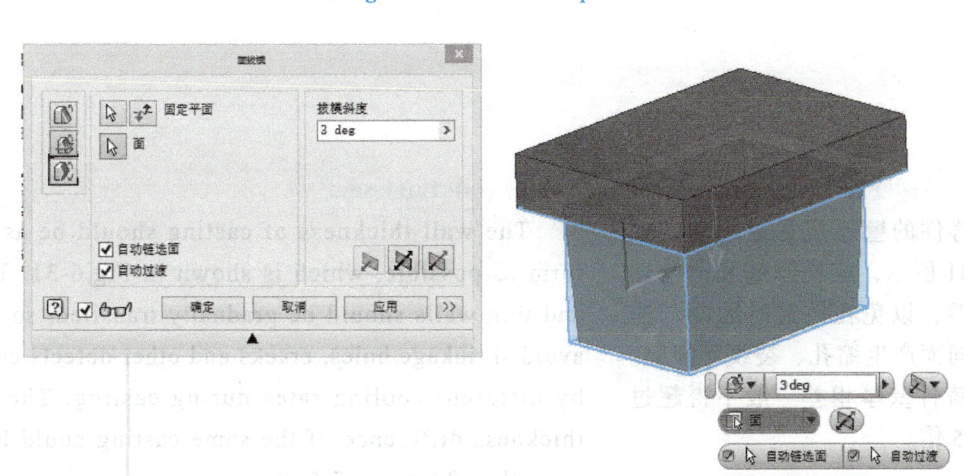

图 6-29 确定拔模斜度

Fig.6-29 Determine draft slope

通过预览，单击“确定”按钮，完成拔模操作，如图 6-30 所示。

After preview, click OK and then complete the draft operation, as shown in Fig.6-30.

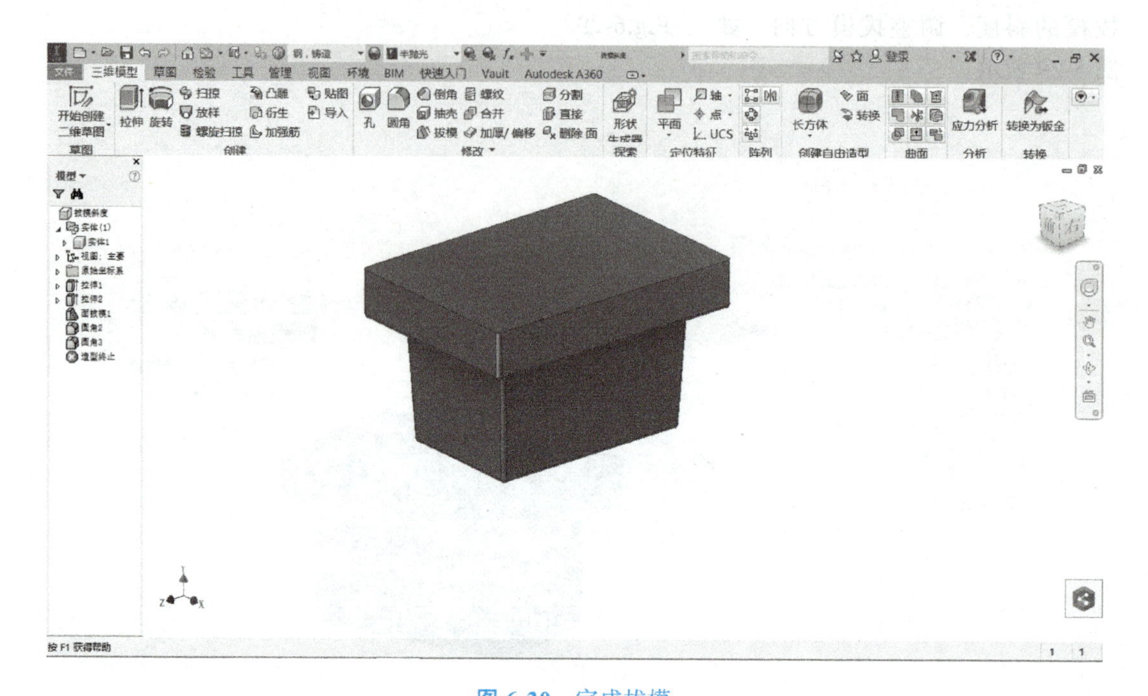

图 6-30　完成拔模

Fig.6-30　Finish draft operation

2. 铸件壁厚

铸件的壁厚应尽量均匀，如图 6-31 所示，应使厚壁和薄壁逐渐过渡，以免在铸造时因冷却速度不同而产生缩孔、裂纹等缺陷。同一铸件壁厚相差一般不得超过 2～2.5 倍。

2. Wall thickness

The wall thickness of casting should be as uniform as possible, which is shown in Fig.6-31. Thick and thin walls should be gradually transited, so as to avoid shrinkage holes, cracks and other defects caused by different cooling rates during casting. The wall thickness difference of the same casting could be no more than 2 times ~ 2.5 times.

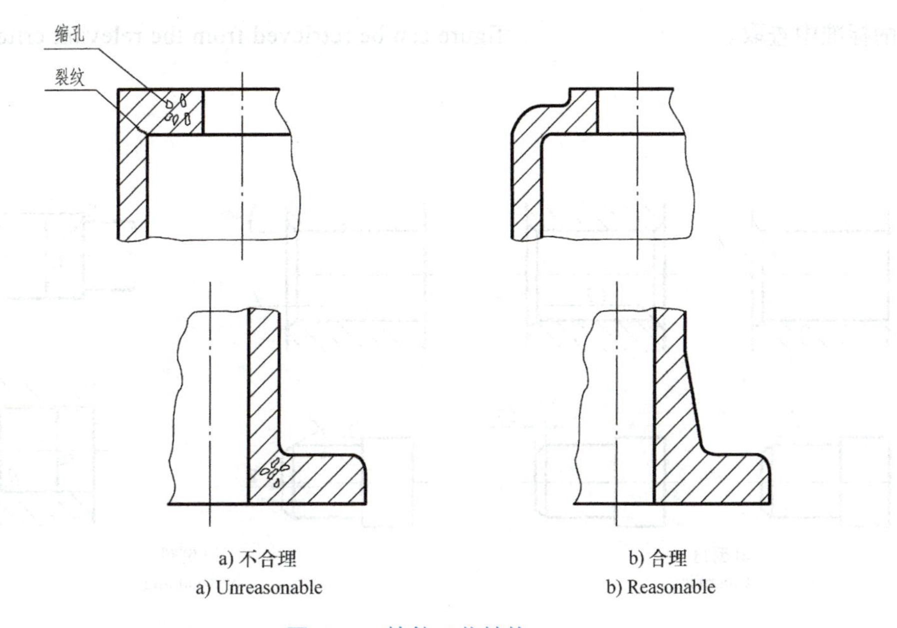

a) 不合理
a) Unreasonable

b) 合理
b) Reasonable

图 6-31　铸件工艺结构
Fig.6-31　Casting structure

6.4.2　机械加工零件的工艺结构

零件的加工面是指切削加工得到的表面，即通过车、钻、铣、刨或镗等去除材料的方法加工形成的表面。

1. 倒角和圆角

为了便于装配及去除零件的毛刺，常在轴、孔的端部加工出倒角，常见倒角为 45°，也有 30° 或 60° 的倒角。阶梯轴轴肩的根部为避免应力集中，常在轴肩根部加工成圆角过渡，称为倒圆。标注方法如图 6-32 所示。

2. 退刀槽和砂轮越程槽

在切削过程中，为保证加工质量，使刀具易于退刀，并在装配时容易与有关零件靠紧，需要在加工面末端加工出退刀槽或砂轮越程槽，退刀槽和越程槽的结构及尺寸如图 6-33 所示。图中的数据可从

6.4.2　Process Structure of Machined Parts

The machined surface of the part refers to the surfaces produced by the cutting process, that is, by means of turning, drilling, milling, planning or boring, which is used to remove the material.

1. Chamfer and fillet

The chamfer is processed at the end of the shaft and hole to facilitate assembly and removal of burrs. Common chamfer angle is 45°, 30° and 60°. In order to avoid stress concentration, the roots of the stepped shaft shoulder are usually transition into rounded corners, which is called rounding. The dimension labeling method is shown in Fig.6-32.

2. Escape and grinding undercut

In cutting, undercut or traveling limit undercut needs to be produced at the end of the machining surface, in order to easily withdraw the cutter or grinding wheel, and easy to be close to the relevant parts during assemble. The structure and dimension of the undercut and grinding undercut are shown in Fig.6-33. The data in the

相关的标准中查取。

figure can be retrieved from the relevant criteria.

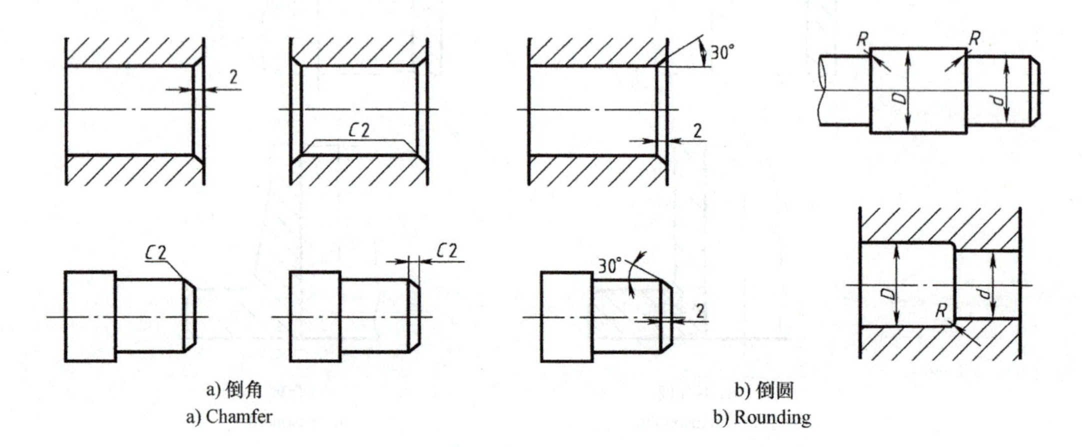

图 6-32 倒角和倒圆

Fig.6-32 Chamfer and rounding

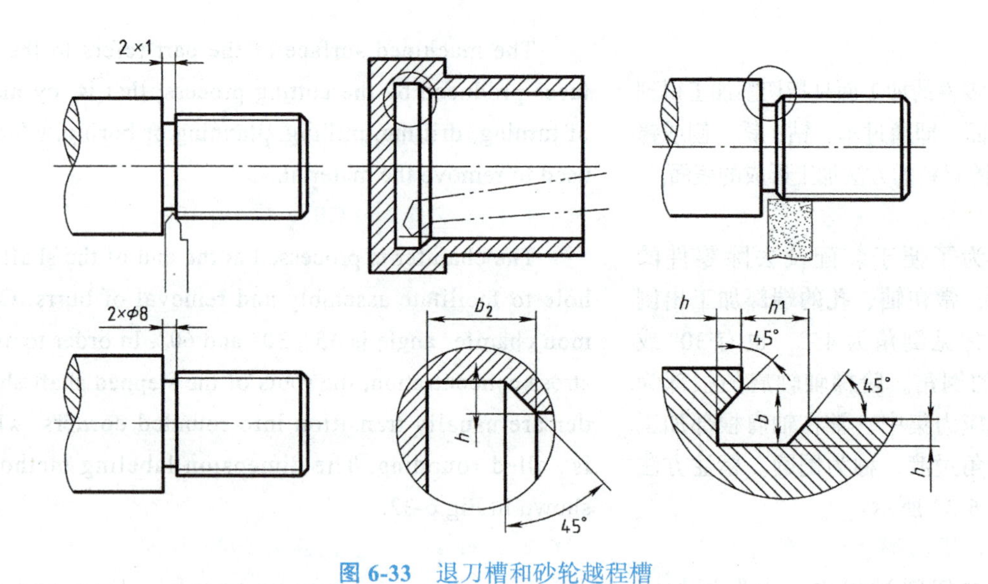

图 6-33 退刀槽和砂轮越程槽

Fig.6-33 Escape and grinding undercut

3. 凸台与凹坑

零件上与其他零件接触的表面，一般都要加工。为了减小加工面积，保证两零件表面接触良好，常在两接触表面处制出凸台和凹坑，结构和尺寸标注如图 6-34 所示。

3. Boss club and recessed surface

Usually, all surface on the part which connect other parts need to be machined. The boss club or recessed surface is added to keep good contact on the surfaces of two parts to reduce the machining area. The structure and dimension are marked as shown in Fig.6-34.

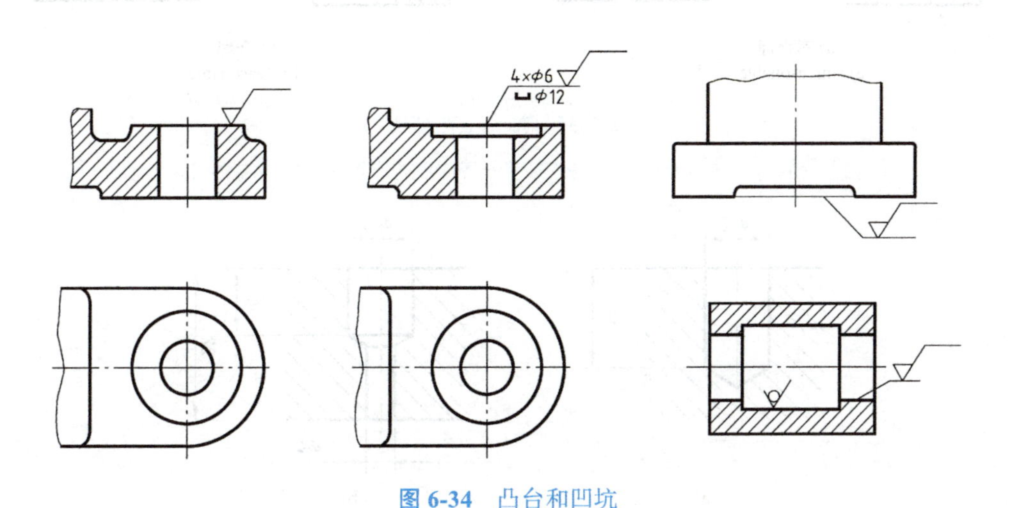

图 6-34　凸台和凹坑

Fig.6-34　Boss clubs and recessed surfaces

4. 钻孔结构

钻孔时，要求钻头垂直于孔的表面，以保证钻孔准确和避免钻头折断；在与孔轴线倾斜的表面上，应先制成与钻头垂直的凸台或凹坑，如图 6-35 所示。钻削加工的不通孔，钻头顶角会在钻孔底部留下一个大约 120° 的锥顶角，画图时，按 120° 画出锥角，不需标注尺寸，钻孔深度尺寸不包括锥角，如图 6-36a 所示；在钻阶梯孔的过渡处也存在 120° 锥角的圆台，圆台孔深不包括锥角，如图 6-36b 所示。

4. Structure of drill

The drill should be perpendicular to the surface to be drilled, in order to ensure accurate drilling and to avoid broken bits. On a surface tilted against the axis of the hole, a convex or sinkhole perpendicular to the drill bit should first be made, as shown in Fig.6-35. The top angle of the drill will leave a conical top corner of about 120° at the bottom of the hole when drilling the blind hole. When drawing, the taper angle according to 120° draw, do not need to dimension, drilling depth size does not include cone angle, as shown in Fig.6-36a. there is a 120° cone angle of the round table in the drill step hole in the transition, the boss hole depth does not include taper angle, as shown in Fig.6-36b.

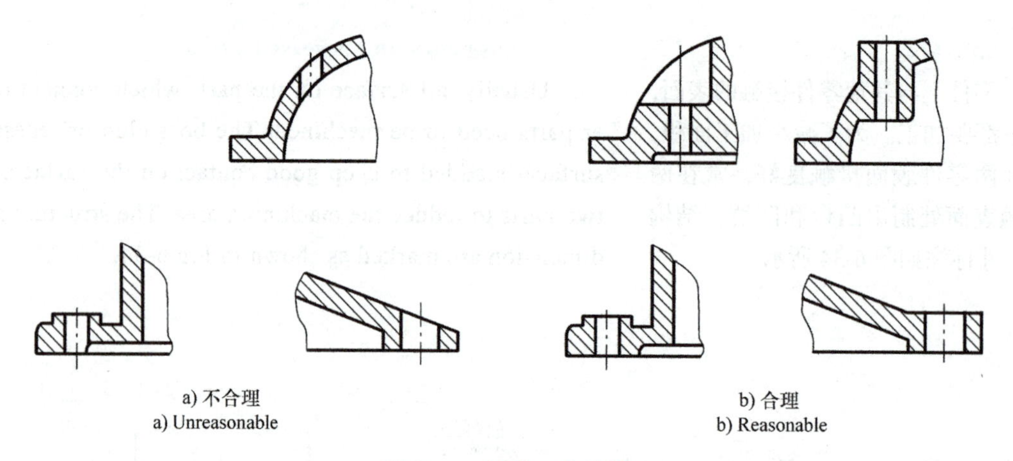

图 6-35　钻孔工艺结构

Fig.6-35　Processing structure of the drill

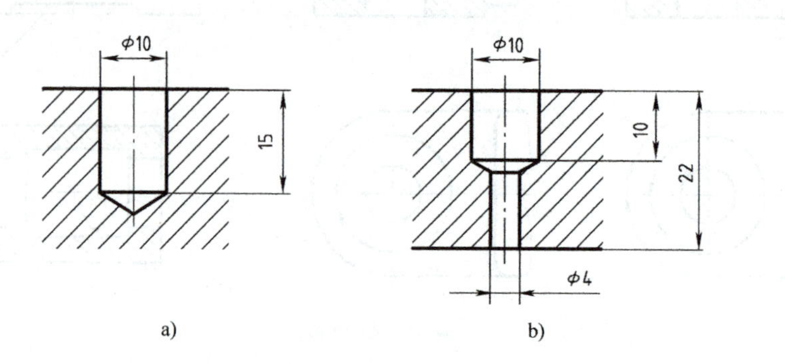

图 6-36　钻孔结构

Fig.6-36　Structure of the drill hole

6.5　读零件图

工程技术人员在对零件设计、生产加工以及技术改造过程中都需要读零件图，通过读图全面了解零件的结构形状，了解零件的制造方法和技术要求等。下面以手压阀阀体零件图为例，介绍读零件图的一般方法和步骤，如图 6-37 所示。

6.5　Interpreting Detail Drawings

The following is an example of a hand pressure valve body part diagram, which describes the general method and procedure for reading part drawings, as shown in Fig.6-37. Engineering designers need to read the parts drawing in the process of part design, production processing and technical evolution. Through understanding the drawing read, the structure of the parts, the manufacturing methods and technical requirements of the parts. The following is an example of a hand valve body drawing as an example to introduce the general methods and steps for reading the part drawing, as shown in Fig. 6-37.

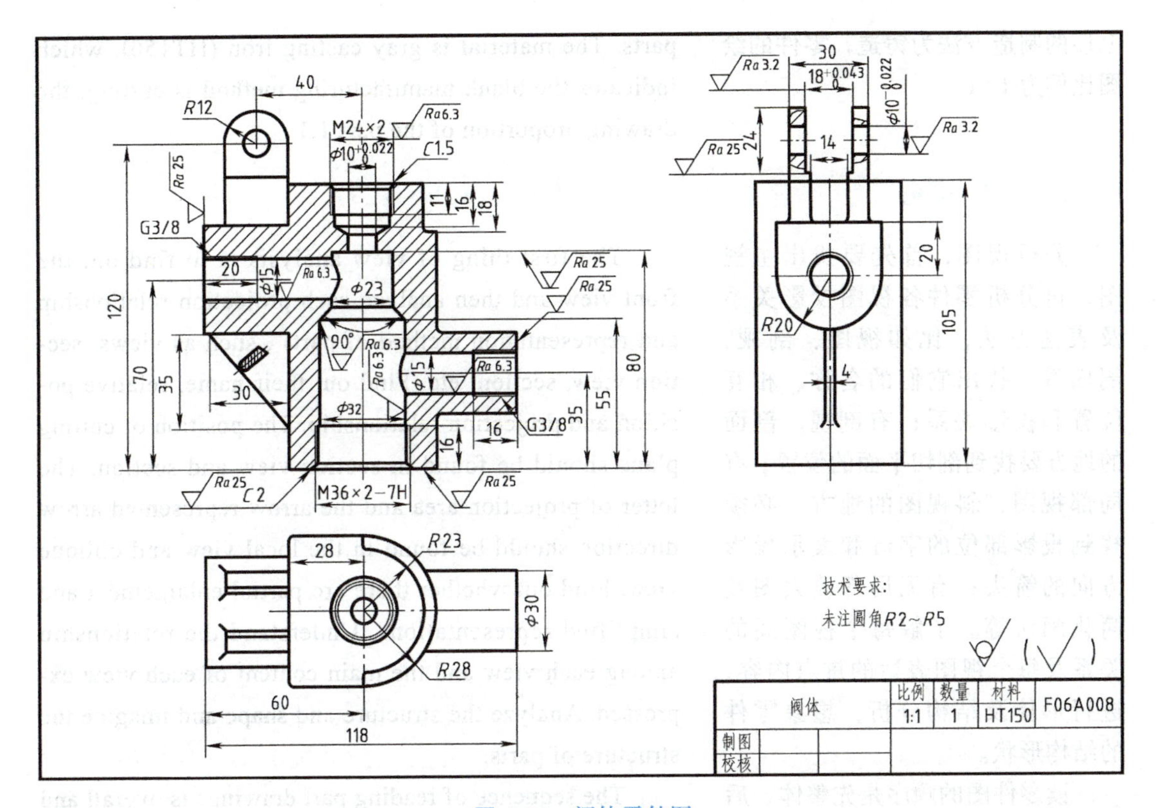

图 6-37　阀体零件图

Fig.6-37　Drawing of a valve

6.5.1　概述

从标题栏中可以了解零件的名称、材料、数量和比例等内容。由零件名称可判断零件属于哪一类零件，由材料大致了解其加工方法，根据比例可估计零件的实际大小。阅读零件图时，最好结合零件在机器或部件中的位置、功能以及与其他零件的装配关系。

图 6-37 所示是手压阀的阀体零件图。由图可知，零件的名称是阀体，阀体的作用是支撑和包容其他零件，属于箱体类零件。零件的材料为灰铸铁，牌号 HT150，说明

6.5.1　Summary

Understand the name, material, quantity and scale by the title block. The name of part determines the type of the part. The material indicates its processing method and the actual size of the part can be estimated according to the ratio.

When reading the part drawing, it is better to combine the position and function of the part in the machine or component and the assembly relationship with other parts.

Fig.6-37 shows the valve drawing of manual pressure valve. According to the figure, the name of the part is the valve. The function of the valve is to support and contain other parts , so a valve belongs to the box-type

毛坯的制造方法为铸造，零件的绘图比例为1:1。

parts. The material is gray casting iron (HT150), which indicates the blank manufacturing method is casting, the drawing proportion of the part 1:1.

6.5.2 分析视图

分析视图，首先要找出主视图，再分析零件各视图投影关系及表达方法，比如视图、剖视、剖面等。找出它们的名称、相互位置和投影关系；有剖视、剖面的地方要找到剖切平面的位置；有局部视图、斜视图的地方，必须找到投影部位的字母和表示投影方向的箭头；有无局部放大图及简化画法等。了解每个视图间的关系及每个视图表达的重点内容，进行形体及结构分析，想象零件的结构形状。

读零件图的顺序是先整体、后局部；先主体结构、后局部结构；先读懂简单结构、再分析复杂结构。从基本视图看出零件的大体内外形状；结合局部视图、斜视图以及剖视图等表达方法，读懂零件的局部或斜面的形状；同时，也从设计和加工方面的要求，了解零件结构的作用。

阀体零件图采用了三个基本视图。主视图是全剖视图，表达阀体比较复杂的内腔结构形状，内部用于容纳其他零件。左视图采用局部剖视图，主要表达左侧U形柱体的外形、筋板厚度和安装手柄的孔的结构。俯视图采用基本视图的表达方法，主要表达了半圆柱的阀体外形。

将三个视图综合起来想象阀体的结构形状，并仔细看懂各部分的局部结构，结合零件在手压阀中的

6.5.2 View Analysis

The first thing of view analysis is to find out the front view and then analyze each projection relationship and representation method of parts , such as views, section view, section, etc. Find out their name, relative position and projection relationship. The position of cutting plane should be found in section view and section. The letter of projection area and the arrow represented arrow direction should be found in the local view and oblique view. Find out whether there are partial enlargement and simplified representations. Understand the relationship among each view and the main content of each view expressed. Analyze the structure and shape and imagine the structure of parts.

The sequence of reading part drawings is overall and then partial. Firstly, understand simple structure and then analyze complex structure. Understand the overall shape of the parts from the basic view. Furthermore, combine the representation methods of local view, oblique view and section view to understand the shape of the parts in partial or oblique plane. At the same time, the function of the structure of part could be understood in the term of design and processing.

The valve body diagram could be represented by three basic views. The front view is a full section view, which expresses the complex structure of the inner chamber and the inner part is used to accommodate other parts. The left view is local section view, which mainly expresses the shape of the left U-shaped cylinder, the thickness of the rib plate and the structure of the hole in which the handle is mounted. The top view is based on the expression of the basic view, which mainly represents the shape of the valve body of the half cylinder.

Combining the three views to imagine the structure and shape of the valve body and understand the local

作用，可看出阀体主要由以下几部分组成，如图 6-38 所示。

左侧是截面为 U 形的柱体，内部有螺纹孔。

左上方有两个一端是圆柱面的立板，上面有两个圆柱孔。

左下方有筋板支承。

中间立柱左端是平面，右端是两段直径不同的圆柱面，立柱中有孔。

右下方是水平放置的圆柱，内部有螺纹孔。

structure carefully. Combined with the role of parts in the hand valve to understand the composition of valve body , which consists of the following parts, as shown in Fig. 6-38.

The left side is the U-type cylinder and there are threaded hole.

On the upper left side are two vertical plates with a cylindrical plane and two cylindrical holes on the upper.

The bottom left is supported by rib plate.

The left side of the intermediate column is a plane. The right end is a cylinder with two different diameters, which has the hole in the column.

The right of bottom is a horizontal cylinder with threaded holes inside.

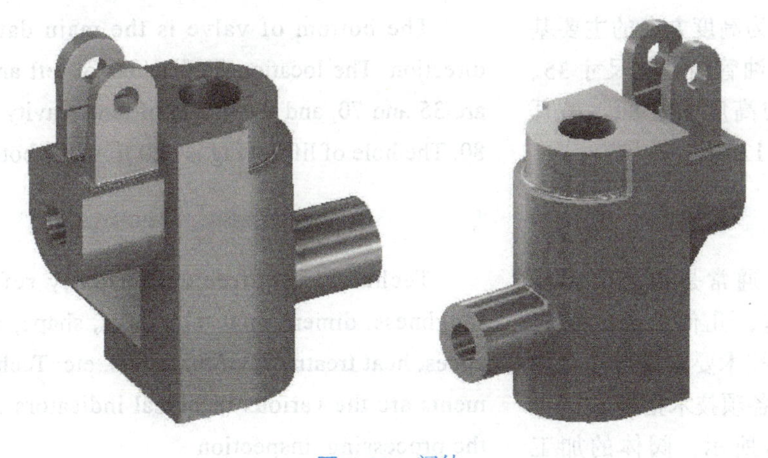

图 6-38 阀体

Fig.6-38 Valve

6.5.3 分析尺寸

分析尺寸主要是根据零件的结构特点找出尺寸基准，了解零件各部分的定形、定位尺寸和零件的总体尺寸，明确尺寸种类和标注形式；结合极限偏差及公差带代号和表面结构代号看尺寸，进一步想象

6.5.3 Analyzing Dimensions

Dimension analysis is to find out the dimension datum according to the structural characteristics of parts. Understand the shape, the positioning dimension and the overall size of parts to define the type of dimension and the form of marking. Combine the limit deviation and the tolerance with code and the surface structure code to

零件的空间形状。

阀体中间立柱轴线为长度方向的主要基准，以确定安装手柄的距离 40，注出内腔的尺寸 ϕ23、ϕ32，阀体螺孔是管螺纹 G3/8，同时注出铅垂孔轴线到左端面的距离 60。长度方向的辅助基准是 U 形的柱体端面，以此确定阀体的总长 118。

阀体前后对称面为宽度方向的主要基准，在左视图上注出阀体立板的定位尺寸 18、30。两立板之间要安装手柄，因此标注了尺寸公差 $18^{+0.043}_{0}$。俯视图标注半圆柱体外形尺寸 *R*23、*R*28、圆柱的外形尺寸 ϕ30 。

阀体底面为高度方向的主要基准，确定左右油管的定位尺寸 35、70，内腔孔的高度 55、80。吊耳的孔距离底面 120 等。

see the dimensions. Then imagine the spatial shape of the part in the further.

The middle axis of the valve body is the main reference of the length direction to determine the distance of the installing handle 40. The dimensions of the inner cavity are ϕ23 and ϕ32. The valve body screw hole is G3/8. At the same time, the distance between the axis of the vertical hole and the left end surface is 60. The auxiliary reference for the length direction is the U type cylinder end face, so as to determine the total length of the valve body 118.

The main reference for the width of the valve body is the front and rear symmetry planes. Location dimensions 18 and 30 are on the left side view. There is a handle to be installed between the two plates, so dimensioning tolerance is $18^{+0.043}_{0}$. The outline dimensions of the half cylinder are shown in the top view, *R*23, *R*28 and cylindrical size 30.

The bottom of valve is the main datum of height direction. The location dimensions of left and right tubing are 35 and 70, and the height of inner cavity hole is 55 and 80. The hole of lifting lug is 120 from the bottom.

6.5.4 了解技术要求

技术要求通常是指表面粗糙度、尺寸公差、几何公差，材料的热处理等。技术要求是零件在加工、检验时的各项技术指标。

如图 6-38 所示，阀体的加工面是安装手柄的立板结构，之间的距离 $18^{+0.043}_{0}$ 与孔的尺寸 $\phi10^{+0.022}_{0}$，*Ra* 值一般为 3.2μm；阀体上端螺孔 ϕ10 与阀杆有配合关系，内孔表面粗糙度 *Ra* 的上限值为 6.3μm，零件上其他加工表面粗糙度 *Ra* 为 25μm。

6.5.4 Understanding Technical Requirements

Technical requirements usually refer to surface roughness, dimensional tolerances, shape, position tolerances, heat treatment of materials, etc. Technical requirements are the various technical indicators of the parts in the processing, inspection.

As shown in Fig.6-38, the machining surface of the valve is the vertical plate structure of the mounting handle. the interval dimension $18^{+0.043}_{0}$ and the hole dimension $\phi10^{+0.022}_{0}$, *Ra* is 3.2μm. The upper screw hole ϕ10 has relationship with the valve stem, the surface roughness of inner hole Ra_{max} is 6.3μm. Other machined surface roughness *Ra* is 25μm on the valve.

6.5.5 综合分析

读懂零件图以分析视图、想象形状为核心，把读懂的零件结构

6.5.5 Comprehensive Analysis

The core of understanding part drawings is to analyze and visualize shapes. Integrate the structures,

形状、尺寸标注和技术要求等内容综合起来，这样就能读懂一张零件图。对于比较复杂的零件图，还需参考有关的技术资料，如装配图、相关零件的零件图及说明书等。

shapes, dimensions, and the technical requirements to figure out the complete drawing of the part. For more complex parts, it is necessary to refer to the relevant technical material, such as assembly drawings, the relevant parts drawings and instruction of part.

6.6　绘制零件图

6.6　Drawing Detail Drawings

用软件绘图的过程（以图 6-11 为例）见微课视频。

The process of drawing (take the Fig.6-11 as an example) with software is shown in the micro lesson.

6.6-1

6.6-2

6.6-3

6.6-4

6.6-5

6.6-6

6.6-7

[本章习题]

[Chapter exercises]

1. 根据轴（图 6-39）、叉架（图 6-40）、箱体（图 6-41）立体图，构建物体的三维模型，采用适当的表达方法，完成工程图的绘制。

1. According to the axis (Fig.6-39), fork (Fig.6-40), box (Fig.6-41) stereogram, build the 3D model of the object, and use appropriate representation method to draw the engineering drawing.

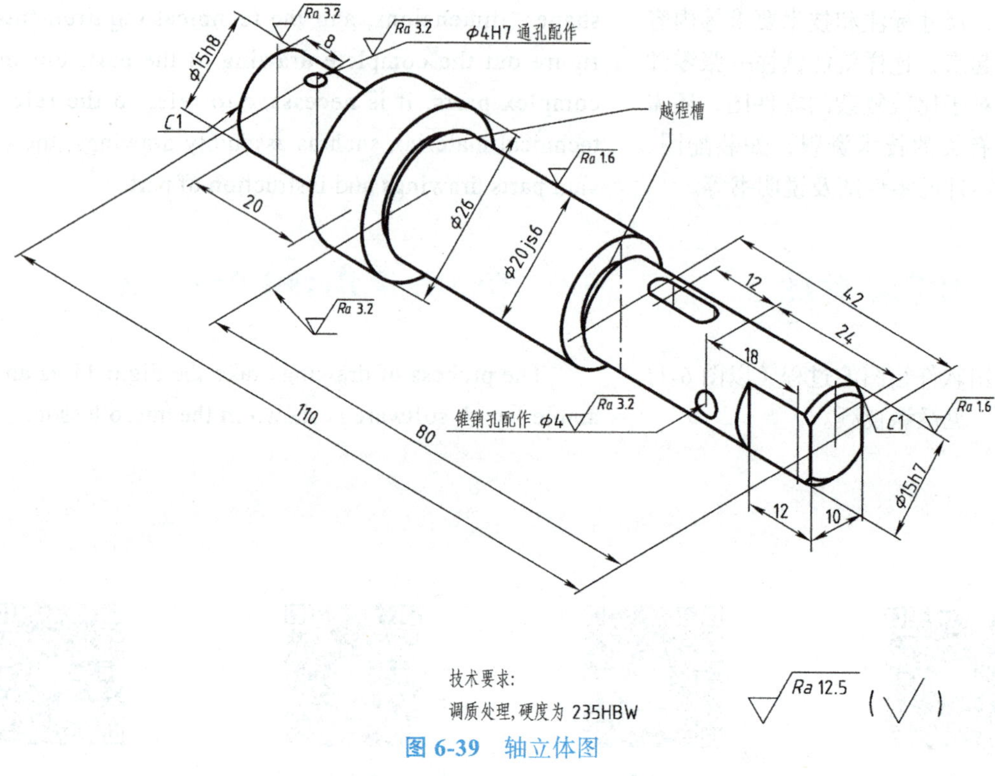

图 6-39 轴立体图

Fig.6-39 Shaft stereogram

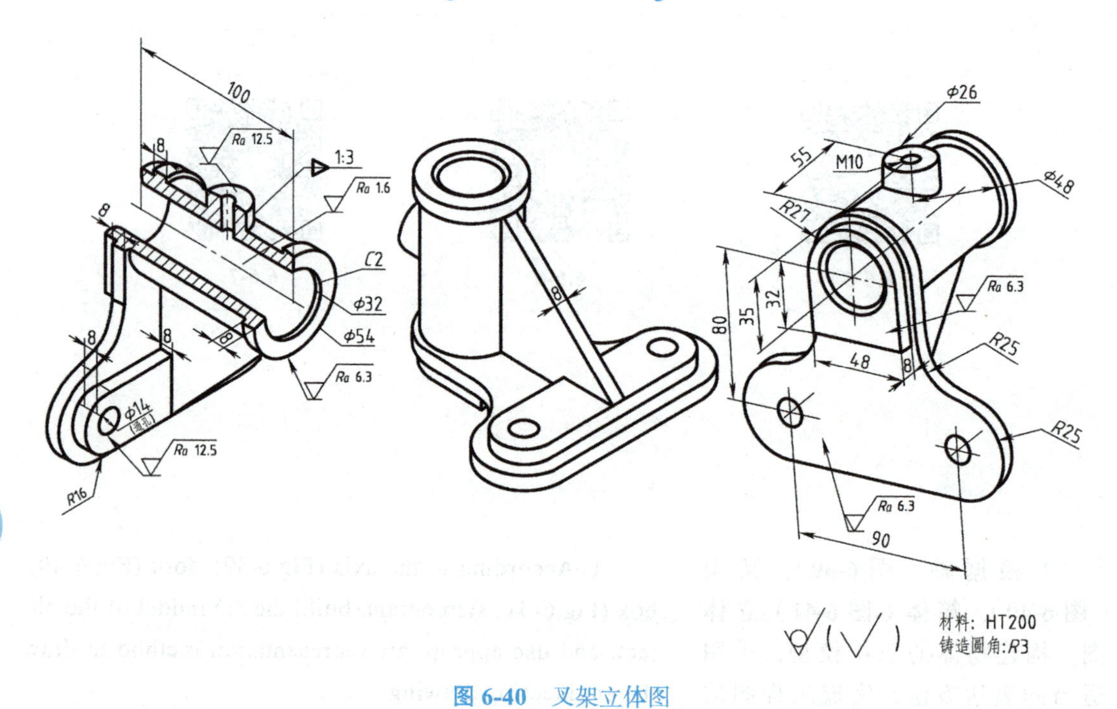

图 6-40 叉架立体图

Fig.6-40 Frame stereogram

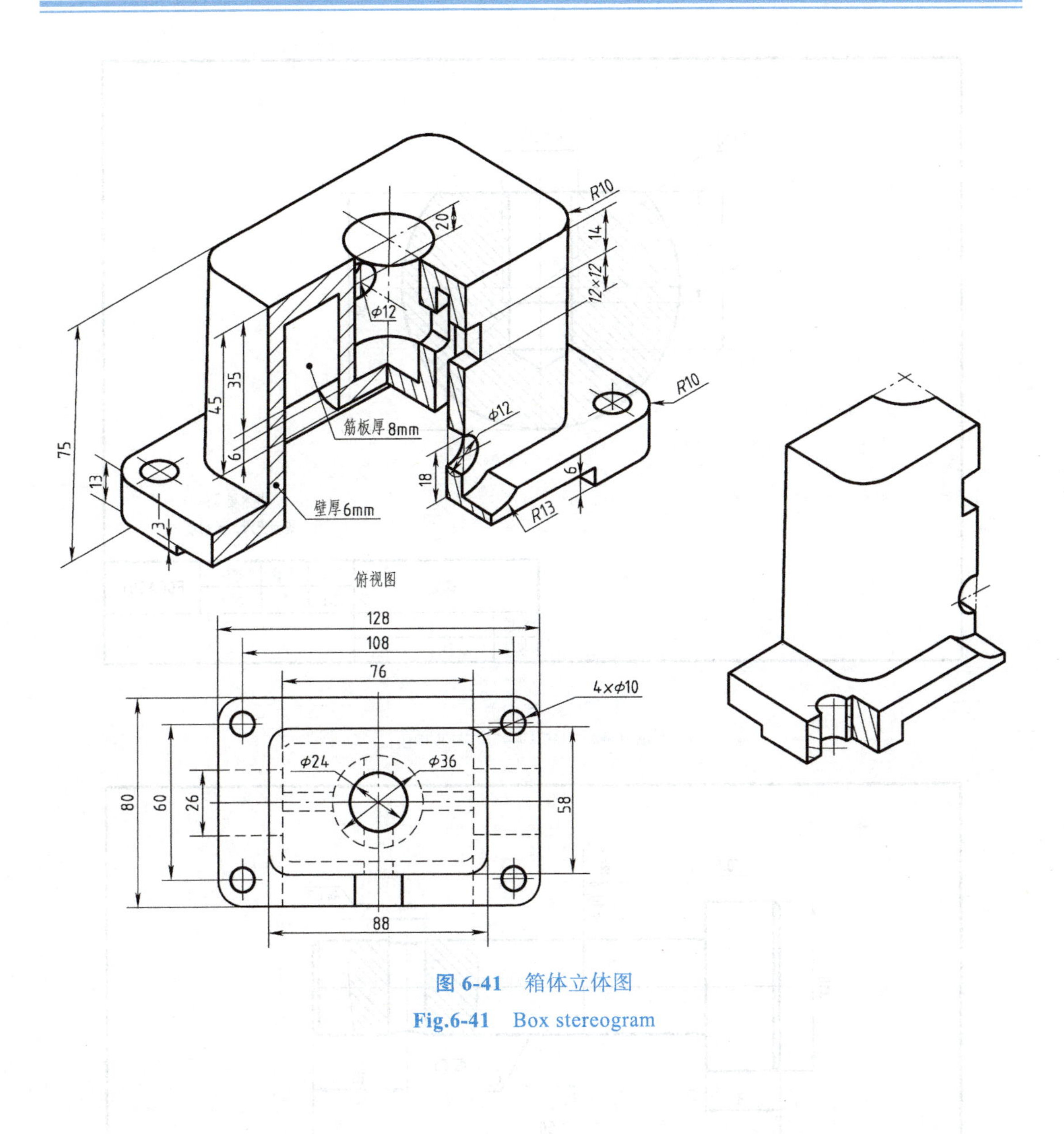

图 6-41　箱体立体图

Fig.6-41　Box stereogram

2. 绘制手压阀中的零件图（见图 6-42~ 图 6-46）。

2. Draw the parts in the hand pressure valve (as shown in Fig.6-42~Fig.6-46) .

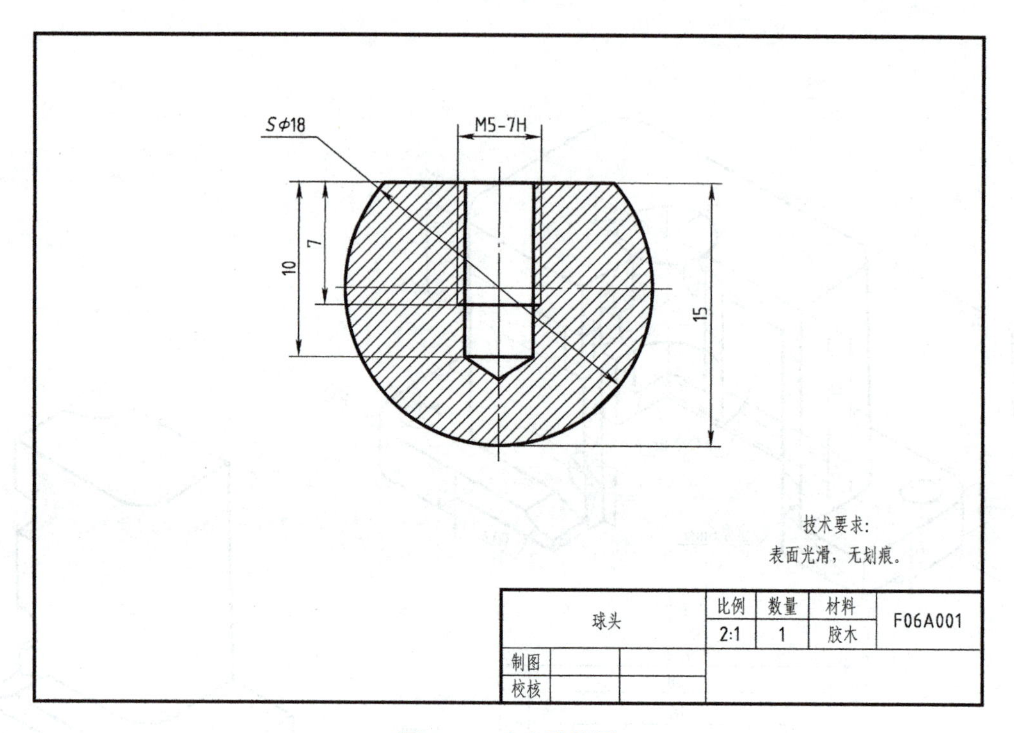

图 6-42 球头零件图

Fig.6-42 Ball head drawing

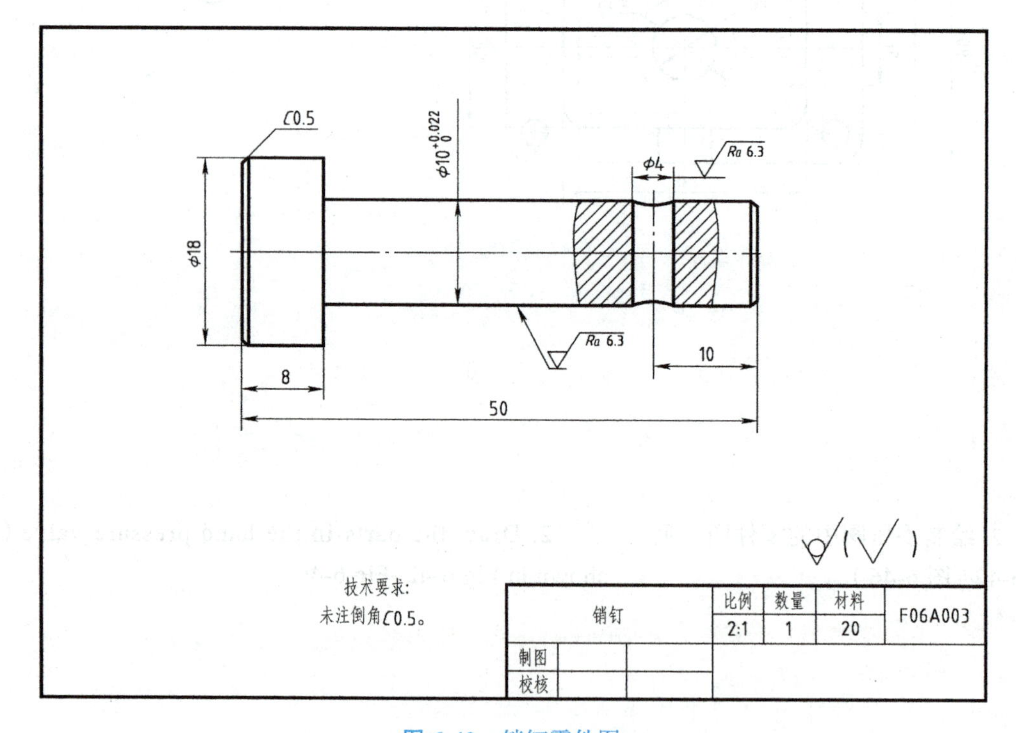

图 6-43 销钉零件图

Fig.6-43 Pin drawing

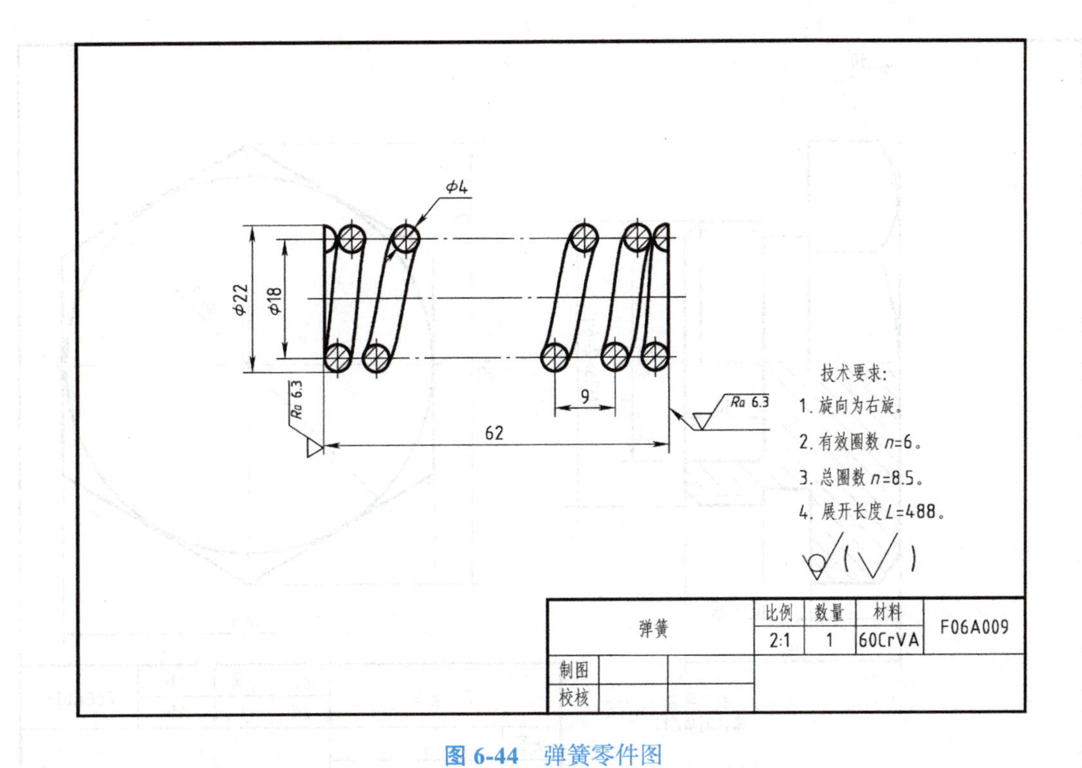

图 6-44　弹簧零件图

Fig.6-44　Spring drawing

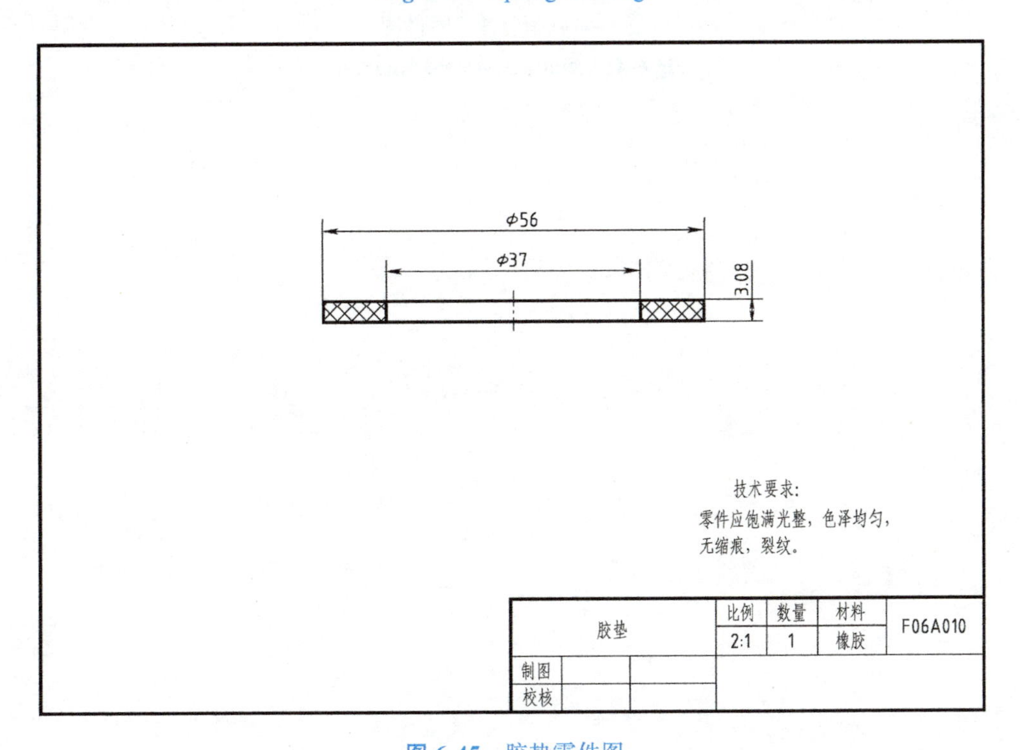

图 6-45　胶垫零件图

Fig. 6-45　Rubber mat drawing

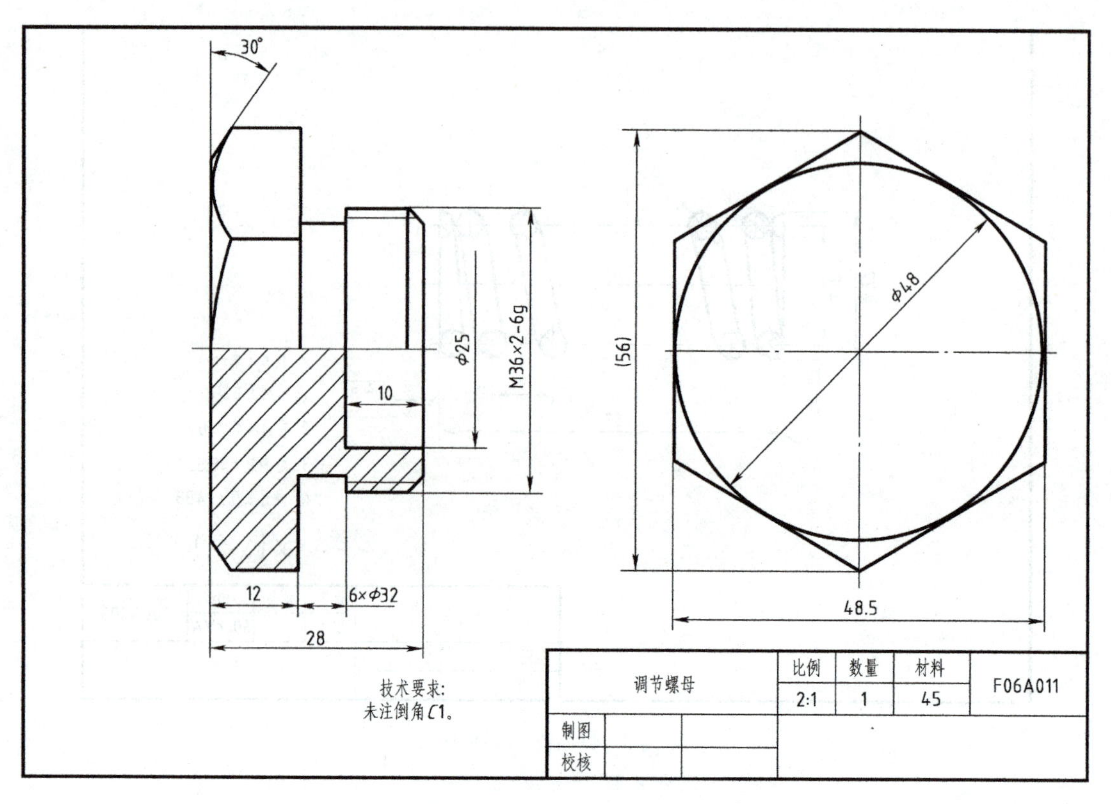

图 6-46 调节螺母零件图

Fig.6-46 Nut adjustment drawing

第 7 章　装配图的识读与绘制

Chapter 7　The Reading and Drawing of the Assembly Drawings

7.1　装配图的内容

机器或部件是由若干个零件按一定的关系和技术要求组装而成的，表达机器或部件的图样称为装配图。

本节以拆卸器为例，介绍装配图的内容。拆卸器分解图如图 7-1 所示，拆卸器装配图如图 7-2 所示。从图 7-2 中可以看出，一张完整的装配图包括以下几个方面的内容。

（1）一组视图　表示部件各零件间的相对位置关系、相互连接方式和装配关系，表达主要零件的结构特征以及机器或部件的工作原理。

（2）必要的尺寸　装配图只需要标注机器或部件的规格性能、装配、安装尺寸，总体尺寸和一些重要尺寸。

（3）技术要求　用符号或文字说明部件在装配、检验和使用过程中必须满足的条件。

（4）零件序号、明细栏　在装配图上，必须对每种零件和标准件进行编号，并在明细栏中依次标出零件的序号、名称、数量和材料等有关事项。

（5）标题栏　在标题栏中，需要注明装配体的名称、图号和绘图比例，以及图样责任者的签名等内容。

7.1　Contents of Assembly Drawings

A machine or subassembly is assembled from numbers of parts based on the relationship and technical requirements, and a drawing for expressing a machine or subassembly is called an assembly drawing.

7.1-1

This section takes the puller as an example to introduce the content of the assembly drawing. The exploded view of the puller is shown in Fig.7-1, and the assembly drawing of the puller is shown in Fig.7-2. As we can see in the Fig.7-2, a complete assembly diagram includes the followings:

7.1-2

(1) A set of views.　They represent the relative position relationship, the connection relationship and the assembly relati onship among the parts and illustrate the structural characteristics of the main parts as well as the working principle of the machine or subassembly.

(2) Necessary dimensions.　In assembly drawings, only the specification performance, as assembly, installation size, overall size and some important sizes of the machine or subassembly need to be dimensioned.

(3) Technical requirements.　Use symbols or words to describe the condition that must be met during assembly, inspection and use.

(4) Part number, Item lists.　Each part and standard part should be numbered in the drawing, and the order number, name, quantity and material of the part should be marked accordingly in the parts list.

(5) Title block.　In the title block, you need to indicate the name of the assembly, drawing number and the drawings scale, as well as signature of the drawings responsible personnel and so on.

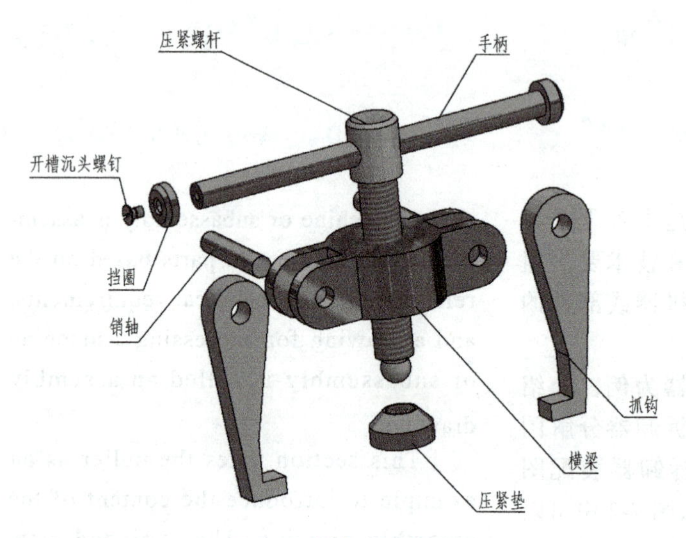

图 7-1 拆卸器分解图

Fig.7-1 Exploded view of the puller

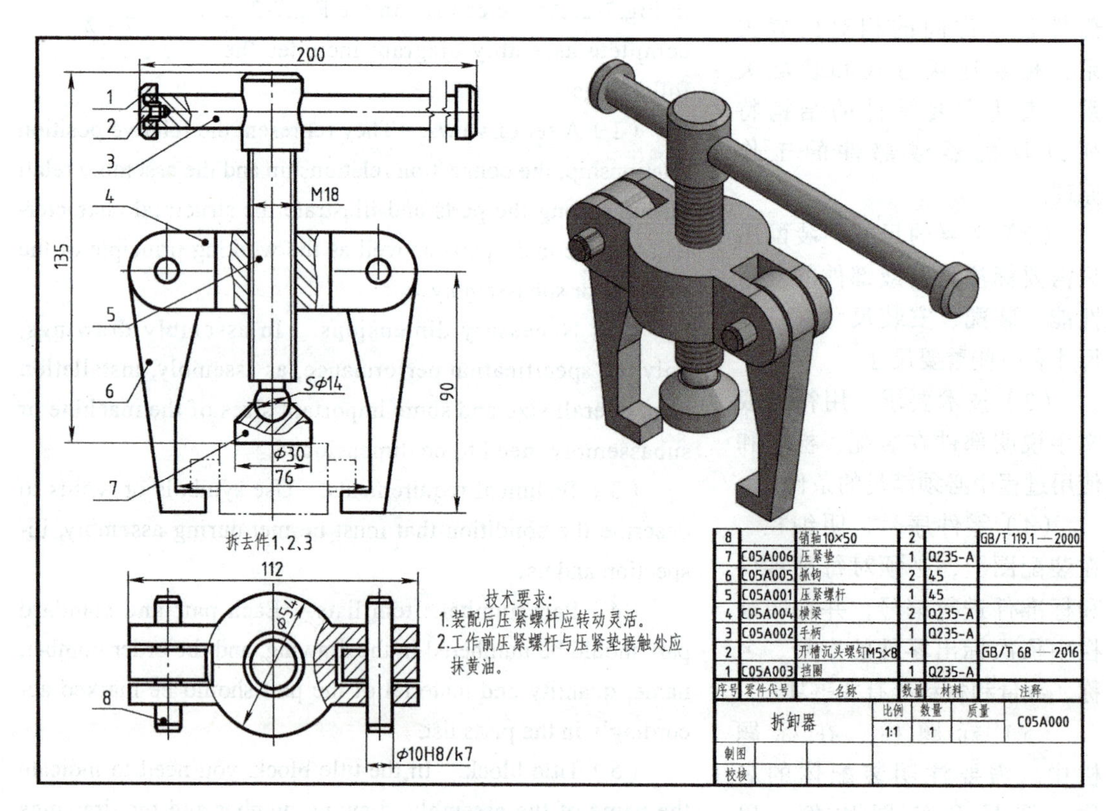

序号	零件代号	名称	数量	材料	注释
8		销轴10×50	2		GB/T 119.1—2000
7	C05A006	压紧垫	1	Q235-A	
6	C05A005	抓钩	2	45	
5	C05A001	压紧螺杆	1	45	
4	C05A004	横梁	2	Q235-A	
3	C05A002	手柄	1	Q235-A	
2		开槽沉头螺钉M5×8	1		GB/T 68 — 2016
1	C05A003	挡圈	1	Q235-A	

拆卸器	比例	数量	质量	C05A000
	1:1	1		
制图				
校核				

图 7-2 拆卸器装配图

Fig.7-2 Assembly drawing of the puller

7.2　装配图的尺寸标注、技术要求和零件编号

7.2　Dimensioning of Assembly Drawings, Technical Requirements and Part Numbers

7.2.1　装配图的尺寸标注

装配图和零件图在生产中的作用不同，因此标注尺寸的要求也不同。装配图中应标注一些必要的尺寸，这些尺寸按作用不同，可以分为以下几类。

（1）性能尺寸（规格尺寸）　表示机器或部件的规格、性能的尺寸，是设计和选用机器或部件的主要依据。如图 7-2 所示抓钩的中心高度 90。

（2）外形尺寸　表示机器或部件的总长、总宽、总高的尺寸。它们反映机器或部件所占空间的大小，可供包装、运输、安装和厂房设计时参考。如图 7-2 所示的 112、200。

（3）装配尺寸　表示部件中各零件间装配关系的尺寸，包括配合尺寸和主要零件间相对位置尺寸。配合尺寸表示零件间的配合性质和公差等级的尺寸。相对位置尺寸表示装配时需要保证的零件间相互位置尺寸，如重要的间隙、距离等。如图 7-2 所示的 M18、ϕ10H8/k7。

（4）安装尺寸　表示将部件安装到机器上或将机器安装到机座上需要的尺寸。

（5）其他重要尺寸　其他重要尺寸指根据装配体的特点和需要，必须标注的尺寸。如运动零件的活动范围，非标准零件上的螺纹尺寸，其他一些设计、加工时必须保证的尺寸等。

7.2.1　Dimensioning of Assembly Drawings

The function of assembly and part drawings is different in production. So the dimensioning requirements are also different. The assembly drawing should be marked with the necessary dimensions by their function which could be grouped into the following categories.

(1) Characteristic dimensions (specification dimensions). The dimensions that represent the specification and characteristic of a machine or subassembly are the main basis of the design and choosing of machines or subassembly. As shown in Fig.7-2, the center height of the hook is 90.

(2) External dimensions. The dimensions represent the general length, width, and height of a machine or subassembly. They reflect the size of the space occupied by machines or subassembly for the reference in packaging, transportation, installation and plant design, which have been shown in Fig.7-2, 112, 200.

(3) Fitting dimensions. The dimensions that represent the assembly relationship between parts in subassembly, including the fitting dimensions and the relative position dimensions between the main parts. The fitting dimensions represent the fitting properties of the parts and the dimensions of the tolerance grades. Relative position size indicates the dimensions of the parts that need to be guaranteed when assembling, such as the important gap and distance, etc, as shown in Fig.7-2 (M18, ϕ10H8/k7).

(4) Installation dimensions. The dimensions represent the size which is to install the subassembly on a machine or a machine on a frame.

(5) Other important dimensions. Other important dimensions refer to the dimensions that must be marked according to the characteristics and needs of the assembly, Such as the motion range of moving parts, the thread size of non-standard parts, other dimensions that must be guaranteed when designing and processing, etc.

7.2.2　装配图的技术要求

装配图的技术要求包括以下内容。

（1）装配要求　装配过程中的注意事项和装配后应满足的要求等。

（2）检验、试验的条件和要求　机器或部件装配后对基本性能的检验、试验方法及技术指标等要求与说明。

（3）其他要求　包括部件的性能、规格参数、包装、运输及使用时的注意事项等。

7.2.2　Technical Requirements of Assembly Drawings

7.2.2

The technical requirements for assembly drawings include the following.

（1）Assembly requirements.　The attention items in the assembly process and the requirements to be satisfied after assembly.

（2）Conditions and requirements for inspection and testing.　Requirements and instructions for inspection, testing methods and technical specifications of basic properties after assembly of machines or subassembly.

（3）Other requirements.　The other requirements include the performance, specifications, packaging, transportation and the use notice of the subassembly.

7.2.3　装配图的零件序号和明细栏

1. 装配图的零件序号

（1）基本要求　为了便于看图和图样管理，对部件中每种零件或组件进行编注序号，并根据零件序号绘制相应的明细栏。

1）装配图中的所有零件，均应按顺序编写序号，相同零件只编写一个序号。

2）序号应该按顺时针或逆时针方向顺序排列。

3）一组紧固件或装配关系明显的零件组，可采用公共的指引线，如图 7-3a 所示。

4）序号应填写在指引线一端的横线上或圆圈内，指引线的另一端由零件的可见轮廓线内引出，并在引出位置用圆点表示。如遇零件很薄或涂黑剖面，不宜画圆点，可用箭头代替，如图 7-3b 所示。

7.2.3　Part Numbers and Part Lists of Assembly Drawings

1. Part Numbers of Assembly Drawings

7.2.3-1

（1）Basic requirements.　In order to facilitate drawing reading and pattern management, each part or components of the assembly is numbered and the corresponding item list is drawn according to the part number.

1）All parts of assembly drawings should be numbered in sequence and the same part has only one serial number.

2）Serial numbers should be in a clockwise or counter-clockwise order.

3）A set of fasteners or parts group with obvious assembly relationship can be used in public guidelines, as shown in Fig.7-3a.

4）The serial number should be filled in the one end of the guideline on the horizontal line or in circle and another end of the guideline is drawn from the visible contour of the part and is represented by a dot at the lead-out position. As shown in Fig.7-3b, the part is appropriate to draw the arrow instead of the dot if the part is very thin or painted black section.

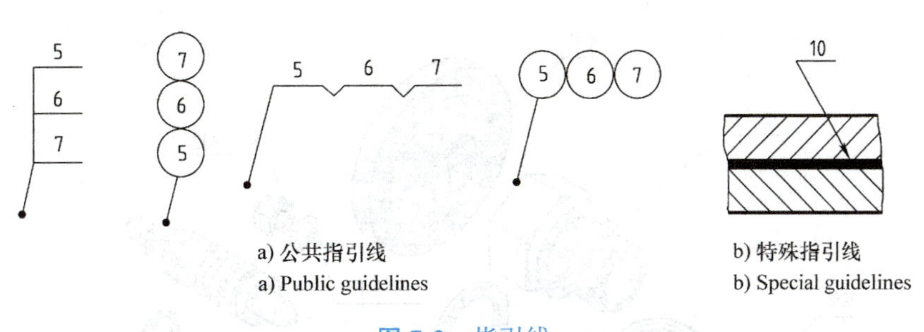

a) 公共指引线
a) Public guidelines

b) 特殊指引线
b) Special guidelines

图 7-3　指引线
Fig.7-3　guidlines

（2）Inventor 的零件序号创建　在 Inventor 中，装配图的零件引出序号有两种方式：引出序号和自动引出序号，如图 7-4 所示。引出序号一次只能创建一个零件的引出序号标注。自动引出序号可以创建一个或多个零件的引出序号。

（2）Creation of part number in Inventor.　In Inventor, the two ways to create the part number is “Balloon” and “Auto Balloon”, as shown in Fig.7-4. Balloon can only create one part’s number at a time. However, Auto Balloon can create the part number for one or more parts.

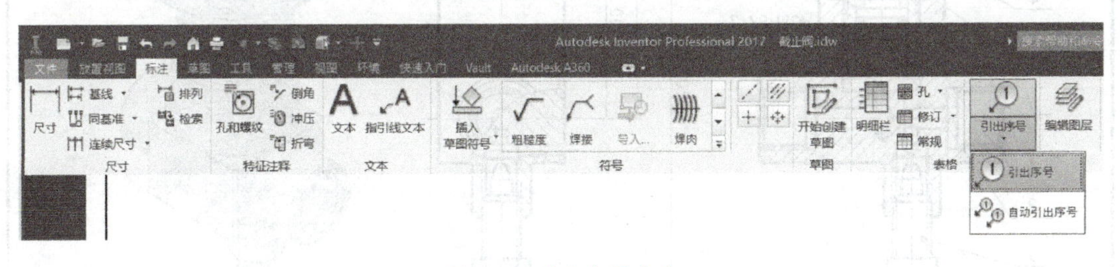

图 7-4　引出序号命令
Fig.7-4　Balloons

以截止阀为例，介绍零件序号的创建。截止阀分解图如图 7-5 所示，截止阀装配图如图 7-6 所示。

创建装配图的引出序号。如果采用引出序号方式，每次只能创建一个零件的序号引出，如图 7-7 所示。如果采用自动

Take the globe valve as an example to introduce the part number creation. Globe valve exploded and assembled view is shown in Fig.7-5 and Fig.7-6 separately.

The balloon is created in the assembly drawing. As shown in Fig.7-7, only one balloon could be created each time if the “Balloon” way is taken. Use the “Auto Balloon” way shown in Fig.7-8, pop-up “Automatic Balloon” dialog box and select the view. Then add the parts that need to be

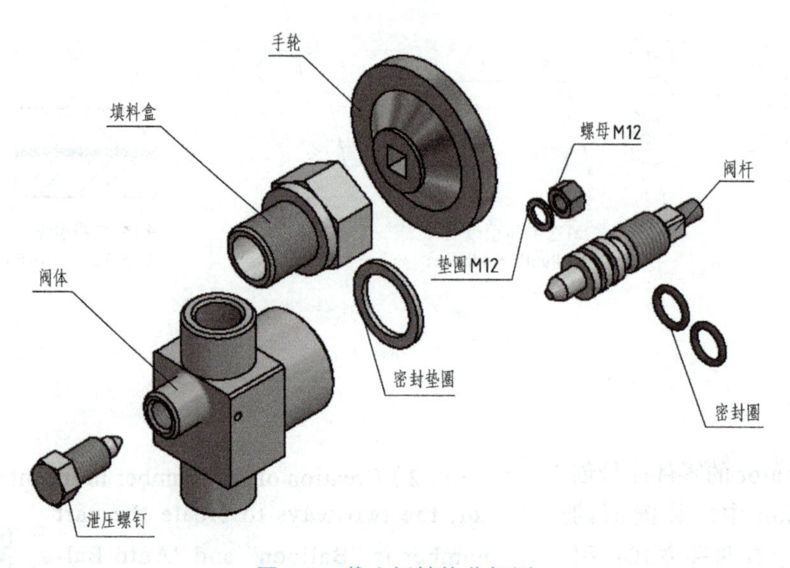

图 7-5 截止阀结构分解图

Fig.7-5 Structural Exploded View of Globe Valve

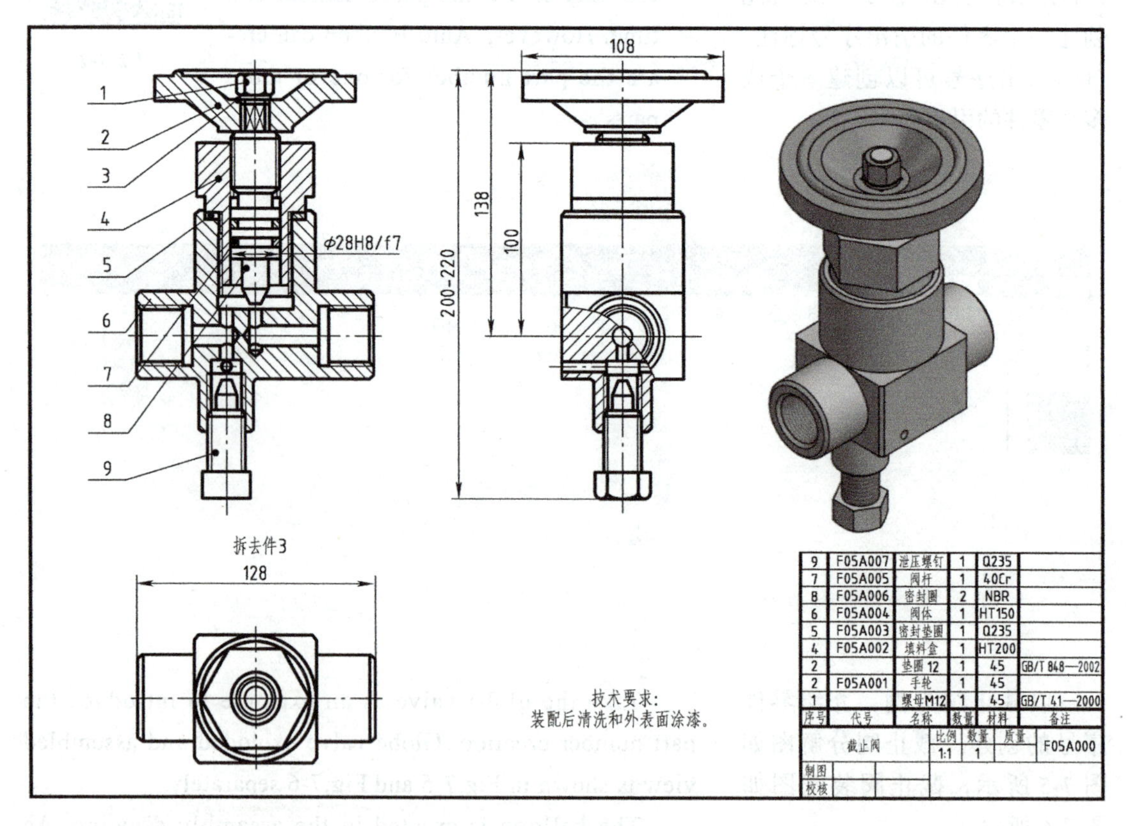

序号	代号	名称	数量	材料	备注
9	F05A007	泄压螺钉	1	Q235	
7	F05A005	阀杆	1	40Cr	
8	F05A006	密封圈	2	NBR	
6	F05A004	阀体	1	HT150	
5	F05A003	密封垫圈	1	Q235	
4	F05A002	填料盒	1	HT200	
2		垫圈 12	1	45	GB/T 848—2002
2	F05A001	手轮	1	45	
1		螺母M12	1	45	GB/T 41—2000

截止阀	比例	数量	质量	F05A000
	1:1	1		
制图				
校核				

图 7-6 截止阀装配图

Fig.7-5 Assembly drawing of globe valve

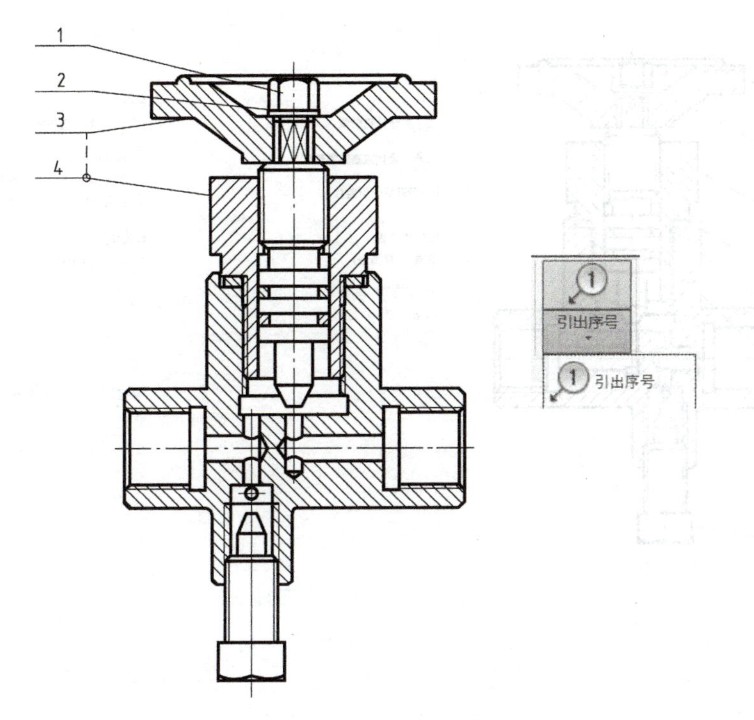

图 7-7　引出序号

Fig.7-7　Balloons

引出序号方式，如图 7-8 所示，弹出“自动引出序号”对话框，选择视图，添加需要引出序号的零件，然后选择放置方式，也可以调整序号之间的间距，即偏移间距。单击“确定”按钮后，将引出箭头修改为国标规定的小圆点，也可以调整零件的引出位置，最后创建完成截止阀的引出序号，如图 7-6 所示。

taken “Balloon” way. The following operation is to select the placement method, and you can also adjust the spacing among the serial numbers, which called offset spacing. After clicking “OK”, the leading arrow will be come small dot specified by the national standard. The lead-out position of the part can also be adjusted. Finally, the serial number of the globe valve is created, as shown in Fig.7-6.

2. 明细栏

（1）基本要求　明细栏应放置在标题栏上方，并与标题栏相连接。当位置不够时，可以将明细栏的一部分移至标题栏左边。

2. Part lists

（1）Basic requirement.　The part lists should be placed above the title block and it is also need to connected with it. When the position is not enough, a portion of the part lists can be moved to the left of title block.

7.2.3-3

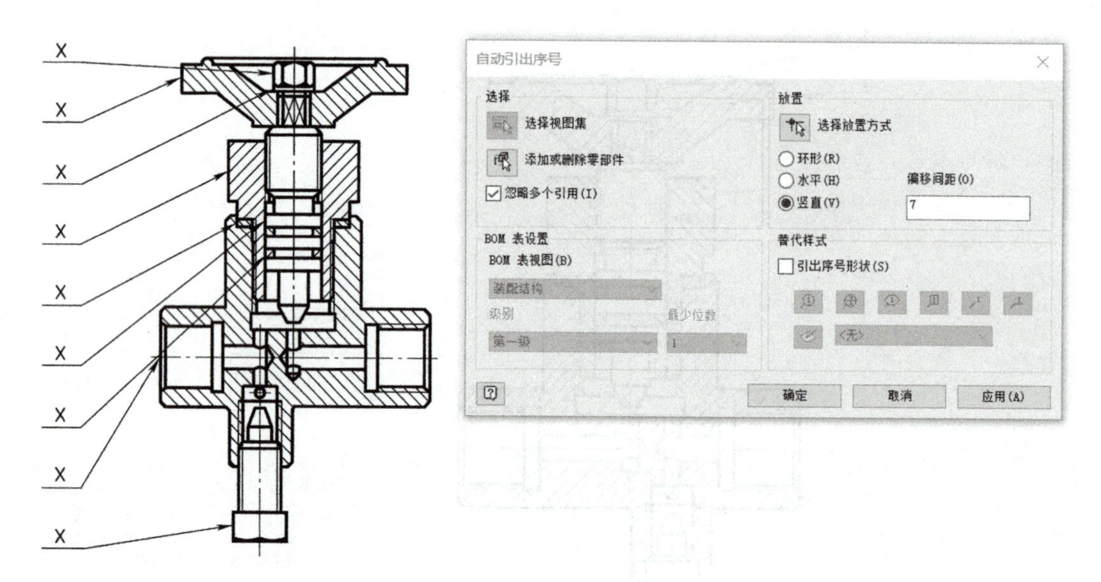

图 7-8 自动引出序号

Fig.7-8 Auto Balloons

在明细栏中，零件序号应自下而上有序填写。标准件应填写其规格和标准号。明细栏的尺寸如图 7-9 所示。

The part number should be written bottom-up in the part list. Standard parts should be filled with their specifications and standard number. The part list size has been shown in Fig.7-9.

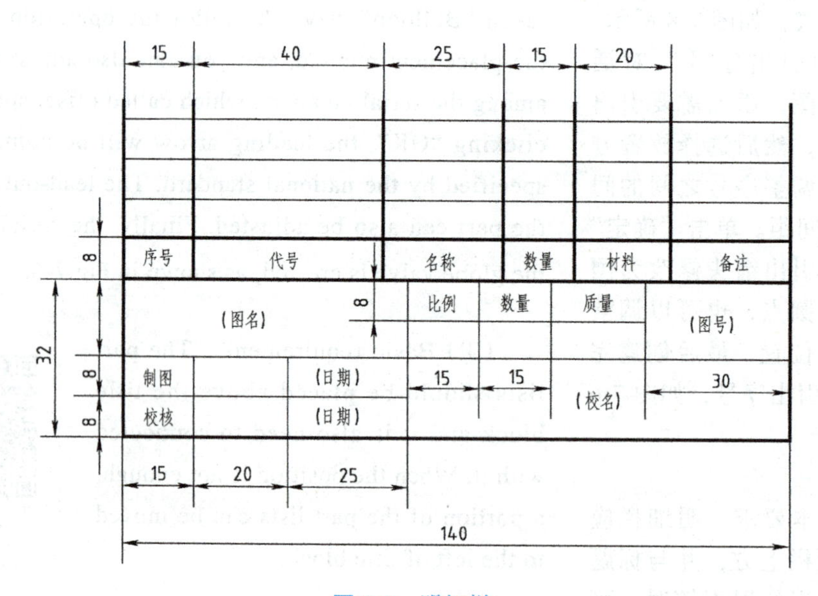

图 7-9 明细栏

Fig.7-9 Part lists

（2）Inventor 的明细栏创建　在 Inventor 中创建明细栏的过程非常简单。在零件序号标注完成后，单击“标注”选项卡中的“明细栏”命令，如图 7-10 所示。以截止阀装配图的明细栏创建为例，启动“明细栏”命令后，选择标有序号的主视图，如图 7-11 所示。将明细栏移置标题栏上方，完成创建，如图 7-6 所示。

（2）Creation of part list in Inventor. It is very simple to create the part list in Inventor. When the part number labeling is finished, click “Part List” command in the callout tab, as shown in Fig.7-10. Take the part list creation of a globe valve assembly drawing as an example. After clicking “parts list” command, the front view can be chosen, as shown in Fig.7-11. The title bar can be moved above the detail bar and part list is created, as shown in Fig.7-6.

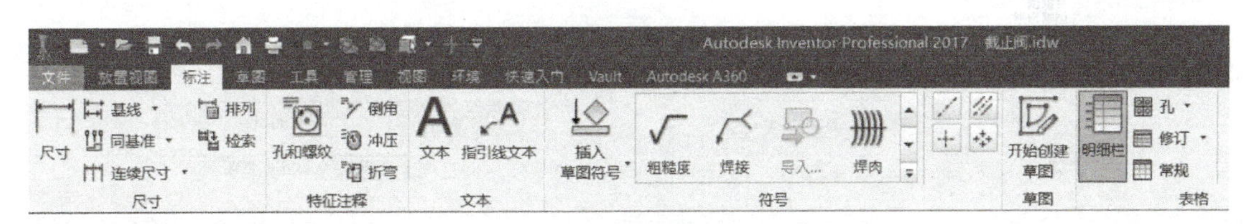

图 7-10　“明细栏”命令

Fig.7-10　Part list

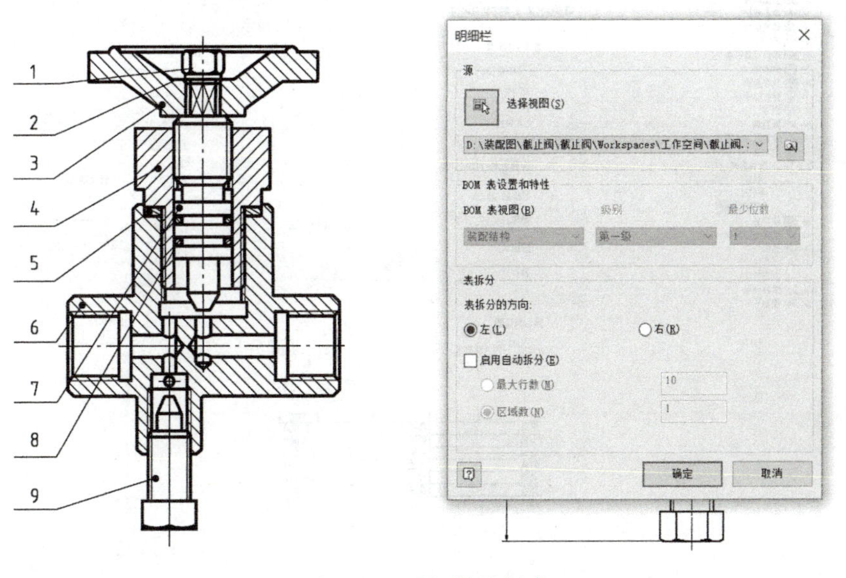

图 7-11　明细栏的创建

Fig.7-11　Creation of part list

明细栏的表头和尺寸同样是在“样式编辑器”中设置。打开“明细栏列选择器”，对明细栏的表头内容进行删减和调整，如图 7-12 所示。明细栏的尺寸设置，如图 7-13 所示。

The header size of part list is also set by “Style and Standard Editor”. Opening the “Style and Standard Editor”, the header contents of part list can be deleted and adjusted, as shown in Fig.7-12. Fig.7-13 shows that the size of part list can be set.

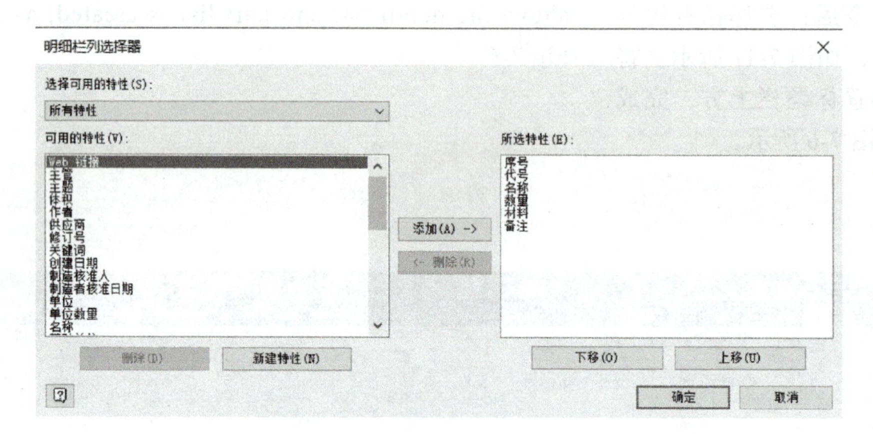

图 7-12 明细栏表头选项的设置

Fig.7-12 Setting of part list header

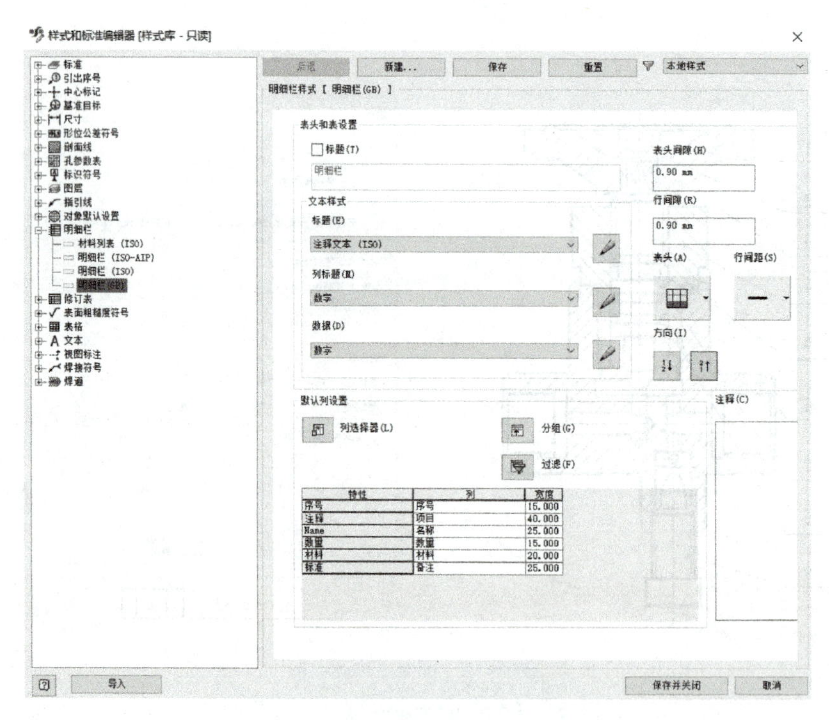

图 7-13 明细栏尺寸的设置

Fig.7-13 Setting of part list size

7.3　装配图的表达方法

相比零件图，装配图表达的重点在于反映机器或部件的工作原理、零件间的装配连接关系和主要零件的结构特征，因此装配图还有一些特殊的表达方法和规定画法。

7.3.1　规定画法

1. 相邻两零件的画法

相邻两个零件的接触面或配合面，规定只画一条线。对于非接触面、非配合表面，即使间隙再小，也必须画两条线。

如图 7-14 所示，轴与连接板下端小孔配合，孔与轴的接触面只画一条线。而轴与连接板上端大孔的公称尺寸不一样，因此画了两条线。

7.3　Representation Methods of Assembly Drawings

Compared with the representation of part drawings, the focus of the assembly drawing is to reflect the working principle of the machine or subassembly, the connection relationship among the parts and the structural features of the main parts, therefore there are some special and conventional representations for the assembly drawing.

7.3.1　Conventional Representation

1. Representation of two contiguous parts

The contact surface or fitting surface of the two contiguous parts should be represented by a line. The non-contact surface or non-fitting surface should be represented as two lines even if the gap is very small.

As shown in Fig.7-14, The shaft is matched with the small hole at the lower end of the connecting plate, and only one line is drawn on the contact surface of the hole and the shaft. While the upper hole's nominal size of the shaft and the connecting plate is different and the two lines is drawn.

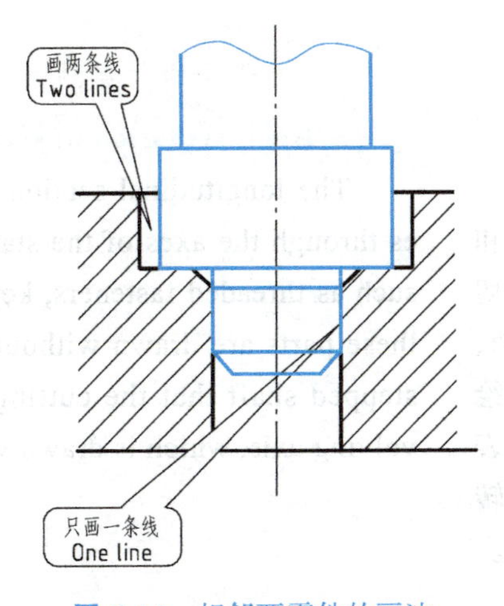

图 7-14　相邻两零件的画法

Fig.7-14　Representation of two contiguous parts

2. 剖面线的画法

相邻两个零件的剖面线倾斜方向应相反，或方向一致，间隔不同，便于装配图中区分不同零件。同一零件在各个视图上的剖面线方向和间隔应一致。

如图 7-15 所示，轴承座和轴承端盖的剖面线方向相反，轴承与轴承座的剖面线间隔不同。

2. Representation of section line

The section lines of two contiguous parts should be in the opposite or same direction and the intervals of two contiguous parts are different, so the different parts in the assembly drawing should be distinguished. The section line direction and interval of the same part on different view should be consistent.

As shown in Fig.7-15, the section line direction of the bearing housing and the bearing-end cover is opposite, and the section line interval between the bearing and the bearing housing is different.

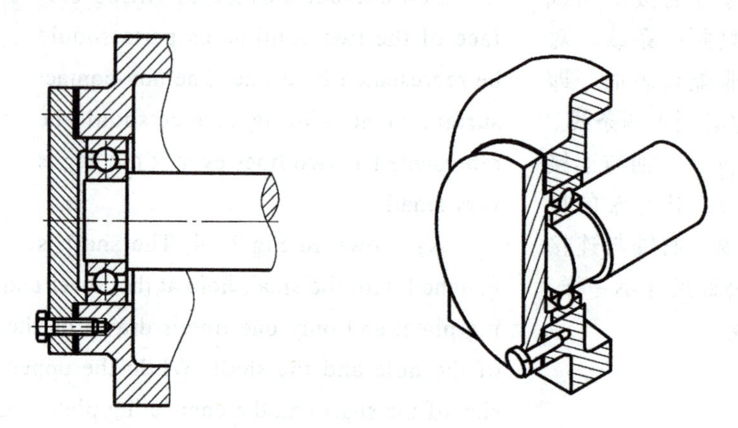

图 7-15　装配图画法的一般规定

Fig.7-15　Conventional representation of assembly drawing

3. 螺纹紧固件及实心件的画法

当剖切平面通过标准件和实心零件的轴线，即纵向剖切时，如螺纹紧固件、键、销、轴、杆等，这些零件按不剖绘制。如图 7-15 所示阶梯轴的表达，剖切平面通过阶梯轴的回转轴线，此时阶梯轴按不剖绘制。

3. Representation of screw fasteners and solid parts

The longitudinal section is that the section plane passes through the axes of the standard parts and the solid part, such as threaded fasteners, keys, pins, shafts, rods, etc., and these parts are drawn without section . Fig.7-15 shows the stepped shaft that the cutting plane passes through its revolving axis, which is drawn without section.

7.3.2　特殊表达方法

1. 拆卸画法

当某些零件遮住了需要表达的其他结构或装配关系，而这些零件在其他视图中已表示清楚时，可假想将其拆去，只画出所要表达部分的视图，并在该视图的上方加注“拆去 × ×”，此方法称为拆卸画法。

如图 7-2 中，拆卸器的俯视图采用拆去手柄等零件表达。

2. 假想画法

（1）当需要表达运动零件的运动范围或极限位置时，可将运动件画在一个极限位置或中间位置上，另一个极限位置用双点画线画出。如图 7-16 所示，双点画线表示运动部位的左侧极限位置。

7.3.2　Special Representation Methods

1. Dismounting representation

When some parts obscure other structures or assembly relationships which need to be expressed and these parts are represented clearly in other views, you can imagine that they are removed and draw the view of the part to be expressed, and add comment “Remove XX” above the view. This is called dismounting drawing.

As shown in Fig.7-2, the top view of the puller is represented with removing the handle.

2. Imagination representation

（1）When the moving range or limit location of the moving part need to be expressed, the moving part could be drawn in one limit location or the middle location, and another limit location is drawn with the double dot-dash line. As shown in Fig.7-16, the double dot-dash line indicates the left limit location of the moving part.

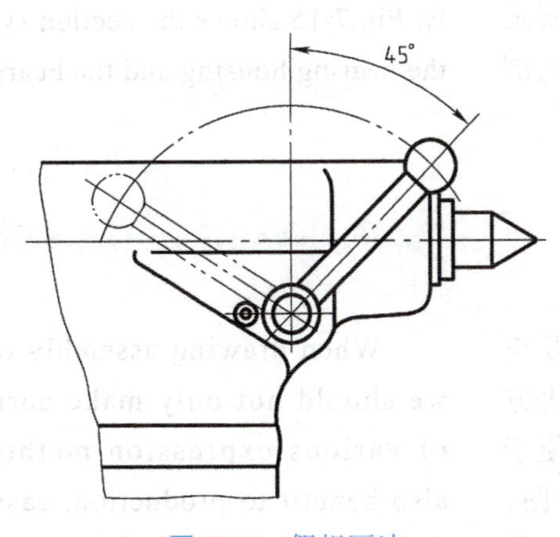

图 7-16　假想画法

Fig.7-16　Imagination representation

（2）当需要表达装配体与相邻机件的装配连接关系时，可用双点画线表示出相邻机件的外形轮廓。如图 7-2 中，拆卸器主视图下端双点画线所示的工件。

（2）When representing assembly connection relationship between the assembly and the contiguous parts, the contour of the contiguous parts can be represented by the double dot dash line. As shown in Fig.7-2, the workpiece displayed at the lower end of the front view of the puller is represented by the double dotted line.

3. 简化画法

3. Simplified representation

如图 7-15 所示，在装配图中有些结构可以采用简化画法。

As shown in Fig.7-15, some structures in the assembly drawing can be simplified.

（1）有若干组相同的螺纹联接件时，可详细画出一组，其余只需在装配图中表示其中心位置。

（1）When there are several groups of threaded connectors, one group can be drawn in detail and the others only need to be shown with the central location in the assembly drawing.

（2）装配图中的滚动轴承按表达需要可采用规定画法或通用画法。

（2）According to expression need, conventional representation or general representation could be used to the rolling bearing of assembly drawing.

（3）零件的工艺结构如圆角、倒角和退刀槽等允许省略不画。

（3）Parts of the processing structure, such as round, chamfer and escape can be omitted.

4. 夸大画法

4. Enlargement representation

对于装配图中较小的间隙、垫片和弹簧等细小部位，允许将其涂黑代替剖面符号或适当加大尺寸画出，如图 7-15 所示轴承座与轴承端盖间垫片的剖面符号。

The small parts (small gaps, gaskets and springs) of the assembly drawing can be represented with black painting to replace the section symbol or size enlargement appropriately. Fig.7-15 shows the section symbol of the gasket between the bearing housing and the bearing end cover.

7.4 装配图的绘制

7.4 Assembly Drawing

绘制装配图时，不但要正确运用装配图的各种表达方法，还要从有利于生产、便于读图出发，恰当地选择视图，将部件或机器的工作原理，各零件间的装配关系及主要零件的基本结构完整、清晰地表达出来。

When drawing assembly drawings, we should not only make correct use of various expression methods, but also benefit to production, easy to read the drawing, choose the view properly. The subassembly or machine working principle, assembly relationship among parts and the major parts structure should be expressed clearly.

7.4.1　部件装配关系分析

机器或部件是由若干零件按一定的关系和技术要求组装而成的，目的是表达机器或部件的工作原理及其装配关系等。因此，在绘制装配图时，首先要弄清部件的用途、工作原理、装配关系、传动路线及其主要零部件的结构。

以手压阀为例，如图 7-17 所示。手压阀主要由阀体、调节螺母、阀杆和手柄等零件组成。当手柄向下压紧阀杆时，阀杆压缩弹簧向下移动，液体入口与出口相通，液体正常流动。松开手柄，由于弹簧弹力作用，阀杆向上压紧阀体，使液体停止流动。

7.4.1　Analysis of Subassembly Assembly Relationship

A machine or subassembly is assembled from numbers of parts according to the relationship and technical requirements. The purpose is to express the working principle and the assembly relationship of a machine or subassembly. Therefore, the usage, working principle, assembly relationship, driving routines and the major parts' structure should be figured out at first during drawing.

Take the hand valve as an example, as shown in Fig.7-17. The hand valve mainly consists of the valve body, regulating nut, valve stem and handle. When handle is pressed downward, the stem compresses spring downward. Then the liquid inlet and outlet are communicated with each other so that liquid could flow normally. The following step is to loosen the handle and the stem compresses spring upward to stop liquid from flow because of spring elasticity.

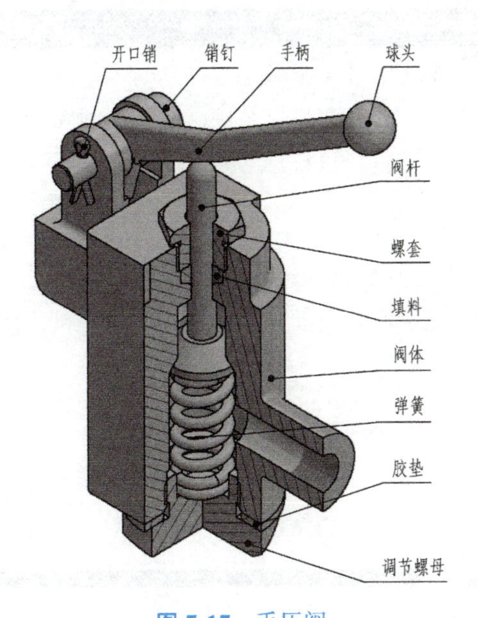

图 7-17　手压阀

Fig.7-17　Hand valve

7.4.2 部件装配

打开“Standard.iam”模板，进入装配环境，如图 7-18 所示。

1. 非标零件添加

通过“放置”命令，依次调入已经建好的非标零件模型，也可以按住 <Ctrl> 键，一次选择多个零件，如图 7-19 所示。

7.4.2 Subassembly Assembly

Open the “Standard.iam” template to enter the assembly environment, as shown in Fig.7-18.

7.4.2-1

1. Adding non-standard parts

Non-standard parts are added in the non-standard model in turn by “Place Component” command. Multiple parts can also be selected by holding down the “Ctrl” key, as shown in Fig.7-19.

7.4.2-2

7.4.2-3

7.4.2-4

7.4.2-5

7.4.2-6

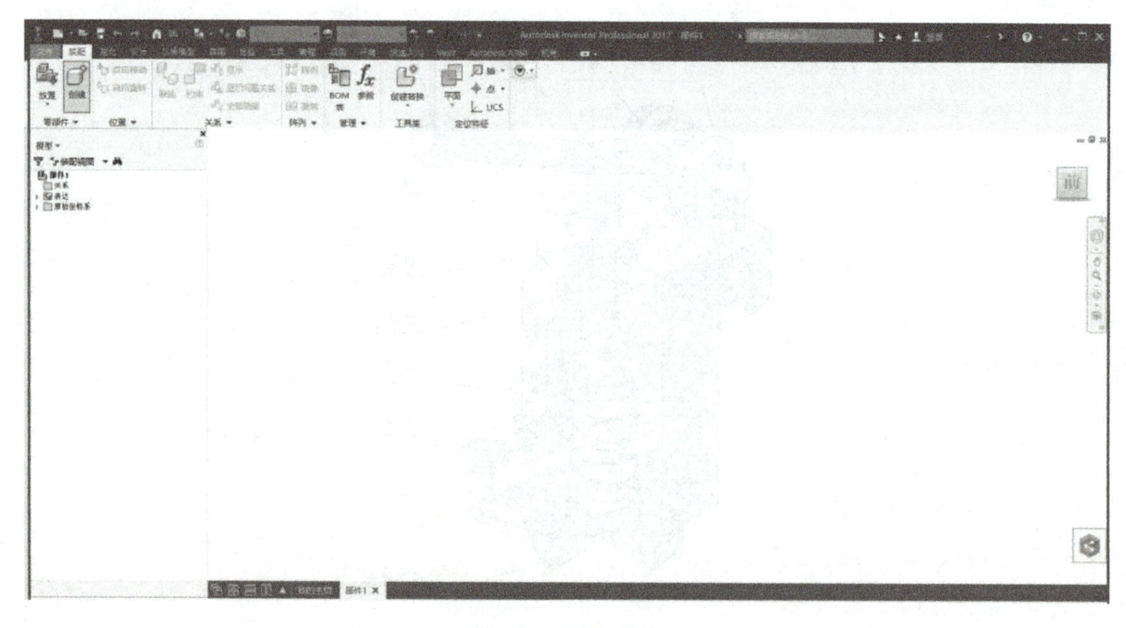

图 7-18 装配环境

Fig.7-18 Assembly environment

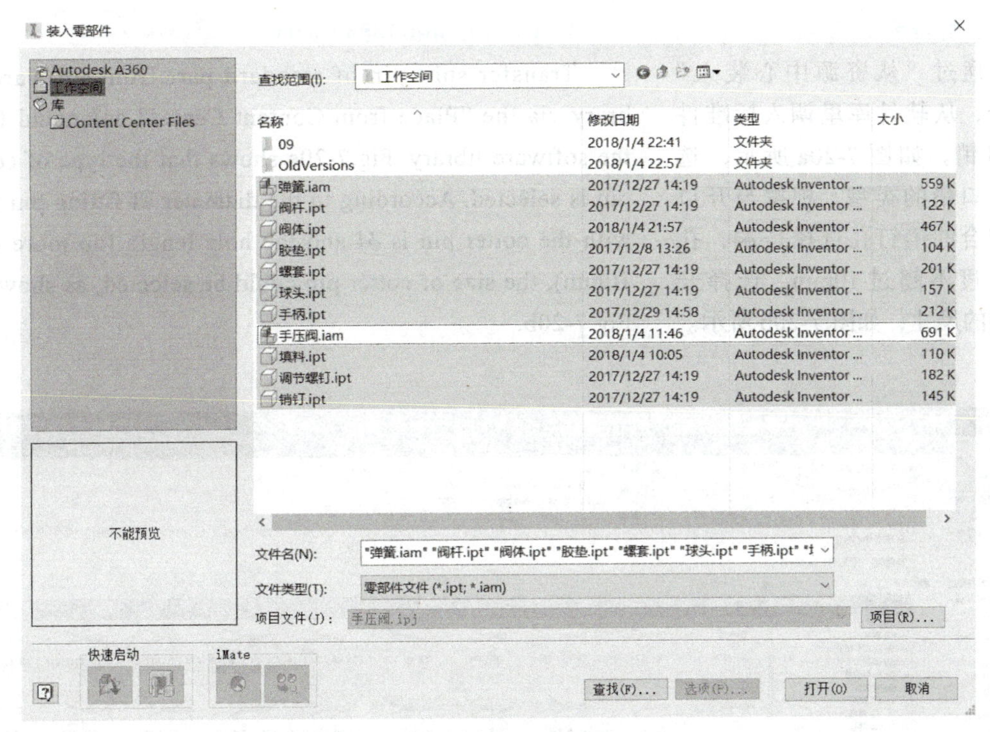

a) 多个零件添加选择

a) Adding multiple parts

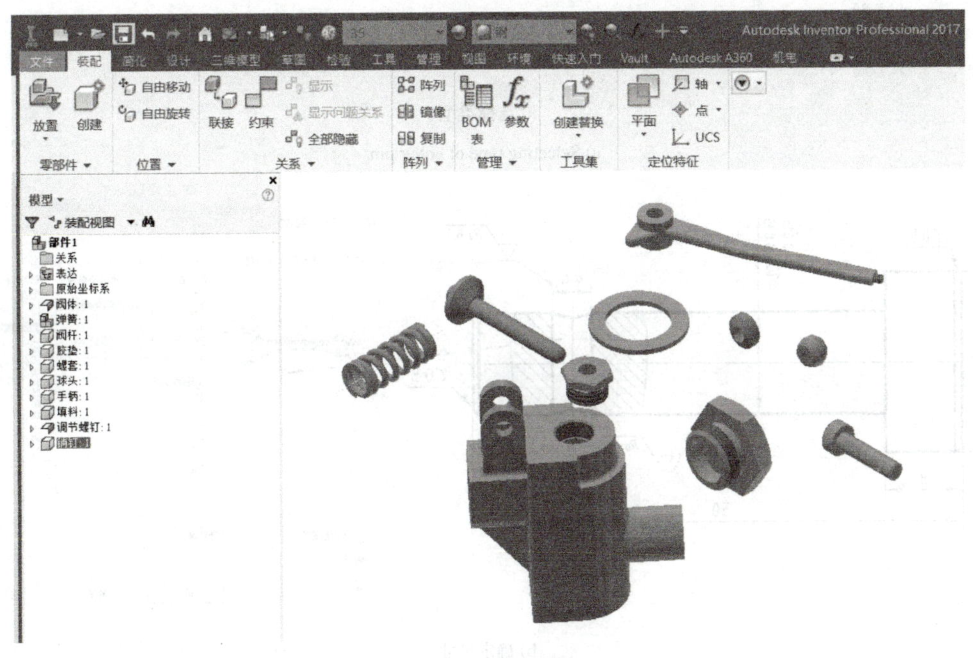

b) 非标零件添加后

b) After adding non-standard parts

图 7-19　非标零件添加

Fig.7-19　Adding non-standard parts

2. 标准件添加

通过“从资源中心装入”命令，从软件库里调入标准件开口销。如图 7-20a 所示，选择开口销的类型。根据与开口销配合的销钉孔直径是 $\phi4$，孔的长度不超过 10mm，选择开口销的尺寸，如图 7-20b 所示。

2. Adding Standard Parts

Transfer split pin of standard part from software library via the “Place from Content Center” command from the software library. Fig.7-20a shows that the type of cotter pin is selected. According to the diameter of fitting pin hole with the cotter pin is $\phi4$ and the hole length (no more than 10mm), the size of cotter pin could be selected, as shown in Fig.7-20b.

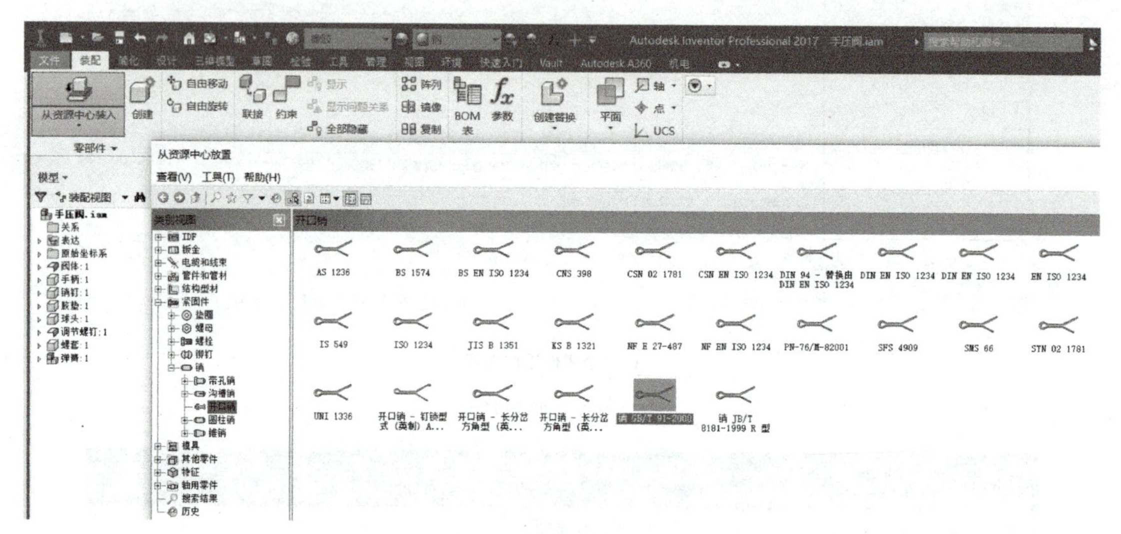

a) 选择类型

a) Selecting type of cotter pin

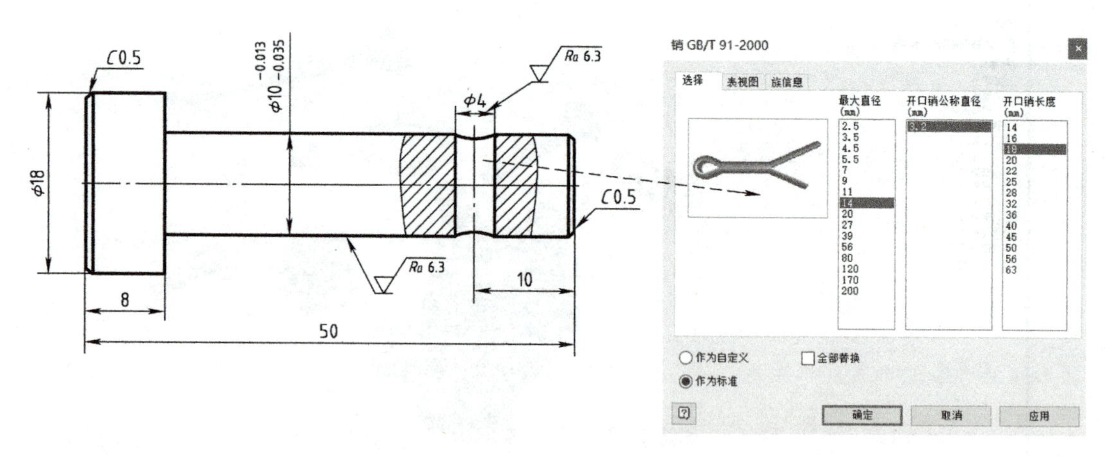

b) 确定尺寸

b) Confirming size

图 7-20 标准件添加

Fig.7-20 Adding standard part

3. 零件位置调整

如图 7-21 所示，通过“自由移动”命令和“自由旋转”命令调整调入的所有零件位置，按照合适的位置进行排列。

3. Part position adjustment

The positions of all parts is adjusted by “Move” and “Rotate” command. Then they will be arrayed based on the appropriate positions, as shown in Fig.7-21.

图 7-21　零件位置调整

Fig.7-21　Adjusting part’s positions

4. 部件装配

在工具面板上单击“约束”命令，在弹出的“放置约束”对话框中选择零件间的约束类型，然后选择需要装配的零件

4. Subassembly assembly

Click the “Constrain” command on the tool panel to pop up “Constrain” dialog, then constrain type can be selected in it. Fig.7-22 shows the handle and valve assembly as an example. The handle and valve body have

进行装配。以手柄和阀体的装配为例，如图 7-22 所示，手柄和阀体之间由于是同轴，单击“插入”约束方式，选择手柄和阀体对应的孔，单击“应用”按钮完成装配，如图 7-23 所示。

the same axis. So click “Insert” constrain type and select the hole corresponding to handle and valve body. Finally, assembly could be completed with clicking “application”, as shown in Fig.7-23.

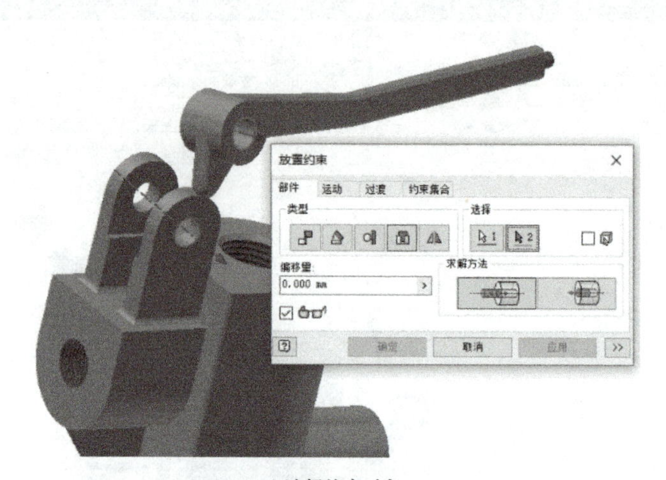

a) 选择约束对象

a) Selecting constraint object

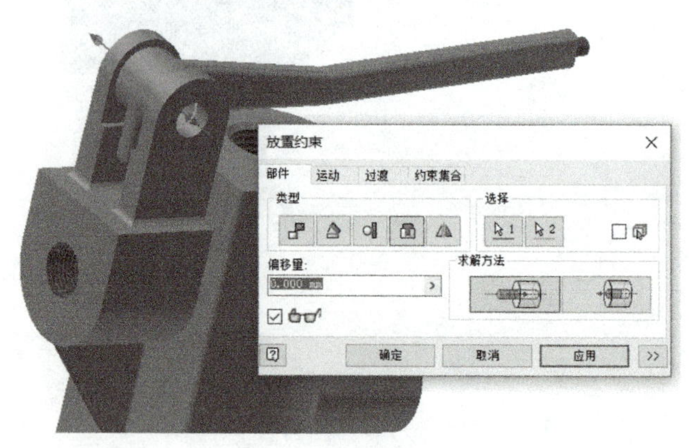

b) 约束完成

b) Constraint completion

图 7-22　手柄与阀体装配

Fig.7-22　Assembly of handle and valve body

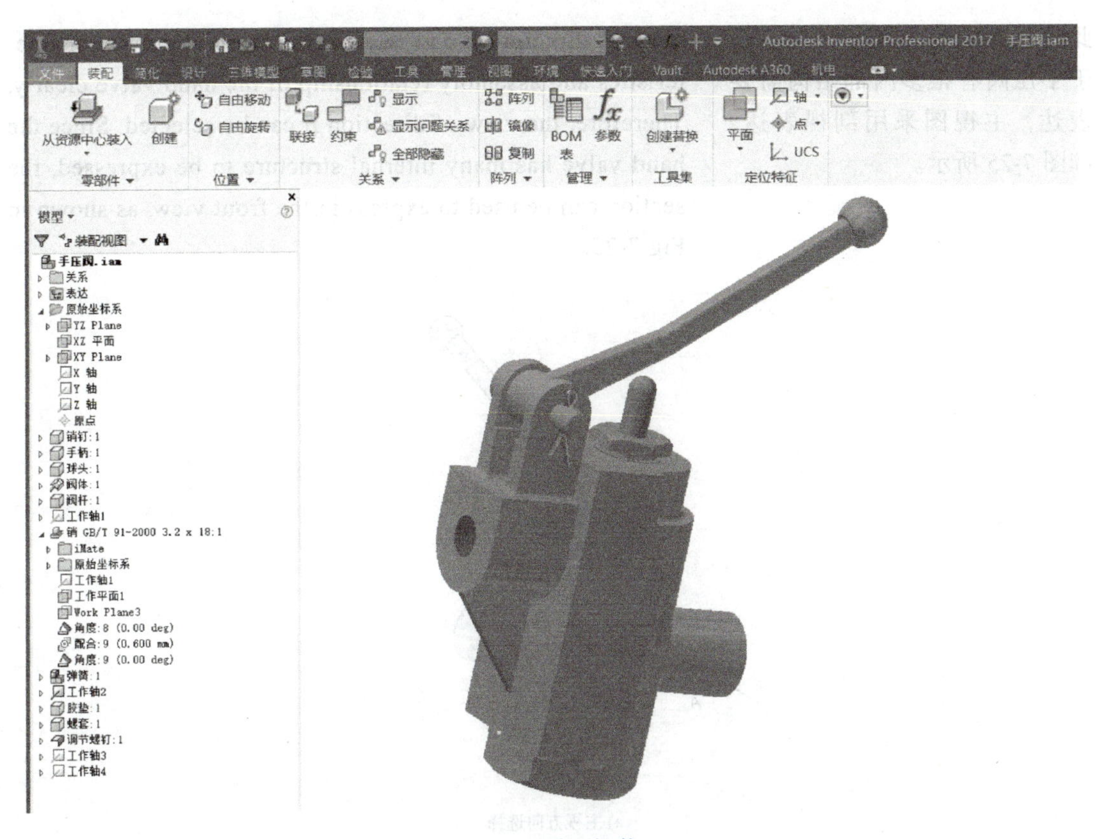

图 7-23　手压阀装配

Fig.7-23　Assembly of handle valve

7.4.3　装配图的表达方案

1. 主视图的选择

部件的放置位置应尽量与工作位置一致，主视图应以表达部件的工作原理和主要装配关系为重点，且采用适当的剖视。

手压阀的放置如图 7-24a 所示。分析 *A* 向和 *B* 向的投影特征，如图 7-24b、c 所示。可以看出，*A* 向有许多结构重叠，*B* 向更能清楚地表达手压阀的主要结构特征和装配关系，因

7.4.3　Representation Method of Assembly Drawing

1. Selecting the front view

The subassembly placed position should be consistent with the working position. The front view should focus on the working principle and major assembly relationship of the expressed part. The appropriate section should be used.

The placement of the hand valve is shown in Fig.7-24a. Fig.7-24b and 7-24c illustrate the analysis of projection characteristics for direction *A* and direction *B*. It is clearly that direction *A* has many overlapping structures, while the

此选择 B 向为主视图方向。由于手压阀有很多内部结构需要表达，主视图采用剖视表达，如图 7-25 所示。

view of direction B can express the main structural characteristics and assembly relationship of the hand valve clearly. Therefore, the view of direction B can be selected. Since the hand valve has many internal structure to be expressed, the section can be used to express in the front view, as shown in Fig.7-25.

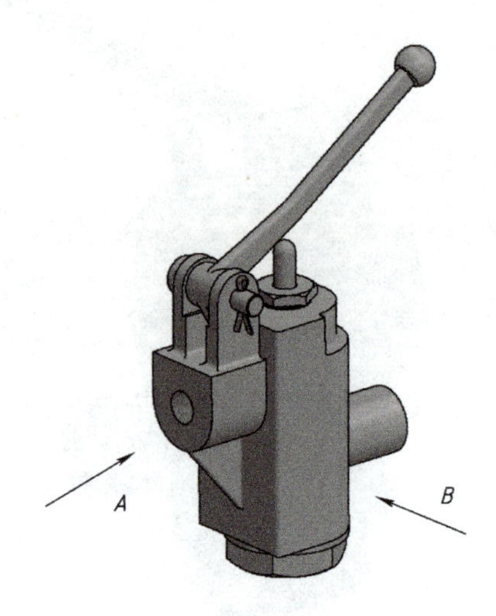

a) 主视方向选择

a) Selecting the front view

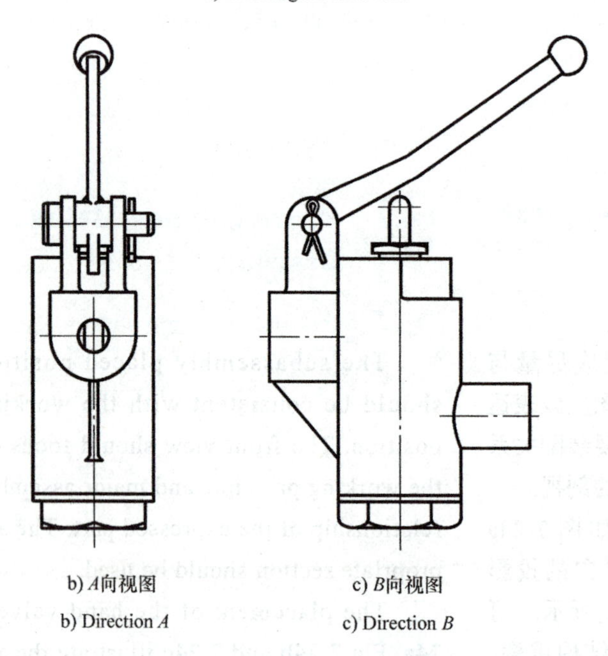

b) A向视图

b) Direction A

c) B向视图

c) Direction B

图 7-24 手压阀主视图的选择

Fig.7-24 Selecting the front view of hand valve

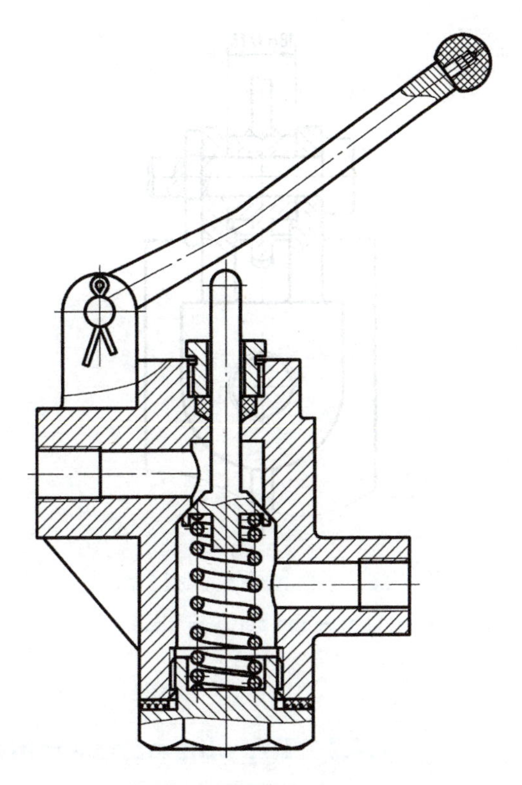

图 7-25　手压阀主视图表达

Fig.7-25　Front view expression of hand valve

2. 其他视图的选择

其他视图一般选取主视图尚未表达清楚的其他装配关系，并选取适当的剖视表达方法。以完整、清晰表达部件的装配关系为前提，部件表达尽量采用最少的视图。

手压阀主视图表达方案确定后，其主要工作原理和配合关系已经表达清楚。但手柄和阀体间的配合连接还没有表达清晰，可以考虑用局部视图表达，如图 7-26 所示。

2. Selecting other views

Generally, other views select the other assembly relations which are not clearly expressed in the front view, and select the appropriate section -view expression methods. The few views should be used to express the subassembly based on the complete and clear expression of the assembly of parts.

The main working principle and fitting relationship of the hand valve are expressed clearly after the expression method of the front view is determined. However, the fitting connection between the handle and the valve body is not expressed clearly. At this time, the local view can be considered to use to express, as shown in Fig.7-26.

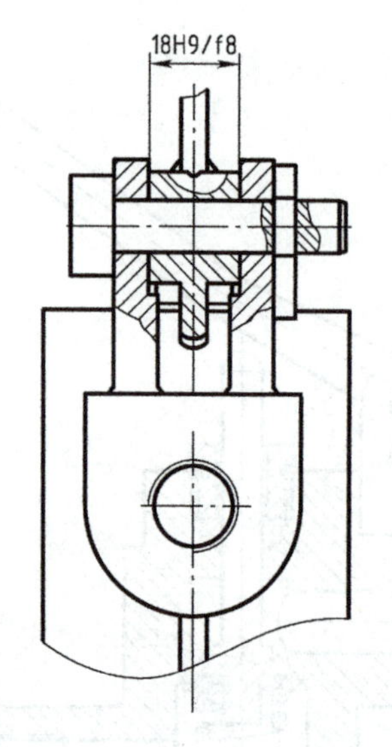

图 7-26　手柄和阀体间的配合表达

Fig.7-26　Fitting expression between handle and valve body

7.4.4　绘制装配图

确定了部件（手压阀）表达方案后，绘制部件装配图还需要考虑以下几个方面。

（1）确定图幅　结合部件（手压阀）整体尺寸大小，布置各视图位置，选择比例，确定图幅大小等。手压阀的总体尺寸高 210，长 118，结合视图布局，选用 A3 图纸幅面。

（2）部件表达　调入部件（手压阀），确定主视图位置。选择合适的方法表达，如局部视图表达。

（3）尺寸标注　标注部件（手压阀）的必要尺寸：配合尺寸、整体尺寸和性能规格尺寸，如图 7-27 所示。

7.4.4　Drawing of Assembly Drawing

After determining the subassembly (hand valve) expression way, the following aspects need to be considered in drawing the subassembly diagram.

7.4.4

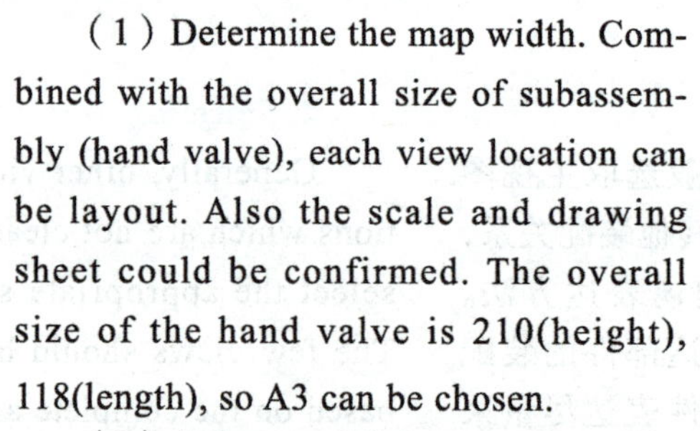

(1) Determine the map width. Combined with the overall size of subassembly (hand valve), each view location can be layout. Also the scale and drawing sheet could be confirmed. The overall size of the hand valve is 210(height), 118(length), so A3 can be chosen.

7.4.4-1

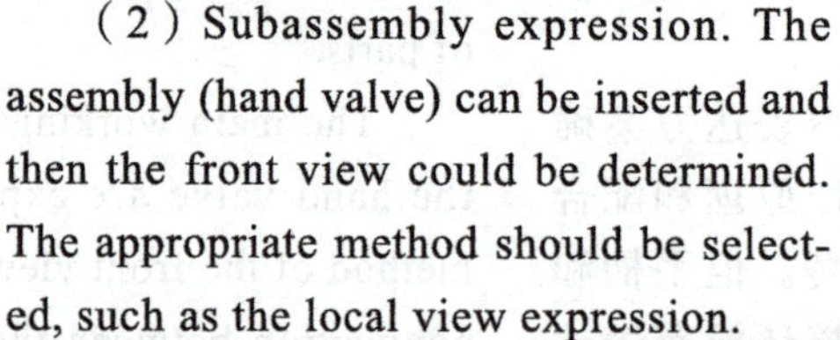

(2) Subassembly expression. The assembly (hand valve) can be inserted and then the front view could be determined. The appropriate method should be selected, such as the local view expression.

7.4.4-2

(3) Dimensioning. Label the necessary dimensions of subassembly (hand pressure valves): fitting dimensions, overall dimensions and performance specifications, as shown in Fig.7-27.

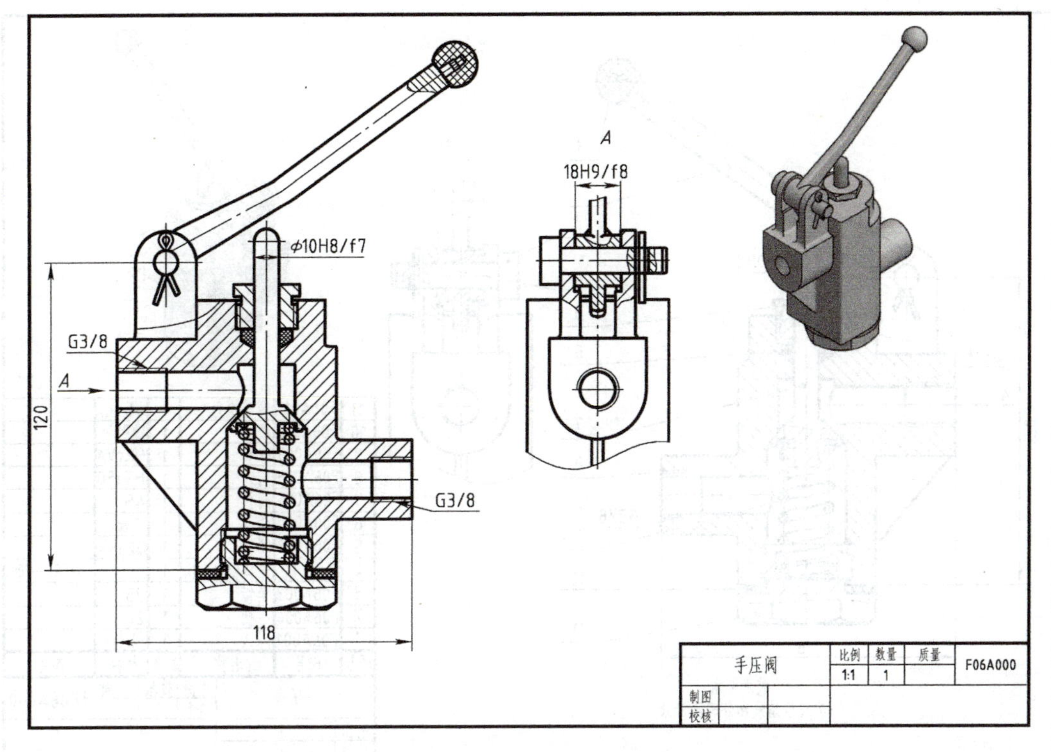

图 7-27　手压阀尺寸标注

Fig.7-27　Dimensions of hand valves

（4）标注序号　标注部件（手压阀）零件序号，单击“自动引出序号”命令，选择视图、零件和放置方式，创建零件的序号标注。

（5）明细栏创建　单击“明细栏”命令，创建部件（手压阀）明细栏。

（6）技术要求编写　根据部件（手压阀）装配过程中的注意事项和装配后应满足的要求等填写技术要求。

（7）审核、检查　对完成的部件（手压阀）装配图进行最后审核，检查标注尺寸是否规范，零件序号是否整齐等问题。检查无误后，完成部件装配图，如图 7-28 所示。

（4）Marking serial number. Click the “Auto Balloon” command and select the way of view, parts and placement to create the parts' balloon.

7.4.4-3

（5）Creation of part list. After clicking “parts list” command, the parts list of subassembly (hand valve) could be created.

（6）Technical requirements writing. Technical requirements should meet the precautions and the requirement of the assembly (hand pressure valves).

7.4.4-4

（7）Checking. The finished subassembly assembly drawing (hand valves) should be final checked whether dimensioning is specification and parts number is neat. Fig.7-28 shows the complete subassembly assembly drawing, which has been inspected.

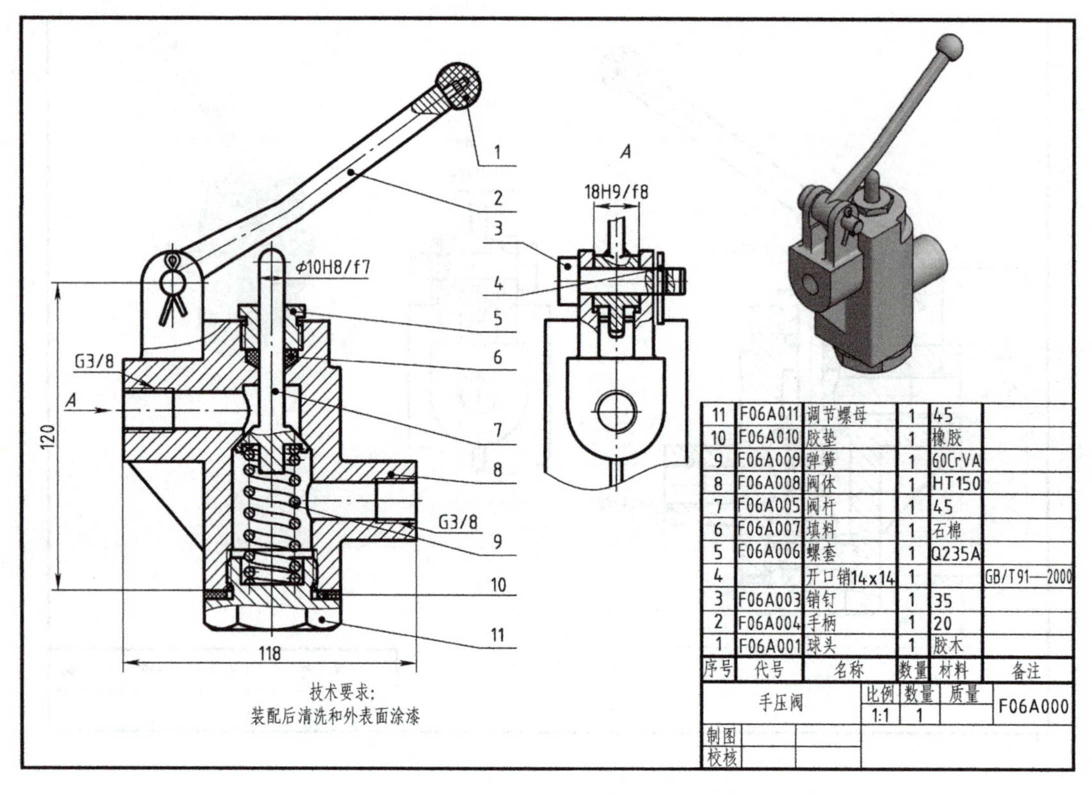

图 7-28 手压阀装配图

Fig.7-28 Assembly drawing of hand valves

7.5 装配结构的合理性

7.5 Rationality of Fitting Structure

为了保证机器或部件能顺利装配，须考虑装配结构的合理性。本节主要介绍一些常见的装配工艺结构。

To ensure the machine or subassembly could be assembled successfully, it is necessary to consider the rationality of the fitting structure. This section introduces the common fitting structures.

7.5.1 相邻两零件的接触面

7.5.1 Contact Surface of Two Adjacent Parts

为了保证装配体某一方向有可靠定位面，相邻零件在某一方向只能有一个接触面，否则会给加工和装配带来困难，如图 7-29 所示。

In order to ensure that the assembly has a reliable locating surface in a direction, the adjacent parts can only have a contact surface in a direction, otherwise it is difficult to process and assemble, as shown in Fig.7-29.

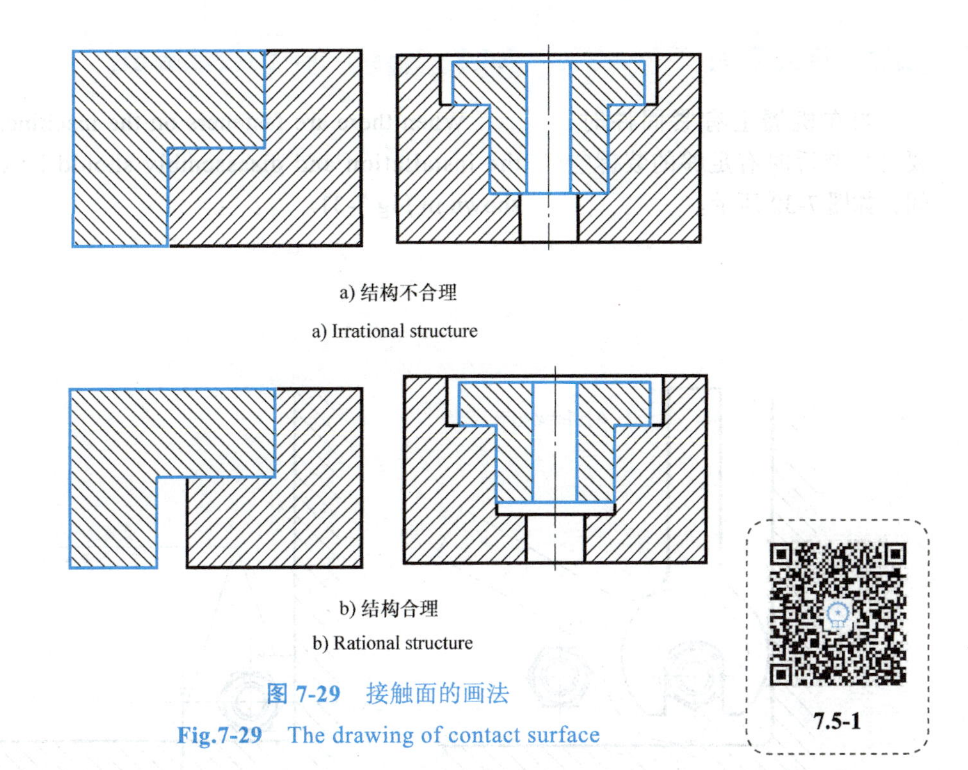

a) 结构不合理

a) Irrational structure

b) 结构合理

b) Rational structure

图 7-29　接触面的画法

Fig.7-29　The drawing of contact surface

7.5.2　轴与孔配合

为了保证轴和孔的配合接触良好，应在孔的接触端面制成倒角或在轴肩根部切槽，如图 7-30 所示。

7.5.2　Fitting of Shafts and Holes

To ensure the good contact of the shaft and the hole, a chamfer should be made at the contact end face of the hole or grooved at the root of the shoulder, as shown in Fig.7-30.

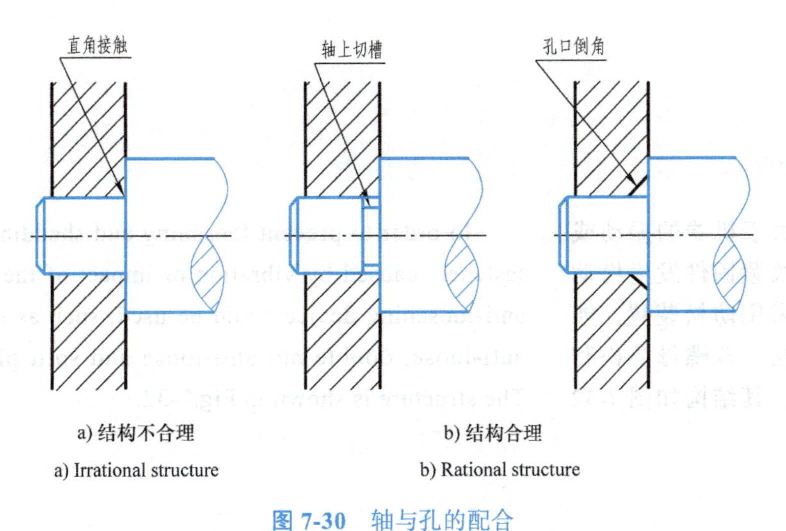

a) 结构不合理

a) Irrational structure

b) 结构合理

b) Rational structure

图 7-30　轴与孔的配合

Fig.7-30　Fitting of shafts and holes

7.5.3 拆装空间

当在机器上有紧固件时，要考虑装拆时有足够的装拆空间，如图 7-32 所示。

7.5.3 Space for Disassembly

When there are fasteners on the machine, the space for the installation and disassembly should be considered, as shown in Fig.7-31.

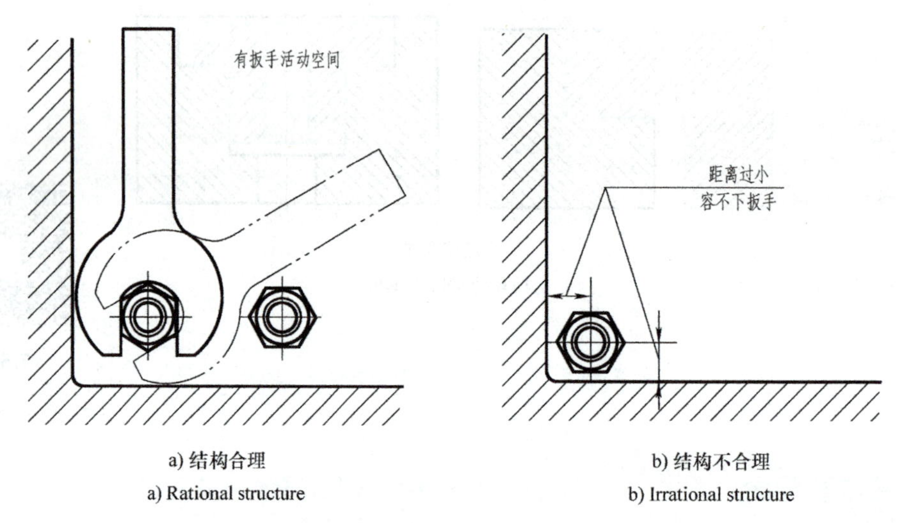

图 7-31 装拆空间

Fig.7-31 Space for disassembly

7.5.4 防松结构

为防止由于机器的振动或冲击引起螺纹紧固件发生松动和脱落，可采用防松装置，如弹簧垫圈防松、双螺母防松和开口销防松，其结构如图 7-32 所示。

7.5.4 Anti-loosening Structure

In order to prevent loosening and shedding of threaded fasteners caused by vibration or impact of the machine, the anti-loosening device could be used, such as spring washer anti-loose, double nut anti-loose and split pin anti-loose. The structure is shown in Fig.7-32.

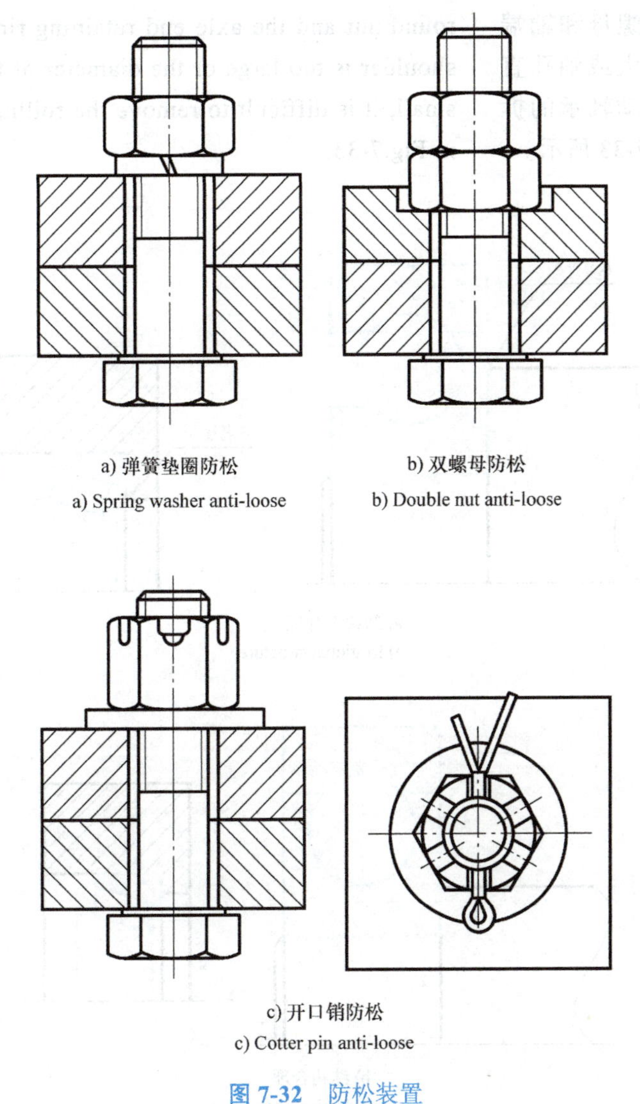

图 7-32 防松装置

Fig.7-32 Anti-loosening structure

7.5.5 轴向固定结构

为了防止滚动轴承产生轴向移动，必须采用一定的结构固定其内、外圈。常用的轴向固定结构形式有轴肩、弹性挡

7.5.5 Axial Fixed Structure

In order to prevent axial movement of rolling bearings, the inner and outer rings must be fixed by the structure. There are many axial fixed structures, such as the shaft shoulder, the elastic retaining ring, the end cover flange, the

圈、端盖凸缘、圆螺母和轴端挡圈等。若轴肩过大或轴孔直径过小，都会给滚动轴承的拆卸带来困难，如图 7-33 所示。

round nut and the axle end retaining ring and so on. If the shoulder is too large or the diameter of the shaft hole is too small, it is difficult to remove the rolling bearing, as shown in Fig.7-33.

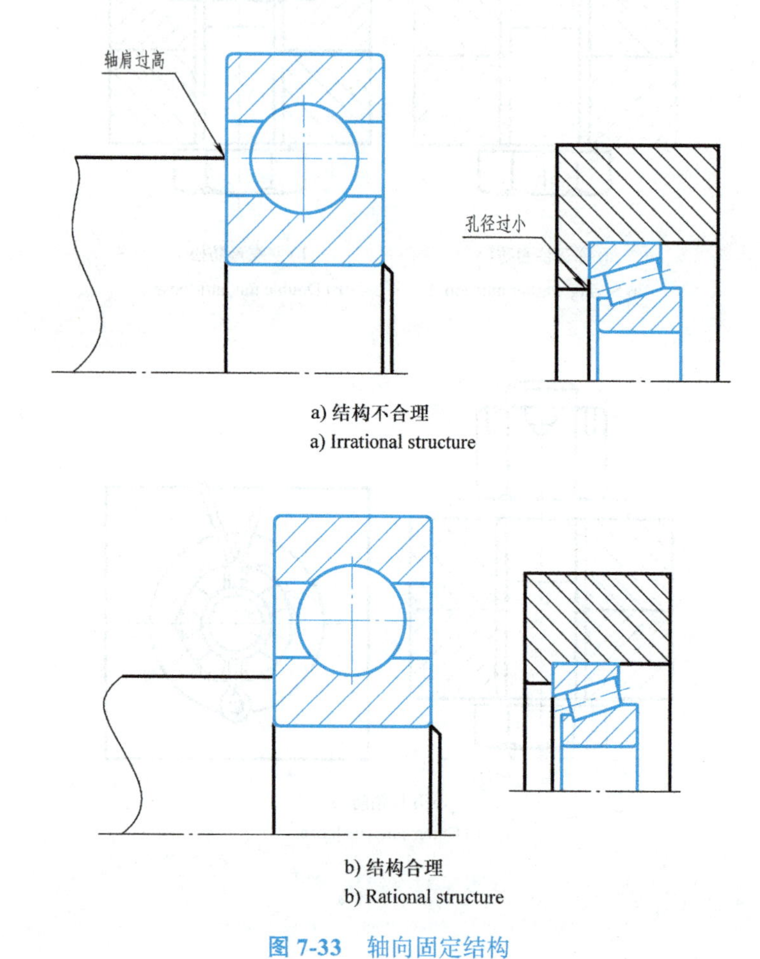

a) 结构不合理
a) Irrational structure

b) 结构合理
b) Rational structure

图 7-33　轴向固定结构
Fig.7-33　Axial fixed structure

7.6　读装配图

7.6　Reading Assembly Drawings

在机器或部件的设计、制造、维修、技术革新和技术交流等生产活动中，常遇到读装配图和拆画零件图的问题。读装配图的目的如下。

In production activities (design, manufacture, maintenance, technological innovation and technical communication, etc.) of machines or subassembly, the problems of reading assembly drawings and extracting detail drawing are often encountered. The following are the purposes of reading assembly drawings.

（1）了解装配体的功用、性能和工作原理。

（1）Learn about the function, performance, and working principle.

（2）读懂各零件间的装配关系和装拆顺序。

（2）Understand the assembly relationships and disassembly sequence among parts.

（3）读懂各零件的主要结构形状和作用等。

（3）Understand the main structure, shape and function of each part.

7.6.1　读装配图的方法和步骤

7.6.1　Methods and Steps of Reading Assembly Drawings

1. 概括了解

读装配图时，首先要看标题栏、明细栏，从中了解该部件的名称、组成该机器或部件的零件名称、数量、材料等。

图 7-34 是夹紧座装配图。

1. General understanding

When reading assembly drawings, the first thing is to read the title bar and the part lists. Some information can be found from them, such as the unit name, part name, quantity, material of parts.

Fig.7-34 reports the assembly draw-

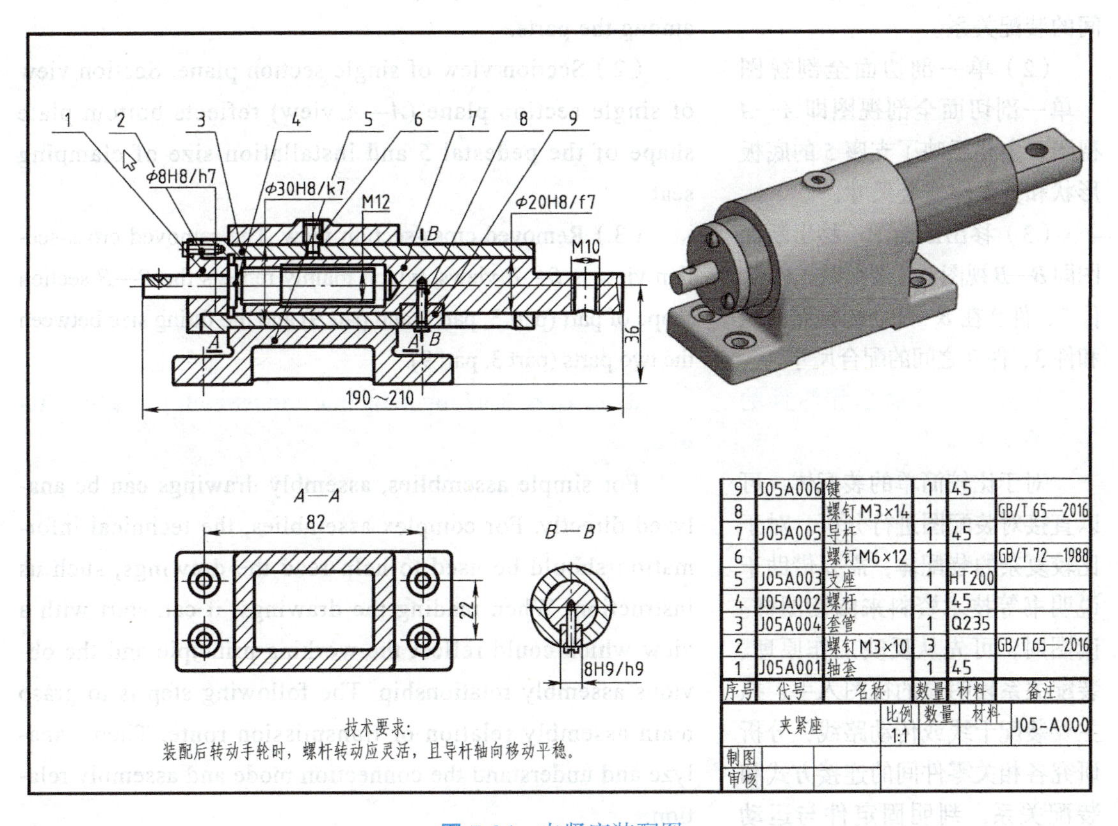

图 7-34　夹紧座装配图

Fig.7-34　Assembly drawing of clamping seat

由标题栏可知该部件名称为夹紧座，对照图上的序号和明细栏，可知它由9种零件组成，其中螺钉2、螺钉6和螺钉8是标准件，其他为非标准件。

ing of clamping seat. From title block, the part name is the clamping seat. According to the serial number and parts list on the comparison diagram, it consists of 9 kinds of parts. screw 2, screw 6 and screw 8 are standard parts others non-standard.

2. 分析视图关系

首先要找到主视图，再根据投影关系识别出其他视图；分析部件的表达方法，从而明确各视图表达的意图和重点。

夹紧座装配图采用了一个基本视图，并采用了局部剖视图、单一剖切面全剖视图、移出断面图等表达方法。

（1）主视图　主视图采用了局部剖视图表达，主要反映了夹紧座的工作原理和零件之间的装配关系。

（2）单一剖切面全剖视图　单一剖切面全剖视图即*A—A*视图，主要反映了支座5的底板形状和夹紧座安装尺寸。

（3）移出断面图　移出断面图即*B—B*视图，主要反映了件3、件7、件9在*B—B*处的截面形状和件3、件9之间的配合尺寸。

2. Analyze relationships of views

The first step is to find the main view and then the other views could be identified according to the projection relationship. The expression method of the subassembly should be analyzed to clarify the purpose and focus of each view.

The clamping seat assembly drawing could be taken in a basic view and several other expressions, such as local section view, section view of single section plane and removed cross-section view.

(1) The front view. The front view adopts the local section view to express, which mainly reflects the working principle of the clamping seat and the assembly relationship among the parts.

(2) Section view of single section plane. Section view of single section plane (*A—A* view) reflects bottom plate shape of the pedestal 5 and installation size of clamping seat.

(3) Removed cross-section view. The removed cross-section view is *B—B* views, which mainly reflects the *B—B* section shape of part (part 3, part 7 and part 9) and the fitting size between the two parts (part 3, part 9).

3. 分析工作原理和零件的装配关系

对于比较简单的装配体，可以直接对装配图进行分析。对于比较复杂的装配体，需要借助于说明书等技术资料来阅读图样。读图时，可先从反映工作原理、装配关系较明显的视图入手，抓主要装配干线或传动路线，分析研究各相关零件间的连接方式和装配关系，判明固定件与运动件，搞清传动路线和工作原理。

3. Analyze working principle and assembly relationship of parts

For simple assemblies, assembly drawings can be analyzed directly. For complex assemblies, the technical information should be used to help read the drawings, such us instruction. When reading the drawings, it can start with a view which could reflect the working principle and the obvious assembly relationship. The following step is to grasp main assembly relation or transmission route. Then, analyze and understand the connection mode and assembly relation.

（1）工作原理　主视图主要反映了夹紧座的工作原理：转动螺杆带动导杆水平移动，从而实现夹紧或松开工件。

（2）装配关系　主视图主要反映了夹紧座的装配关系：套管与支座内孔配合，通过螺钉固定；导杆与套管内孔配合，通过螺钉固定；螺杆与导杆螺纹联接；端盖和套管通过螺钉联接固定。

4. 分析零件的结构及其作用

在弄清上述内容的基础上，还要看懂每一个零件的形状。读图时，借助序号指引的零件上的剖面线，利用同一零件在不同视图上的剖面线方向与间隔一致的规定，对照投影关系以及与相邻零件的装配情况，逐步想象出各零件的主要结构形状。

分析时，一般先从主要零件入手，然后是次要零件。有些零件的具体形状可能表达不够清楚，这时需要根据该零件的作用及与相邻零件的装配关系进行推想，完整构思出零件的结构形状，为拆画零件图作准备。

支座、导杆、套管、螺杆是夹紧座的主要零件，它们在结构和尺寸上都有非常紧密的联系，要读懂装配图，必须看懂它们的结构形状。

导杆、套管、螺杆和轴套属于轴套类零件，根据主视图就可以分析出零件的主要形状。

（1）螺杆　右端有螺纹结构，与导杆螺纹联接，左端有锥孔，用来定位。

(1) Working principle. The front view mainly reflects the working principle of the clamping seat: the guide rod is moved horizontally by rotating worm to clamp or loosen the workpiece.

(2) Assembly relationship. The front view mainly reflects the assembly relationship of the clamping seat: the casing pipe is fitted the internal bore of the pedestal and fixed by screws. The guide rod is fitted the inner hole of the casing pipe through screw fixation. The worm is connected to the guide rod via thread. The end cover and casing pipe are connected with screws.

4. Analyze the structure and function of parts

On the basis of understanding the above contents, the shape of each part should be understood. When reading the drawings, by applying the regulation of "the section lines of the same part on different views have same directions and intervals". the main structural shape of each part is gradually imagined according to the projection relationship and the assembly condition of the adjacent parts.

Normally, the main parts should be analyzed earlier than the secondary part. When the shapes of some parts may not be expressed unclearly, the structural shape of parts could be conceived completely on the basis of reckon of part function and assembly relationship of adjacent parts. It could provide preparation part extraction drawing.

Pedestal, guide rod, casing pipe, worm are the main parts of clamping seat, which have the very closed connection in structure and size. The structural shape should be understood at first in order to read the assembly drawing.

Guide rod, casing pipe, worm and bushing are sleeve parts. According to the front view, the main shape of the parts could be analyzed.

(1) Worm. The right end of the screw has a threaded structure that is connected to the guide rod. The taper hole at the left end of the worm is used to locating.

（2）导杆　右端有垂直螺纹孔，左端内螺纹孔，用于联接螺杆。

（3）套管　套管主要和导杆配合，左端有螺纹孔，与轴套联接。套管设有螺纹孔，用螺钉联接底座。

（4）支座　支座是箱体类零件，由主视图和 *A—A* 剖视图可知，直径 30 的内孔与套管配合。底座有 4 个沉孔，用来安装夹紧座。

每个零件的结构形状都看清楚之后，将各零件联系起来，便可想象出夹紧座的完整结构，如图 7-35、图 7-36 所示。

(2) Guide rod. There is a vertical threaded hole at the right end of the guide rod and a threaded hole at the left-end for connecting the worm.

(3) Casing pipe. The casing pipe is mainly fitted with the guide rod. There is a threaded hole at the left end that connected with the bushing. There is a threaded hole to connect the worm.

(4) Pedestal. The pedestal is a box-type part. According to the front view and *A—A* section view, the inner hole of diameter 30 is fitted with the casing pipe. The base has 4 counter bores that are used to install the clamp seat.

The complete structure of clamping seat could be imagined after the structural shape of each part being understood and connected.

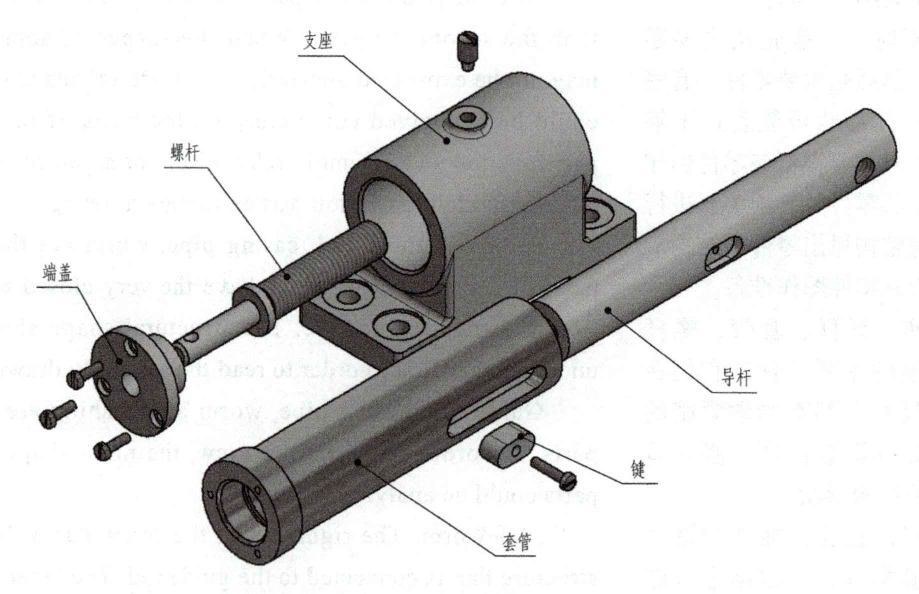

图 7-35　夹紧座结构分解图

Fig.7-35　Exploded view of the clamping seat

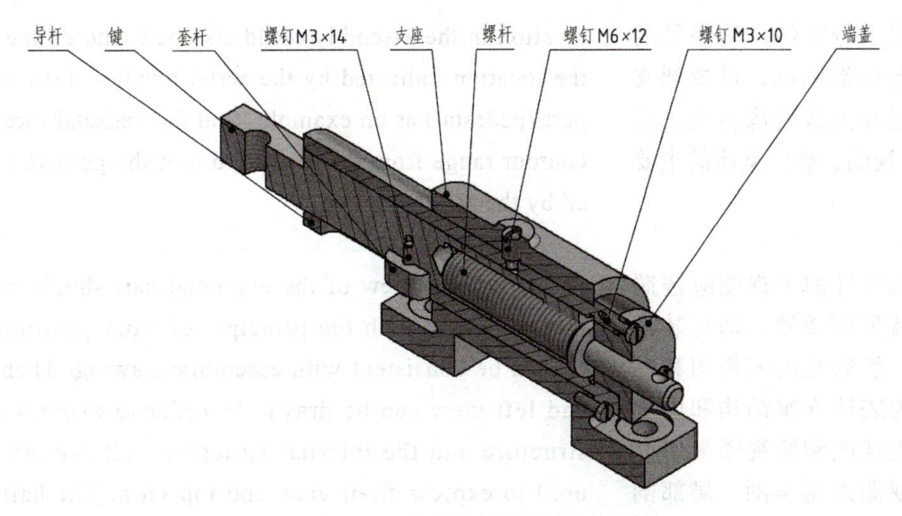

图 7-36　夹紧座结构剖视图

Fig.7-36　Sectioning structure of the clamping seat

5. 总结归纳

最后，总结归纳工作原理、传动关系、各零件间的装配关系技术要求、尺寸等，综合分析总体结构，从而对装配体做全面了解。

5. Summary

Finally, summarize the working principle, transmission relationship, the assembly relationship among parts, dimension and the overall structure in order to get a comprehensive understanding of the assembly.

7.6.2　由装配图拆画零件图

由装配图拆画零件图，是机械设计和机械维修时经常遇到的问题。在识读装配图的教学过程中拆画零件图，常常是检查是否真正读懂装配图的手段。

下面以夹紧座为例，介绍拆画零件图的方法。

1. 读懂零件形状

根据零件序号和明细栏，找到要拆画零件的序号、名称，再根据序号指引线所指的部位，找到该零件在装配图中的位置。

7.6.2　Extracting Part Drawing from Assembly Drawing

Extracting part drawing from assembly drawings is a common job in mechanical design and maintenance. Extracting part drawing could examine whether the assembly drawings are really understood during the process of reading assembly drawings.

Let's take a clamping seat as an example to introduce the method of drawing part.

1. Comprehending part structure

According to the balloon and the part lists, the serial number, name of the extracted part could be found. The part

如支座是5号零件，从序号的指引线起始端圆点，可找到支座的位置和大致轮廓范围。结合 *A—A* 视图，确定支座的主要形状。

2. 确定零件表达方案

拆画零件的主视图应按照工作位置原则放置，即与装配图一致。然后画出俯视图和左视图。为表达支座结构和内部结构，主视图和俯视图采用全剖，左视图采用半剖、局部剖表达，如图7-37所示。

3. 标注尺寸

如果是装配上标注出的尺寸，可以照抄。未标注出的尺寸从图上测量，按所采用比例

location in the assembly could also be found on the basis of the location indicated by the serial number. Take the No. 5 part (pedestal) as an example, find the pedestal location and contour range from the original dot of the guideline indicated by the serial number.

2. Determine the part representation project

The front view of the extracted part should be placed in accordance with the principle of work position, which should be consistent with assembly drawing. Then the top and left view can be drawn. In order to express the base structure and the internal structure, full section view is used to express front view and top view. The half section and local section are used to express left view, as shown in Fig.7-37.

3. Dimensions

If the dimensions are marked on the assembly, they could be copied. If not, the dimensions should be measured

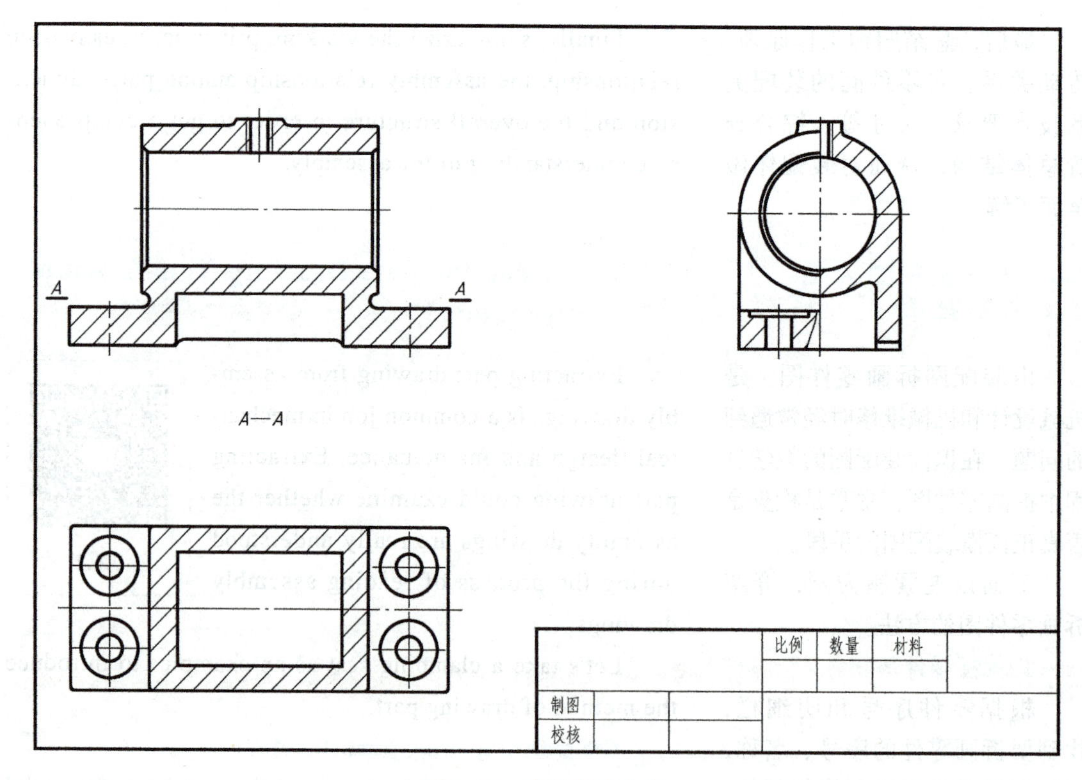

图7-37 支座表达方案

Fig.7-37 Pedestal representation project

寸从图上测量，按所采用比例换算。

from the assembly drawing and converted according to the proportions.

4. 技术要求

补充表面粗糙度等。

4. Technical requirements

Things like surface roughness, etc. need to be added

5. 填写标题栏

根据装配图中的明细栏，绘图比例和绘图者姓名等，在零件图的标题栏中填写零件的名称、材料、数量等，并填写绘图比例和绘图者姓名等。

5. Filling in the title bar

According to the part list in the assembly drawing, the scale, quantity, material, name of the parts, and draftsman's name of assembly drawing. The title block of the part drawing should be filled with the name, material, quantity, scale and the name of the draftsman.

6. 检查校对

检查校对是拆画零件图的最后一步。首先查看零件是否表达清楚，投影关系是否正确。然后校对尺寸是否有遗漏，相互配合的相关尺寸是否一致，以及技术要求与标题栏等内容是否完整。支座的零件图如图 7-38 所示。

6. Checking and proofreading

Checking and proofreading is the last step of extracting the drawing of part. Firstly, it should be checked whether the parts are clearly expressed and the projection relationship is correct. Then you must check if some sizes are omitted and the relevant dimensions of the coordination are consistent or not. The technical requirements, title block and other content should also be confirmed . The part drawing of the pedestal is shown in Fig.7-38.

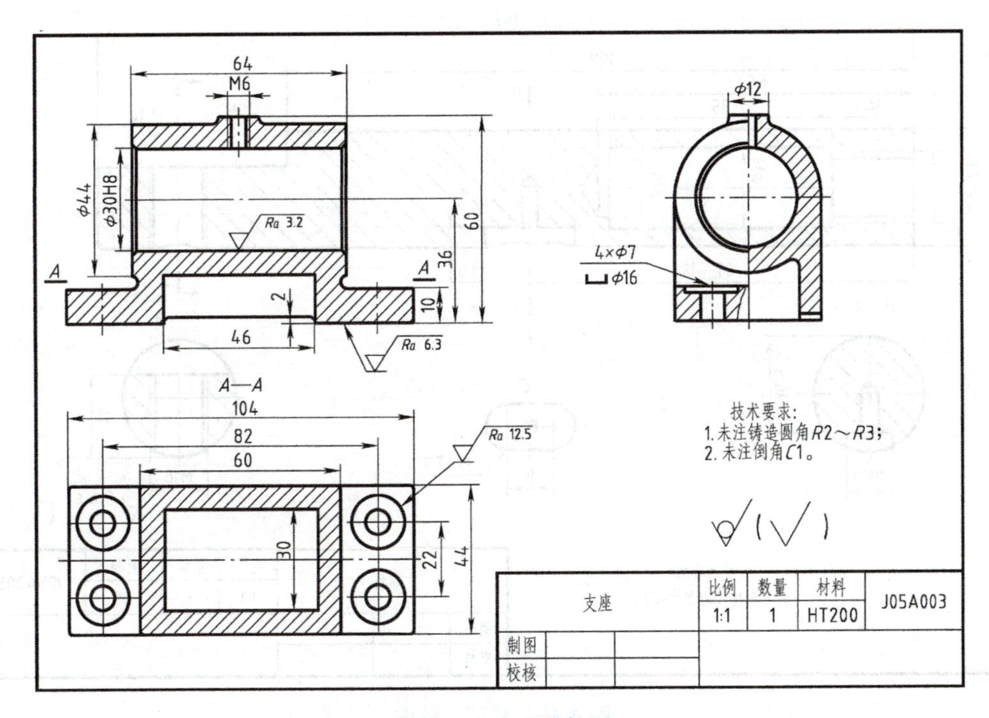

图 7-38　支座零件图

Fig.7-38　Pedestal detail drawing

[本章习题] [Chapter exercises]

根据夹紧座装配图（图7-34）、夹紧座结构分解图（图7-35）、支座零件图（图7-38）、导杆零件图（图7-39）、键零件图（图7-40）、套管零件图（图7-41）、轴套零件图（图7-42）、螺杆零件图（图7-43），对夹紧座进行零件建模、部件装配以及生成夹紧座装配工程图。

According to assembly drawing (Fig.7-34) and exploded view (Fig.7-35) of clamping seat, pedestal part drawing (Fig.7-38), guided rod (Fig.7-39), key (Fig.7-40), casing pipe (Fig.7-41), shaft sleeve (Fig.7-42) and worm (Fig.7-43), create part model, subassembly assembly, and draw assembly drawings for the clamping seat.

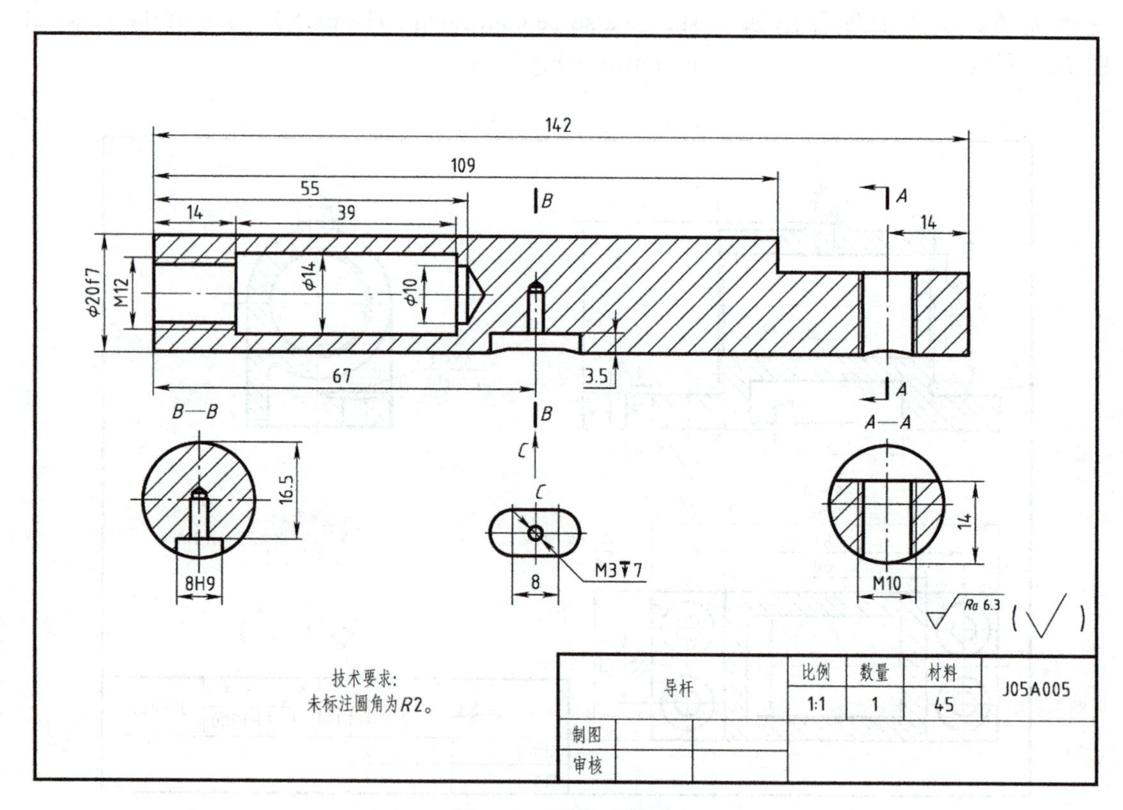

图 7-39 导杆零件图

Fig.7-39 Guide rod detail drawing

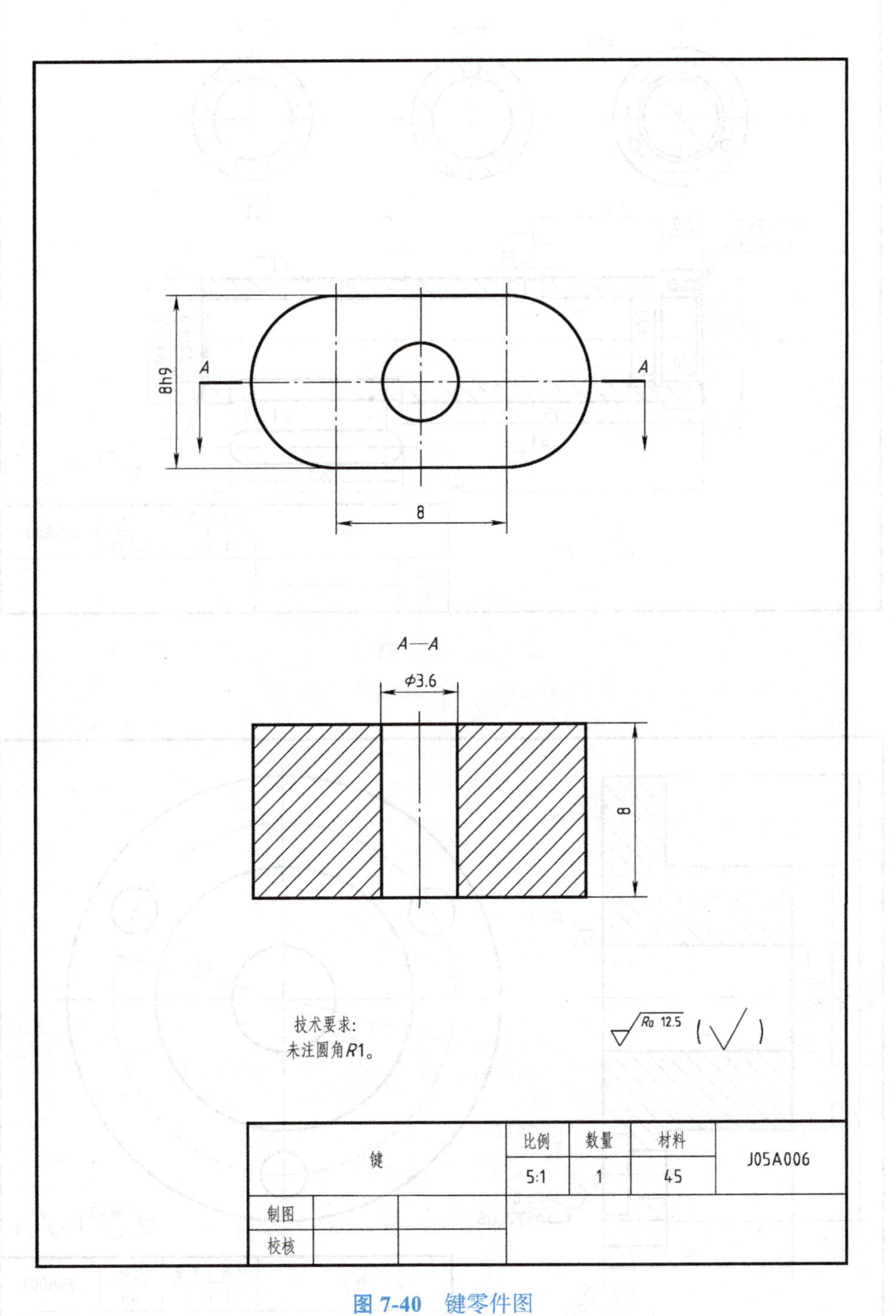

图 7-40　键零件图

Fig.7-40　Key detail drawing

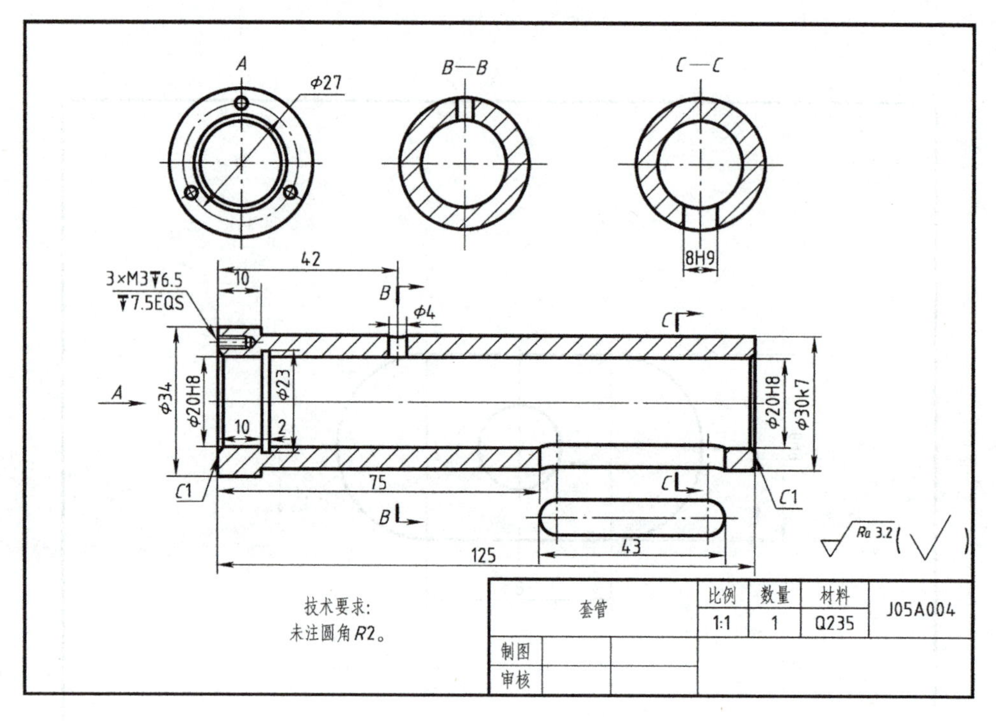

图 7-41 套管零件图

Fig.7-41 Casing pipe detail drawing

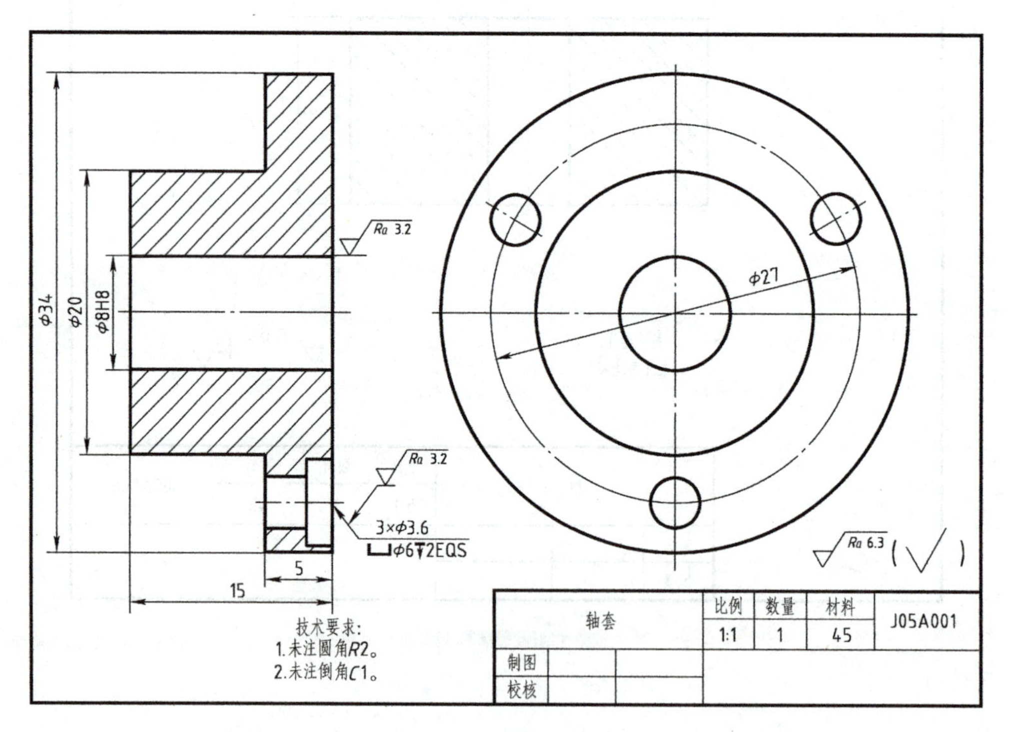

图 7-42 轴套零件图

Fig.7-42 Shaft sleeve detail drawing

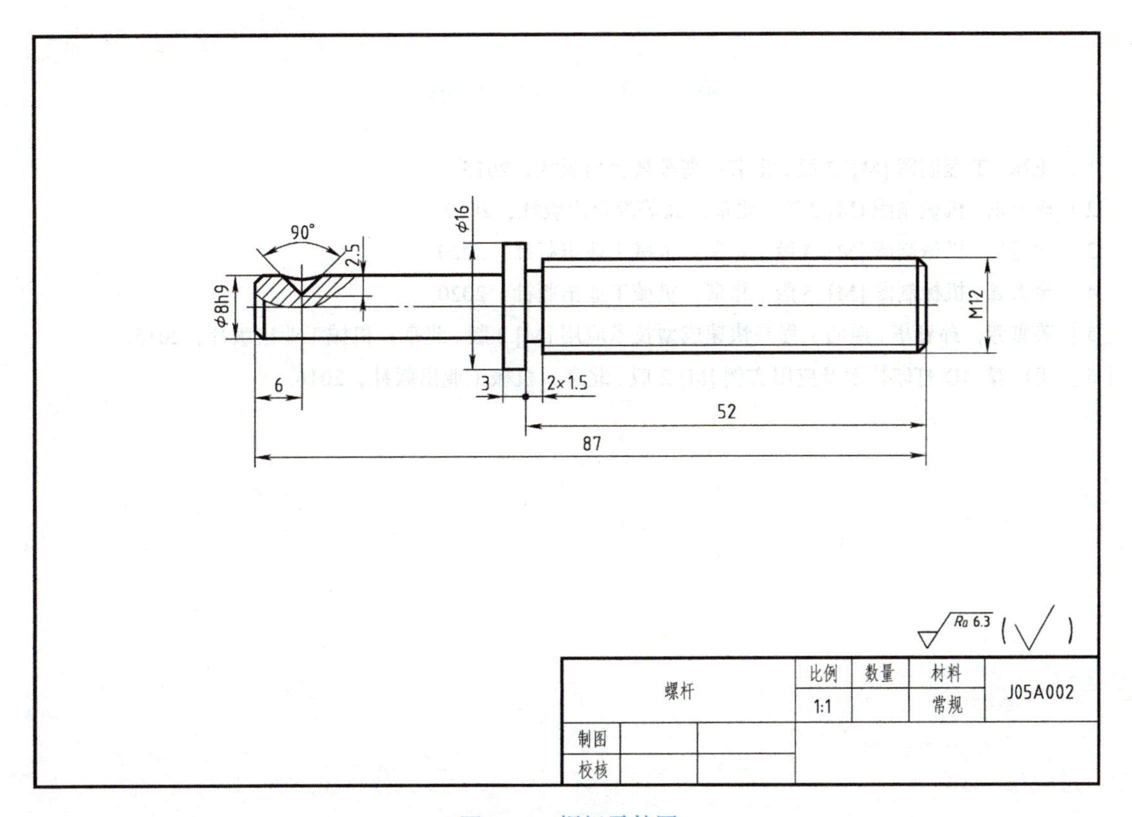

图 7-43 螺杆零件图

Fig.7-43 Worm detail drawing

参考文献　References

[1] 王冰 . 工程制图 [M]. 2 版 . 北京：高等教育出版社，2015.

[2] 唐卫东 . 机械制图 [M].2 版 . 北京：高等教育出版社，2019.

[3] 胡建生 . 机械制图 [M]. 4 版 . 北京：机械工业出版社，2020.

[4] 金大鹰 . 机械制图 [M]. 5 版 . 北京：机械工业出版社，2020.

[5] 陈雪芳，孙春华 . 逆向工程与快速成型技术应用 [M].2 版 . 北京：机械工业出版社，2015.

[6] 王广春 .3D 打印技术及应用实例 [M].2 版 . 北京：机械工业出版社，2016.